8702-2854

D0122198

DATE DUE	
APR 17 2011	

IMPACTS OF FOREST HARVESTING
ON LONG-TERM SITE PRODUCTIVITY

IMPACTS OF FOREST HARVESTING ON LONG–TERM SITE PRODUCTIVITY

Edited by

W.J. DYCK

New Zealand Forest Research Institute
Rotorua
New Zealand

and

D.W. COLE

College of Forest Resources
University of Washington
Seattle

and

N.B. COMERFORD

Department of Soil and Water Sciences
University of Florida
Gainesville
USA

CHAPMAN & HALL

London · Glasgow · Weinheim · New York · Tokyo · Melbourne · Madras

NORTHWEST COMMUNITY
COLLEGE

Published by Chapman & Hall, 2-6 Boundary Row, London SE1 8HN, UK

Chapman & Hall, 2-6 Boundary Row, London SE1 8HN, UK

Blackie Academic & Professional, Wester Cleddens Road, Bishopbriggs, Glasgow G64 2NZ, UK

Chapman & Hall GmbH, Pappelallee 3, 69469 Weinheim, Germany

Chapman & Hall USA, One Penn Plaza, 41st Floor, New York NY 10119, USA

Chapman & Hall Japan, ITP-Japan, Kyowa Building, 3F, 2-2-1 Hirakawacho, Chiyoda-ku, Tokyo 102, Japan

Chapman & Hall Australia, Thomas Nelson Australia, 102 Dodds Street, South Melbourne, Victoria 3205, Australia

Chapman & Hall India, R. Seshadri, 32 Second Main Road, CIT East, Madras 600 035, India

First edition 1994

© 1994 Chapman & Hall

Printed in Great Britain by St Edmundsbury Press Ltd, Bury St Edmunds, Suffolk

ISBN   0 412 58390 9

Apart from any fair dealing for the purposes of research or private study, or criticism or review, as permitted under the UK Copyright Designs and Patents Act, 1988, this publication may not be reproduced, stored, or transmitted, in any form or by any means, without the prior permission in writing of the publishers, or in the case of reprographic reproduction only in accordance with the terms of the licences issued by the Copyright Licensing Agency in the UK, or in accordance with the terms of licences issued by the appropriate Reproduction Rights Organization outside the UK. Enquiries concerning reproduction outside the terms stated here should be sent to the publishers at the London address printed on this page.
  The publisher makes no representation, express or implied, with regard to the accuracy of the information contained in this book and cannot accept any legal responsibility or liability for any errors or omissions that may be made.

A catalogue record for this book is available from the British Library

∞ Printed on permanent acid-free text paper, manufactured in accordance with ANSI/NISO Z39.48–1992 and ANSI/NISO Z39.48–1984 (Permanence of Paper).

CONTENTS

CONTRIBUTORS

P.N. Beets
New Zealand Forest Research Institute
Private Bag 3020
Rotorua
New Zealand

J.A. Burger
Virginia Polytechnic Institute
and State University
Blacksburg
VA
USA

D.W. Cole
College of Forest Resources
University of Washington
Seattle
WA 98195
USA

N.B. Comerford
Department of Soil and Water Sciences
University of Florida
Gainesville
FL 32611
USA

W.J. Dyck
New Zealand Forest Research Institute
PO Box 31011
Christchurch
New Zealand

J.W. Hornbeck
USDA Forest Service

NE Forest Experimental Station
Durham
NH 03824
USA

D.W. Johnson
Biological Sciences Center
Desert Research Institute
Reno
NV 89512
USA

R.K. Jones
ESRI Canada Limited
Victoria
British Columbia
Canada

J.P. Kimmins
Faculty of Forestry
University of British Columbia
Vancouver
Canada

H. Lundkvist
Swedish University of Agricultural
Sciences
S-750 07 Uppsala
Sweden

J. Manz
Weyerhaeuser Company
Tacoma
WA 98531
USA

D.J. Mead
Lincoln University
Canterbury
New Zealand

R.E. Miller
USDA Forest Service
Pacific Northwest Research Station
3625 93rd Avenue
SW Olympia
WA 98512
USA

L.A. Morris
Warnell School of Forest Resources
University of Georgia Athens
GA 30602
USA

D.G. Neary
USDA Forest Service
Rocky Mountain Station
Flagstaff
AZ 86001
USA

R.F. Powers
USDA Forest Service
PSW Research Station
Redding
CA
USA

M.F. Proe
Macaulay Land Use Research Institute
Aberdeen
AB9 2QJ
UK

H.M. Rauscher
USDA Forest Service
NC Forest Experiment Station
Grand Rapids

MN 55744
USA

M.W. Ritchie
USDA Forest Service
PSW Research Station
Redding
CA
USA

C.T. Smith
New Zealand Forest Research Institute
Ltd
Private Bag 3020
Rotorua
New Zealand

T.A. Terry
Weyerhaeuser Company
Tacoma
WA 98531
USA

H. Van Miegroet
Department of Forest Resources
Utah State University
Logan
UT 84322-5215
USA

J. Yarie
Forest Soils Laboratory
University of Alaska
Fairbanks
AK 99775
USA

D. Zabowski
College of Forest Resources
University of Washington
Seattle
WA 98195
USA

FOREWORD

The International Energy Agency Bioenergy Agreement was initiated as the Forestry Energy Agreement in 1978. It was expanded in 1986 to form the Bioenergy Agreement. Since that time the Agreement has thrived with some fifteen countries (Austria, Belgium, Canada, Denmark, Finland, Italy, Japan, Netherlands, New Zealand, Norway, Sweden, Switzerland, United Kingdom, United States and the CEC) currently being signatories. The objective of the Agreement is to establish increased programme and project cooperation between the participants in the field of bioenergy.

The environmental consequences of intensive forest harvesting have been the subject of intense interest for the Agreement from its initiation. This interest was formulated as a Cooperative Project under the Forestry Energy Agreement in 1984. It developed further under each of the subsequent three-year Tasks of the Bioenergy Agreement (Task III, Activity 3 "Nutritional consequences of intensive forest harvesting on site productivity", Task VI, Activity 6 "Environmental impacts of harvesting" and more recently Task IX, Activity 4 "Environmental impacts of intensive harvesting". The work has been supported by five main countries from within the Bioenergy Agreement: Canada, New Zealand, Sweden, UK, and USA.

The continued work has resulted in a significant network of scientists working together towards a common objective - that of generating a better understanding of the processes involved in nutrient cycling and the development of management regimes which will maintain or enhance long term site productivity.

This book represents the cumulation of several years' work by the editorial team and the individual chapter authors. As such it is a fitting monument to the work of the IEA Bioenergy Agreement and the international cooperation it has fostered over the years.

Jan Erik Mattsson, Paul Mitchell

Operating Agents IEA/BA/Task IX
Biomass Supply from Conventional Forestry
March 1993

PREFACE

The idea for this book originated at a business meeting of the IEA Bioenergy Agreement Project 'Environment Impacts of Intensive Harvesting' held in Rotorua in March 1989. Professor Hamish Kimmins proposed that the information that had been synthesized by collaborators during the project be formally documented in a book for future reference. Since the project began in 1985 there has been an annual workshop of project collaborators, always on a theme related to the overall objective of predicting the impact of intensive harvesting on long-term site productivity. Each workshop has produced a proceedings, consequently some of the chapters closely reflect papers presented at project workshops, whereas others are totally new. It was thought that a book would bring the most important findings together in one volume in a form that was available to a wide audience of forest managers, policy makers, researchers, and students in forestry and land management. Consequently, this book is a synthesis of knowledge on the subject of the impacts of harvesting on long-term site productivity, but, as pointed out in the chapters, is by no means the last word on the subject.

The book discusses the need for long-term site productivity research (Chapter 1) and the methodology that has been used both to investigate the impacts of harvesting and site preparation on site quality (Chapters 2 and 3) and on off-site values (Chapter 4) and to predict the impacts either through the use of computer simulation models (Chapter 6) or classification systems (Chapter 7). Identification of key ecosystem processes to study and model is covered in Chapter 5. In Chapter 8, management systems for sustaining productivity are presented from an operational perspective. Design of long-term site productivity experiments and methodology for investigating important ecosystem processes are reviewed in Chapters 9 and 10 respectively. The final chapter (11) briefly recaps on the contents of the book and makes suggestions for future research in the area of harvesting impacts on long-term site productivity.

The more that 150 collaborators in the project have all shared a common goal, to improve our ability to predict the potential impacts of intensive harvesting on the environment, and particularly on long-term site productivity. A number of international collaborative research projects have stemmed from the IEA project and the annual workshops have provided a focus for researchers, and also forest managers, to discuss concerns and new research ideas. Obviously there is much more research to be done, however, the project has served an important role in bringing researchers together with a common goal and focusing research effort to make the most efficient use of scarce resources and funding.

A number of people have been instrumental in the success of this project, in particular the organizers and hosts of the annual workshops: Mike Messina, Folke Andersson, Helene Lundkvist, TomWilliams, Dale Cole, Hamish Kimmins, Nick Comerford, Bob Powers. Mike Proe, Greg Ruark, Taumey Mahendrappa, and Tat Smith. We gratefully recognise their contributions and also the many people that assisted them. We also acknowledge the contribution the IEA Bioenergy Agreement has made in making this book possible, particularly Paul Mitchell and Jan Erik Mattson, and also the members of the Executive Committee who have supported the project and the book initiative throughout. In particular, John Tustin (EC member for New Zealand) has played an integral role in the intiation of the project and also in realizing the publication of this book.

Finally, and most importantly, we thank Christine Bow (Mees), both for her assistance in the management of the project, but particularly for her invaluable and very substantial effort in preparing this book for publication.

W.J. Dyck
D.W. Cole
N.B. Comerford

CHAPTER 1

REASONS FOR CONCERN OVER IMPACTS OF HARVESTING

D.W. JOHNSON

Biological Sciences Center, Desert Research Institute
Reno, NV 89512, USA

INTRODUCTION

There are a number of effects of forest harvesting that cause concern among various environmental groups and forest managers. These include effects on aesthetics, wildlife habitat, water quantity and quality, susceptibility to fire, and long-term site productivity. The subject of this volume is restricted to only one of these issues, namely, long-term site productivity. This by no means implies that long-term site productivity is the only harvesting effect worthy of consideration; indeed, forest harvesting practices in the western United States are under attack at this moment because of concerns over wildlife habitat. However, taking the assumption that harvesting will continue and likely even intensify in many commercial forests around the world, the topic of harvesting effects upon long-term site productivity is timely and will certainly fill this volume.

HISTORICAL CONCERNS

Concern over the impacts of forest harvesting on long-term site productivity go back at least 100 years. Second-rotation declines in the productivity of Norway spruce plantations were observed in Germany and Switzerland in the late 19th century (Weideman [1] cited in Pritchett and Fisher [2]). Nutritional causes were suspected, but apparently not conclusively proven (see review by Pritchett and Fisher [2]). Ebermeyer's "Complete Treatise on Forest Litter" [3] evaluated the effects of harvesting forest litter on forest productivity. Ebermeyer concluded that nutrient removal - especially calcium -was the primary cause of the productivity decline (see [4]). Ebermeyer's work is credited with stimulating studies of forest nutrient cycling much later in the 1950's and 1960's [4].

The studies by Likens *et al.* [5] at Hubbard Brook, New Hampshire created a major controversy over the policy of clearcutting because of potential effects upon water quality as well as nutrient loss via leaching. Specifically, Likens *et al.* [5] showed major increases in streamwater nitrate concentrations following clearcutting and herbicide treatment at Hubbard Brook, New Hampshire. These results contrasted sharply with the result of Cole and Gessel [6] in nitrogen-poor Douglas-fir forests in Washington which showed little increase in nitrate leaching following harvesting. The Hubbard Brook results created major controversy over the practice of clearcutting and stimulated research programs designed to evaluate the effects of disturbance on nitrate leaching (e.g., [7]). The harvesting - nitrate leaching issue eventually faded as several studies showed that 1) leaching losses of nitrogen and other nutrients following harvesting are normally much less than nutrient export in biomass, and 2) the nitrate response following the clearcutting and herbicide studies at Hubbard Brook was anomalously large compared to most other instances (e.g., [8]), even for other similar northern hardwood forest systems which were harvested and not treated with herbicide [9]. There have even now been cases where nitrate leaching decreases following harvesting in nitrogen-fixing forests [10].

Nutrient removal in biomass re-emerged as a major concern over harvesting - especially whole-tree harvesting (WTH) - in the 1970's [11, 12, 13], and remains a significant concern to this date. The particular concern with WTH is that the nutrient export per unit biomass increases dramatically with little gain in useable product in many cases. With most commercial forests being deficient in nitrogen or phosphorus, one might expect that these would be the nutrients most affected by WTH. However, nutrient budget analyses indicate that calcium is the nutrient most significantly depleted in most sites [14, 15, 16, 17], as was the case in Ebermeyer's early study. This is an especially curious result in view of the fact that cases of actual calcium deficiency in forests are virtually unheard of. This raises an interesting question - are we facing impending calcium deficiencies in many of these intensively-harvested forests, or are nutrient budget analyses faulted? In the case of calcium, as well as all other nutrients besides nitrogen, inputs to the soil available nutrient pool via weathering remain very uncertain, and may be greatly underestimated in many cases. However, the analyses of Federer *et al.* [17] suggest that even high rates of weathering will not preclude the development of calcium deficiencies in many forest ecosystems, in that calcium depletion is indicated even when soil total calcium pools are included in their analyses. Furthermore, studies in several forest ecosystems have shown significant depletion of exchangeable calcium due to calcium uptake [15, 18, 19].

FACTORS AFFECTING NUTRIENT REMOVAL IN HARVESTING

Several studies have documented the large effects of species composition on the rate of nutrient uptake and, therefore, the rate of nutrient export via WTH (e.g., [15, 16, 20, 21]). Early concerns that afforestation with conifers would lead to soil deterioration (e.g., [22]) stemmed from the belief that hardwoods improved sites because of their high rates of nutrient uptake and cycling whereas conifers caused soil acidification through podzolization. As noted by Stone [23], the early perception that hardwoods improve soils whereas conifers degrade them was based upon field observations that may reflect species site preferences more than species effects upon site quality. Indeed, nutrient budget analyses indicate that hardwoods will usually deplete soil nutrients more quickly than conifers because of the very fact that they take up *and* sequester greater amounts of nutrients [15, 20]. As shown by Alban [18], species that take up and recycle large amounts of nutrients tend to deplete subsoils and enrich litter and surface soils whereas species which take up only low amounts of nutrients cause less subsoil depletion but tend to create acid, nutrient-poor litter and surface soils. The effects of nutrient cycling on soil nutrient distribution are often not included in evaluations of WTH effects on site fertility; usually, the effects of WTH are calculated from the net increment of cations in vegetation, the recycled component being considered as having no net effect. On a whole-soil basis, this may be true, but the distribution of exchangeable cations among horizons may change considerably as a result of recycling, and such changes may have profound effects upon the chemical composition of leachate leaving the soil.

Evaluations of WTH seldom include leaching, even though leaching may be a major cause of nutrient export of base cations. For example, in comparing leaching and WTH effects in loblolly pine and mixed oak forests in eastern Tennessee, Johnson and Todd [24] found that calcium export was four times greater in the mixed oak than in the loblolly pine stand, a result expected from the literature. However, calcium leaching was considerably greater in the loblolly stand than in the mixed oak stand because of differences in soil exchangeable calcium status (which were in turn a result of differences in calcium uptake), and the overall calcium exports via leaching and harvesting at the two sites were almost identical.

The interactions between uptake and leaching of base cations deserve special consideration, especially in managed forests where intensive harvesting occurs. Cation leaching can be described as a function of purely physical and chemical processes (even though these processes are affected by other biological processes). Specifically, the leaching of a particular cation is a function of the concentration of total anions in solution, the ratio of that cation to other cations on cation exchange sites, and the selectivity of that cation for exchange sites. High base saturation soils can be expected to have high rates of base cation leaching balanced by bicarbonate, a major natural leaching agent in many forest soils [25]. Bicarbonate concentration is controlled by partial pressure of CO_2 in the soil atmosphere and soil solution pH. Soil solution pH, in turn, is regulated by base saturation and the level of mineral acid anions in solution [26]. Thus, one would expect to see a relationship between soil exchangeable base cation pools and leaching rates. As shown in Figure 1.1(b), this hypothesis is supported for the forest ecosystems described above.

In contrast to the leaching processes, uptake is driven by biological requirements, which vary considerably among nutrients, species, and with stand age and growing

Harvesting vs Exchangeable Cations

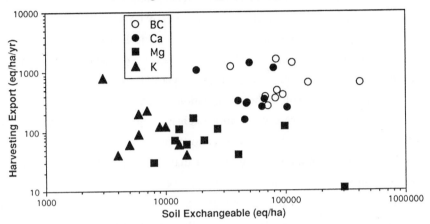

Leaching vs Exchangeable Cations

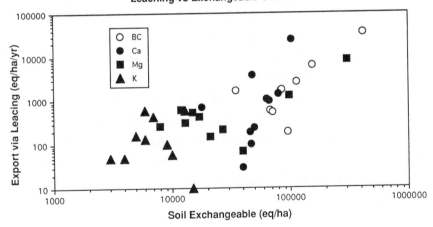

FIGURE 1.1 (a) Base cation export via harvesting *v.* exchangeable cation content and(b) base cation export via leaching *v.* exchangeable cation content in a variety of forest sites (data from [16]). (BC = Total base cations).

conditions, but are not controlled by soil exchangeable reserves. Soil exchangeable reserves may have some effect upon uptake, but by no means regulate it. Thus, one would not expect to see any particular relationship between exchangeable base cations and base cation uptake. This hypothesis is supported by the comparison of exchangeable base cations and cation export via WTH in Figure 1.1(a).

Tree uptake can result in the conservation of cations in short supply by depleting soil exchangeable reserves. Cation exchange equations dictate that a decrease in the level of a given exchangeable cation (for example, by uptake) will cause a relative decrease in the concentration of that cation and an increase in the concentrations of all other cations in soil solution (given constant anion concentrations). Thus, the depletion of a particular cation from subsoils will cause a reduction in the leaching rate of that cation and, other things being equal, an increase in the leaching rates of other cations. In short, cations in most demand by the forest are effectively conserved by depletion of soil exchangeable pools. This depletion by uptake could progress well beyond the levels possible through leaching alone, since plants selectively take up nutrients on an individual basis rather than on the basis of their availability on exchange sites, as is the case with leaching. Indeed, uptake can deplete soil pools to the extent that the system begins to accumulate one or more base cations, even with elevated inputs of H^+ from the atmosphere (some case studies of this effect are reviewed by Johnson and Richter [27]. This effect will occur whether the cation being depleted is sequestered in biomass (the only effect of uptake incorporated in most calculations of uptake effects), or recycled and accumulated in forest floor or upper soil horizons, as has been observed by Alban [18] and Johnson and Todd [24]. Thus, it is possible for recycling of cations from subsurface to surface soils to act as a very effective method of cation nutrient conservation.

EFFECTS ON SOIL ORGANIC MATTER

While nutrient budget analyses indicate that calcium depletion is often the most significant nutritional effect of intensive harvesting, effects of harvesting on soil organic matter and carbon-nitrogen interactions may, in the long run, be most important for long-term site productivity. There are several reasons for making such a statement. First, calcium, like other cation nutrients and phosphorus, can be very readily and effectively added back to the site in the form of inorganic fertilizer if either nutrient budget data or foliar analysis indicate potential deficiency. This is not the case with organic matter. Secondly, despite the indication of potential for calcium depletion from nutrient budget data, nitrogen and phosphorus remain the primary limiting nutrients in most intensively-managed forests. Organic matter will have a direct effect upon the availabilities of these nutrients, especially with respect to nitrogen. The degree of biomass removal will have a direct effect upon the C:N ratio of residues left to decompose on site; a large amount of woody residue may result in nitrogen deficiency due to incorporation of nitrogen into soil organic matter, and complete removal of biomass may result in nitrogen deficiency by removing a pool of readily-mineralizable nitrogen (in foliage and small limbs) for the new plantation. These issues are especially important to consider in future research on harvesting effects, because they are not readily evaluated with simple nutrient budget analyses. Several models of harvesting effects emphasize nitrogen and carbon-nitrogen interactions (see review by Johnson and Dale [28]), and one of the major goals of the IEA Bioenergy Project was to evaluate these models. This effort should obviously continue, along with a basic research program to gain further insights into factors controlling carbon and nitrogen retention in soils.

Soil organic matter is known to have a major influence upon site productivity because of its effects upon physical (bulk density, water holding capacity), biological (microbial populations), and chemical (cation exchange capacity) properties of soils [29]. Recently, there have also been claims as to the benefits of dead wood to forest ecosystems as well [30]. As with all nutrients, however, too much carbon in soils can have negative effects, also, especially with regard to nitrogen immobilization [31].

In addition to issues regarding site productivity, the global carbon - climate change issue has greatly heightened interest in the effects of land management practices on terrestrial carbon balances. Of particular interest is the "missing sink" of carbon, which arises from the difference in CO_2 release by fossil fuels (approximately 6 x 10^{15} g) and the annual CO_2 increase in the atmosphere (approximately 3.4 x 10^{15} g) [32]. One of the major problems in identifying this sink is that background or "natural" pools and fluxes are one or two orders of magnitude larger than both the release from fossil fuels and the observed increases in the atmosphere [33].

In that the soil contains a very large reserve of carbon in organic matter (approximately 1500-1600 x 10^{15} g, as compared to 500-800 x 10^{15} g in terrestrial vegetation and 750 x 10^{15} g in the atmosphere), there is justifiable concern over the soil carbon balance and the potential effects of human activities upon this balance. Early estimates and models of the effects of forest harvesting upon soil organic matter assumed that large (35-50%) losses of soil carbon occur following forest harvesting (e.g., [34]). Recent reviews of the literature on harvesting effects do not support these early assumptions, however. Harvesting may cause either increases or decreases in soil carbon, depending largely upon how much residue is left on site and how it is treated [35, 36]. It appears as if harvesting followed by reforestation has little or no effect upon soil carbon reserves on a global scale, however, in that the overall average effects of harvesting on soil carbon are small [35, 36]. Harvesting followed by conversion to agriculture result in a large loss of soil carbon in most cases, however [37].

When considering the effects of intensive harvesting upon site nutrient status, it is critical to separate the effects of organic matter from nutrients. The studies of intensive harvesting effects on red alder growth at Pack Forest, Washington USA, illustrate this. In these studies, forest residues were removed from experimental plots after harvesting at three levels: 1) normal bole-only harvest, 2) total above-ground tree removal, and 3) total tree plus forest floor removal. After 10 years, growth followed the patterns of residue removal: greatest in bole-only, least in to-tal tree plus forest floor. This might have led to the conclusion that organic matter was the major factor in site productivity; however, a nitrogen fertilization trial on a paired set of these plots showed that the reduced growth in the intensively-harvested plots could be completely reversed by nitrogen fertilization [38].

RESEARCH NEEDS

A coordinated, multi-site study of the effects of harvesting site productivity employing strict, common sampling protocols is badly needed. Although there are numerous very good site-specific studies of the effects of intensive harvesting in the literature, the lack of common experimental design and protocols precludes the kind of generalization needed to predict intensive harvesting effects. It is particularly important to conduct specific tests of model predictions, without the benefit of "post-mortem" calibrations. There is little doubt that models are useful in researching the effects of forest management on site productivity, but their predictive value is, as yet, unproven and their use in assessment and policy making is as yet unwarranted.

ACKNOWLEDGEMENTS

Research supported by the National Council of the Paper Industry for Air and Stream Improvement, Inc and the Nevada Agricultural Experiment Station, College of Agriculture, University of Nevada, Reno.

LITERATURE CITED

1. Weideman, E. Zuwacksrückgang und Wuchstockungen der Fichte in den mittleren und den unteren Hohenlagen der Sachsischen Staatsforstern. Tharandt, 1923. (Translation 302, U.S. Dept. of Agriculture, 1936).

2. Pritchett, W.L. and Fisher, R.F. Properties and Management of Forest Soils. John Wiley and Sons, New York, 1987.

3. Ebermeyer, E. "Die gesamte Lehre der Waldstreu mit Rücksicht auf die chemische Statik des Walbaues". Berlin, 1876.

4. Tamm, C.O. Nutrient cycling and productivity of forest ecosystems. In: *Impact of Intensive Harvesting on Forest Nutrient Cycling*. (Ed.) A.L. Leaf. State University of New York, Syracuse, 1979.

5. Likens, G.E., Bormann, F.H., and Johnson, N.M. Nitrification: Importance to nutrient losses from a cut-over forest ecosystem. *Science*, 1969, **163**, 1205-1206.

6. Cole, D.W. and Gessel, S.P. Movement of elements through forest soil as influenced by tree removal and fertilizer additions. In: *Forest Soil Relationships in North America*. (Ed.) C.T. Youngberg. Oregon State University Press, Corvallis, 1965, pp. 95-104.

7. Vitousek, P.M., Gosz, J.R., Grier, C.C., Melillo, J.M., Reiners, W.A., and Todd, R.L. Nitrate losses from disturbed ecosystems. *Science*, 1979, **204**, 469-474.

8. Mann, L.K., Johnson, D.W., West, D.C., Cole, D.W., Hornbeck, J.W., Martin, C.W., Riekerk, H., Smith, C.T., Swank, W.T., Tritton, L.M., and Van Lear, D.H. Effects of whole-tree and stem-only clearcutting on postharvest hydrologic losses, nutrient capital, and regrowth. *Forest Science*, 1988, **42**, 412-428.

9. Likens, G.E., Bormann, F.H., Johnson, N.M., and Pierce, R.S. Recovery of a deforested ecosystem. *Science*, 1978, **199**, 492-496.

10. Van Miegroet, H., Cole, D.W., and Homann, P.S. The effect of alder forest cover and alder forest conversion on site fertility and productivity. In: *Sustained Productivity of Forest Soils*. Proceedings of the 7th North American Forest Soils Conference. (Eds.) S.P. Gessel, D.S. Lacate, G.F. Weetman, and R.F. Powers. University of British Columbia, Faculty of Forestry Publication, Vancouver, Canada, 1989, pp. 333-354.

11. Weetman, G.L. and Webber, B. The influence of wood harvesting on the nutrient status of two spruce stands. *Canadian Journal of Forestry Research*, 1972, **2**, 351-369.

12. Boyle, J.R., Phillips, J.J., and Ek, A.R. "Whole-tree" harvesting: Nutrient budget evaluation. *Journal of Forestry*, 1973, **71**, 760-762.

13. Leaf, A.L. (Ed.) *Impact of Intensive Harvesting on Forest Nutrient Cycling*. State University of New York, Syracuse, 1979.

14. Johnson, D.W. The effects of harvesting intensity on nutrient depletion in forests. In: *IUFRO Symposium on Forest Site and Continuous Productivity*. (Eds.) R. Ballard and S.P. Gessel. USDA Forest Service General Technical Report PNW-163, Portland, Oregon., 1983, pp. 157-166.

15. Johnson, D.W., Henderson, G.S., and Todd, D.E. Changes in nutrient distribution in forests and soils of Walker Branch Watershed, Tennessee, over an eleven-year period. *Biogeochemistry*, 1988, **5**, 275-293.

16. Johnson, D.W., Kelly, J.M., Swank, W.T., Cole, D.W., Van Miegroet, H., Hornbeck, J.W., Pierce, R.S., and Van Lear, D. The effects of leaching and whole-tree harvesting on cation budgets of several forests. *Journal of Environmental Quality*, 1988, **17**, 418-424.

17. Federer, C.A., Hornbeck, J.W., Tritton, L.M., Martin, C.W., Pierce, R.S., and Smith, C.T. Long-term depletion of calcium and other nutrients in eastern U.S. Forests. *Environmental Management*, 1989, **13**, 593-601.

18. Alban, D.H. Effects of nutrient accumulation by aspen, spruce, and pine on soil properties. *Soil Science Society of America Journal*, 1982, **46**, 853-861.

19. Binkley, D., Valentine, D., Wells, C., and Valentine, U. An empirical model of the factors contributing to 20-yr decrease in soil pH in an old-field plantation of loblolly pine. *Biogeochemistry*, 1989, **8**, 39-54.

20. Marion, G.K. Biomass and nutrient removal in long-rotation stands. In: *Impact of Intensive Harvesting on Forest Nutrient Cycling.* (Ed.) A.L. Leaf. State University of New York, Syracuse, 1979, pp. 98-110.

21. Cole, D.W. and Rapp, M. Elemental cycling in forest ecosystems. In: *Dynamic Properties of Forest Ecosystems.* (Ed.) D.E. Reichle. Cambridge University Press, London, 1981, pp. 341-409.

22. Rennie, P.J. The uptake of nutrients by mature forest growth. *Plant and Soil*, 1955, **7**, 49-95.

23. Stone, E.L. Effects of species on nutrient cycles and soil change. *Phil. Trans. R. Soc. Lond. Ser.*, 1975, **B271**, 149-162.

24. Johnson, D.W. and Todd, D.E. Nutrient export by leaching and whole-tree harvesting in a loblolly pine and mixed oak forest. *Plant and Soil*, 1987, **102**, 99-109.

25. Johnson, D.W., Cole, D.W., Gessel, S.P., Singer, M.J., and Minden, R.V. Carbonic acid leaching in a tropical, temperate, subalpine, and northern forest soil. *Arctic and Alpine Research*, 1977, **9**, 329-343.

26. Reuss, J.O. and Johnson, D.W. Acid deposition and the acidification of soil and water. Ecological Studies No. 59. Springer-Verlag, New York, 1986, 118 p.

27. Johnson, D.W. and Richter, D.D. Effects of atmospheric deposition on forest nutrient cycles. *TAPPI Journal*, 1984, **67**, 81-85.

28. Johnson, D.W. and Dale, V.H. Nitrogen cycling models and their application

to forest harvesting. In: Proceedings, IEA/BA Task II Workshop, Production, Technology, Economics and Nutrient Cycling, Kingston, Canada, May 20-23, 1986. IEA/ENFOR/ OMNR Joint Report, Canadian Forestry Service, 1987, pp. 27-36.

29. Chen, Y. and Aviad, T. Effects of humic substances on plant growth. In: *Humic Substances in Soil and Crop Sciences: Selected Readings*. (Eds.) P. McCarthy, C.E. Clapp, R.L. Malcolm, and P.R. Bloom. American Society of Agronomy, Madison, WI, 1990, pp. 161-186.

30. Maser, C., Tarrant, R.F., Trappe, J.M., and Franklin, J.F. (Eds.). *From the Forest to the Sea: A Story of Fallen Trees*. General Technical Report PN W-GTR-229, US Dept of Agriculture, Forest Service, Portland, OR, 1988.

31. Bollen, W.B. Soil microbes. In: *Environmental Effects of Forest Residues Management*. (Ed.) O.P. Cramer. USDA Forest Service General Technical Report PNW-24. Pacific Northwest Forest and Range Experiment Station, Portland, Oregon, 1974, pp. B1-B41.

32. Lugo, A.E. The search for carbon sinks in the tropics. *Water, Air, and Soil Pollution*, 1992, **64**, 3-9.

33. Post, W.M., Peng, T-H, Emmanuel, W.R., King, A.W., Dale, V.H., and DeAngelis, D.L. The global carbon cycle. *American Scientist*, 1990, **78**, 310-326.

34. Houghton, R.A. and Woodwell, G.M. Global climatic change. *Scientific American*, 1989, **260**, 36-44.

35. Detwiler, R.P. Land use change and the global carbon cycle: the role of tropical soils. *Biogeochemistry*, 1986, **2**, 67-93.

36. Johnson, D.W. The effects of forest management on soil carbon storage. *Water, Air, and Soil Pollution*, 1992, **64**, 83-120.

37. Mann, L.K. Changes in soil carbon storage after cultivation. *Soil Science*, 1986 **142**, 279-288.

38. Compton, J.E. and Cole, D.W. Impact of harvest intensity on growth and nutrition of successive rotations of Douglas-fir. In: *Long-term Field Trials to Assess Environmental Impacts of Harvesting*. Proceedings IEA/BE T6/A6 Workshop, Florida, USA, February 1990. (Eds.) W.J. Dyck and C.A. Mees. IEA/BE T6/A6 Report No 5. Forest Research Institute, Rotorua, New Zealand, FRI Bulletin No 161, 1991, pp. 151-161.

CHAPTER 2

STRATEGIES FOR DETERMINING CONSEQUENCES OF HARVESTING AND ASSOCIATED PRACTICES ON LONG-TERM PRODUCTIVITY

W.J. DYCK

New Zealand Forest Research Institute
PO Box 31011, Christchurch, New Zealand

D.W. COLE

College of Forest Resources, University of Washington
Seattle, WA 98195, USA

INTRODUCTION

Forest harvesting removes nutrients in biomass and, along with site preparation operations, may remove or displace nutrients contained in logging slash and the forest floor. Considerable soil disturbance may also occur resulting in nutrient loss and often soil compaction. Sites vary in their resilience to disturbance, depending not only on inherent site factors, including climate and soil properties (i.e., site quality), but also on the intensity of the forestry operation and site conditions at the time of disturbance.

Emphasis on sustainability of forest production is increasing as world demand for wood expands (currently increasing at approximately 80 million m^3/yr) and greater areas of natural forest are being placed under increasing environmental constraints. Land currently designated for "production" will have an expanding role to play in supplying wood fiber as large forested areas are reserved for non-fiber production uses.

For forest managers and policy makers to be confident that production forestry is sustainable requires that the impacts of forestry practices on site productivity be fully understood. Of primary concern is the impact of harvesting and site preparation operations on nutrient removal and soil physical properties that regulate tree growth. However, both understanding the present and predicting the future requires that any successful research strategy takes an holistic approach and considers wood production in relation to both the site and to external forces.

Site Quality and Productivity

Forest managers are primarily interested in maintaining or improving forest productivity, which generally equates to wood production. Productivity is determined by climate and inherent soil properties (site quality) but is also affected by management (Figure 2.1) [1]. Use of improved tree breeds, weed control, and fertilization all improve productivity, whereas inappropriate harvesting and site preparation operations that remove nutrients or seriously compact the soil may have a negative impact on productivity, and also on site quality.

Whereas productivity can be readily manipulated by management, the challenge for forestry researchers is to determine if site quality has changed and to be able to confidently predict likely impacts at other sites. Although seemingly straightforward, in reality it is difficult to determine if a change in site quality has

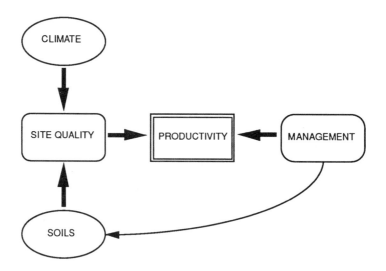

FIGURE 2.1 The relationship between productivity, site quality, and management practices (from [1]).

occurred because of the time frame involved and the potential for confounding effects (Chapter 3). It is even more difficult to make predictions of future impacts, and existing computer simulation models are inadequate for making predictions beyond general trends (Chapter 6).

Research Strategies

Considerable research has been initiated to address concerns over long-term site productivity, however, for various reasons not all attempts have been successful, and many projects have produced misleading results. Shortcomings of many studies have been failure to account for possible confounding influences, and inappropriate design -many studies were designed for other purposes (e.g., watershed studies).

Earlier predictions of future productivity declines resulted from nutrient budget studies that simply estimated the total amount of nutrients contained in an ecosystem and natural nutrient inputs and compared this to the amount calculated to be removed at harvest. Despite predictions of future reductions in productivity for a number of regions, there is little evidence that productivity decline has occurred as a consequence of biomass removal alone [2, 3]. Moreover, with a few exceptions such as "second-rotation decline" in South Australia [4], there is little evidence that productivity decline has occurred as a consequence of even fairly major site disturbances, such as slash burning on nutrient-poor soils. It is difficult to determine if a decline in site quality has occurred at some of these sites but has simply not been detected because insensitive research methodologies have been used [1]. Where growth reductions have been attributed to biomass removal only, it has been because productivity losses were very large and therefore obvious [5], or because well-designed trials were established that could detect relatively small changes in growth between treatments [6].

Above-ground productivity is generally used as an indicator of site quality, and measurements of wood production provide a convenient, as well as relevant indicator of the productive capacity of a site. However, wood production reflects more than just site quality because it is also influenced by management practices (Figure 2.1). In intensively managed plantations, faster growing breeds of trees are planted at the start of each rotation, and herbicides and starter doses of fertilizers are routinely applied to give seedlings an initial boost in growth. Thus, changes in productivity may not be reflecting changes in site quality. Furthermore, fluctuations in climate may obscure comparisons of growth from one rotation to the next.

Strategies for determining the consequences of harvesting and associated practices on long-term site productivity (site quality) can be considered in three categories: chronosequence studies, retrospective studies, and long-term field trials. Each strategy has its relative merits and these must be considered before any strategy is applied. Failure to realize pitfalls may result in misleading conclusions as well as wasted resources.

CHRONOSEQUENCE STUDIES

Concept of Chronosequences

Chronosequences, as a research approach, can provide the investigator with a rapid means of establishing the nature and direction of long-term changes in ecosystem properties as a result of management practices without having to wait the necessary time for the changes to take place.

A chronosequence represents an ecological time series where, to the best judgment of the investigator, the differences observed between the various ecosystems that comprise the sequence have been brought about by differences in age or time and not by environmental differences within the time series. It is assumed in this technique that all ecosystems within the sequence were identical at time zero and since then have not been selectively affected by biological factors that changed the pattern of ecosystem development, such as diseases, pathogens, or the activities of man such as harvesting, burning, and fertilization. It is also assumed that climate does not normally change within the time period covered by the chronosequence and is similar for all the sites used for the comparison. If these assumptions are not met then any relationships defined by the chronosequence may be either incorrect or incorrectly defined.

Clearly it is difficult to meet all of these criteria. Furthermore, for chronosequences that span a long period of time, the lack of adequate documentation on climate, diseases, and other factors of the environment makes it improbable if not impossible to know whether the criteria have been met. However, should they be

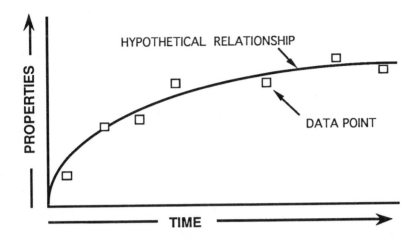

FIGURE 2.2 Hypothetical relationship between ecosystem properties and time as defined through the use of a chronosequence [7].

met, then relationships as illustrated in Figure 2.2 can be developed. For chronosequences that span only a short period of time, the variance associated with the data can result in misleading conclusions, a point that will be discussed in greater depth later in the chapter.

Strengths and Weaknesses

In spite of such uncertainty with the technique, chronosequences play an important role in ecosystem research, and have advantages over alternative research approaches. These advantages include:

1. *Contraction of time*: The most obvious advantage of a chronosequence in ecosystem research is that it allows one to assess ecosystem changes over time without having to wait for the changes to take place. This consideration becomes increasingly critical the longer the time period under study. In some cases the time period required to follow an experiment through to completion, such as a study in soil development or plant succession, would exceed our own tenure as researchers. Using a chronosequence as the basis for following a time series rather than time itself can collapse an experiment into a manageable duration.

2. *Defining the relationship*: As a chronosequence is made up of a number of separate observations representing different points in time, it is possible to define how the system properties changes over time. By knowing the form of this functional relationship, one can speculate about changes that take place with time. Interpolation between data sets allows us to estimate the nature of the relationship at other points in time.

3. *Model testing*: Most ecosystem models are seldom tested with data derived independently from those used in the initial construction of the model. Potentially, chronosequences could play a role in model testing by providing an independent analysis of ecosystem change over time.

The technique is not without its problems, the most critical of which involve basic underlying assumptions built into the concept itself. These assumptions include:

1. *All points within the chronosequence have experienced similar climatic conditions including temperature and precipitation.* For valid comparisons there should not be a shift in the climate over the real time period covered by the chronosequence.

2. *Biotic factors not included in the experimental design (such as a previous pathogen or insect attack), have not selectively affected certain plots within the chronosequence.* If one or more of the plots of the sequence were disturbed by some external influence, this would obviously distort the time series relationship making it appear as if the dependent variable of the experiment, the property under investigation, was behaving erratically within the time series.

3. *The ecosystem properties of each site within the sequence, including soils, slope, aspect, vegetative composition, biomass accumulation, and structure, were reasonably similar to each other at time zero.* This is seldom the case, and there is often a considerable site-to-site variability in the initial conditions. This variability can lead to potentially misleading relationships as

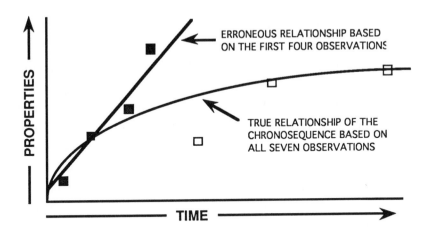

FIGURE 2.3 Effect of insufficient length of the time series on the relationships inferred from the chronosequence data [7].

well as different conclusions from those reached by other long-term studies examining the same relationship.

4. *The sites represent the entire time span of the chronosequence.* This is particularly critical if there is a great deal of variance within the data sets of the chronosequence or if the pattern early in the chronosequence differs significantly from that further on. Data derived from only a short period of the time series can potentially give a very misleading representation of the true chronosequence pattern, as illustrated in Figure 2.3.

5. *There is seldom replication between plots of similar age within the chrono-sequence.* This presents a conceptual problem in the experimental design as to the effect of outlying data points on the shape and the strength of the time-property relationship. This can be tested through influence diagnostics (e.g., Cook's D) which measure the change to the estimates that results from deleting each observation separately [8].

In spite of these inherent difficulties, chronosequences have served a valuable role in ecosystem research and also have application in long-term site productivity research. The successful use of this technique requires that researchers fully understand the shortcomings of the procedure and design the study accordingly. In considering these difficulties, it is apparent that the chronosequence has its greatest utility when changes in relatively simple relationships are being followed for long time periods. The larger the number of plots established within the sequence, the smaller is the likelihood that the investigator will be misled by a single errant observation. If enough sites are sampled, if the initial site-to-site variability is reasonable, and if the sites sampled are representative of the entire time series under consideration, this technique should capture the general pattern of the relationship under study.

Case Examples Involving Chronosequences Studies
Chronosequences have been extensively used to assess long-term changes in ecosystem properties. They have been used to a lesser extent to assess changes brought about by forest management practices including harvesting and associated practices. Examples illustrating both long-term ecological changes as well as changes caused by management practices are discussed below.

Evaluating long-term changes in forest ecosystem properties
1. *Plant Succession*: In that the time frame of plant successional changes is usually very long and the changes quite dramatic, a chronosequence is an ideal means by which studies of this type can be carried out. As discussed above, the longer the time frame and greater the change taking place, the less critical it becomes that the other environmental factors are similar or the sites were all identical at time zero. The early recognition of plant succession by plant ecologists, including the concepts of pioneering and climax species, were clearly the result of observations taken along a time series. More contemporary examples of successional studies include the study by Crocker and Major [9] who used chronosequences very effectively in their study of plant community development on deglaciated surfaces at Glacier Bay, Alaska. Christensen and Peet [10] studied the successional convergence in species

composition for forest stands in the Piedmont region of North Carolina. Heilman [11] used chronosequences to follow changes in sphagnum peat development on sites formerly occupied by productive forests.

2. *Biomass Accumulation*: The basic data used in the construction of most growth and yield tables are essentially derived from chronosequence measurements. Chronosequences have also been effectively used to examine successional changes in both overstorey and understorey species composition and biomass accumulation during stand development. This technique is particularly well suited for this type of analysis in that many of the difficulties with the use of chronosequences as stated earlier are eliminated. At time zero there is no residual accumulation of biomass to confound the time series, the properties are easily measured, and the time series (age of the sites) is easily established.

   The research by MacLean and Wein [12] on jack pine (*Pinus banksiana*) and mixed hardwood forests of New Brunswick; by Ruark and Bockheim [13] for trembling aspen (*Populus tremuloides*) in Wisconsin; by Sprugel [14] for wave-regenerated balsam fir (*Abies balsamea*) on Whiteface Mountain, New York; by Long and Turner [15] on Douglas-fir (*Psuedotsuga menziesii*) forests in Washington; and by Madgwick *et al.* [16] for radiata pine (*Pinus radiata*) in New Zealand are examples of such studies. Figure 4 illustrates foliage and tree biomass accumulation during second growth Douglas-fir stand development, and Figure 2.5 illustrates the shading out of understorey species with the closure of the forest canopy. In both cases the chronosequences have clearly demonstrated the effect of stand age on biomass accumulation and distribution without the need to actually follow these changes over time.

3. *Changes in Mineral Cycling*: Perhaps the most extensive literature on the use of chronosequences in ecosystem research is associated with studies in mineral cycling. The systematic change of ecosystem properties such as nitrogen and organic matter content are far easier to establish with a chronosequence than fluxes associated with these same properties. This is primarily because fluxes are poorly buffered by the system and, thus, can change significantly

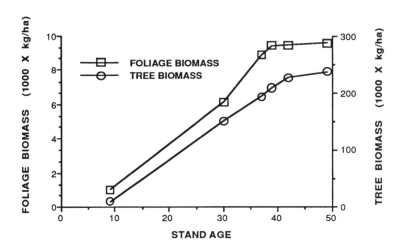

FIGURE 2.4 A chronosequence of foliage and aboveground tree biomass
accumulation for site III Douglas-fir [15].

from year to year, month to month, or day to day. In addition, they do not
necessarily change in the orderly and directional way found in the examples
discussed above. Ovington [17] published perhaps the first article using
chronosequences in mineral cycling studies in an article titled "The
Circulation of Minerals in Plantations of *Pinus sylvestris* L." (Figure 2.6).
Other early examples where chronosequences were used in mineral cycling
studies include those by Switzer and Nelson [18] for loblolly pine (*Pinus
taeda*), and by Forrest and Ovington [19] for radiata pine (*Pinus radiata*).

While the return of nutrients, as derived through a chronosequence,
demonstrates strong and consistent patterns, deriving uptake values is somewhat less consistent and certainly includes far more apparent yearly variation
as can be seen for scots pine [17] and Douglas-fir [20]. This increase in
variation occurs because uptake can not be directly measured - rather it is
calculated from several ecosystem properties all of which are difficult to
derive.

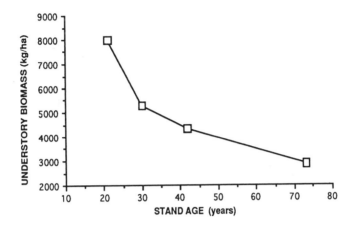

FIGURE 2.5 A chronosequence of understorey biomass accumulation by second growth Douglas-fir [15].

Effect of Management Practices on Ecosystem Changes

Chronosequences have been used far less frequently to assess the effects of management practices on ecosystem properties, probably for several reasons:

1. the spatial variability inherent within the ecosystem is further increased when subjected to the added variability associated with a management practice; and

2. there is difficulty in distinguishing a particular management effect from other factors.

However, because growth and yield tables are essentially constructed from chronosequences, it is important for mensurationists to realize that these models implicitly include the effects of any harvesting and site preparation operations that were applied in the original chronosequence. If site quality has been altered, growth models constructed for first-rotation sites can not be expected to accurately predict second-rotation growth (e.g., South Australia).

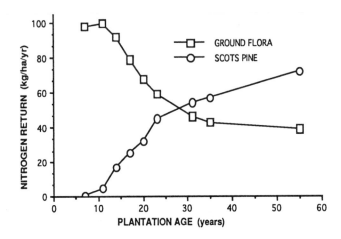

FIGURE 2.6 Annual return of nitrogen by the tree and ground flora in a Scots pine plantation. Adapted from Ovington [17].

RETROSPECTIVE STUDIES

Concept of Retrospective Studies

In a retrospective study, as with a chronosequence study, the investigator is attempting to establish the nature and direction of long-term changes in ecosystem properties and management practices without waiting for the changes to take place. In the case of a retrospective study, this is accomplished by analyzing information from areas or plots that had previously received different treatments to see if the treatments have had an effect on the present situation. The main requirements for a successful retrospective study are that the initial treatments were documented in some way, and that other than the differences caused by the initial treatments, stand development for the areas being studied is not affected by other factors which would confound results and interpretation. This concept, utilizing retrospective observations for determining changes in stand development with time, is illustrated in Figure 2.7.

Ideally, retrospective studies should be conducted on existing experiments for which there is at least initial documentation of site conditions and treatments, but this is not always possible (e.g., [21]). Although the research objectives of the initial experiment and the retrospective study can potentially be very different, the

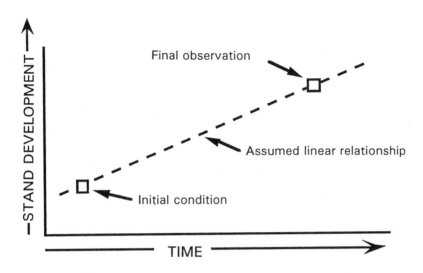

FIGURE 2.7 Changes in stand development established through a retrospective study. The line passing through the two observations represents the hypothetical change of stand conditions over time as defined through the use of a retrospective study.

stronger the similarity in their basic experimental designs, the greater the potential for conclusions reached to be scientifically sound. In addition, it is critical that enough time has passed between the initial measurements and those taken in the retrospective study for the changes to have occurred.

Installation of retrospective studies outside existing experiments is often hampered by a lack of a "control" to compare to the treatment response. The absence of a control greatly increases the uncertainty that results accurately represent the true situation.

Strengths and Weaknesses

The primary strengths of a retrospective study are the same as those discussed for a chronosequence. Long-term changes in various ecosystem properties and management practices can be established without having to wait the necessary time for the changes to take place. This is important when an understanding of the effect of some management practice on the long-term productivity of a forest stand is in question. This consideration becomes critical for longer time periods. Using a retrospective study as the basis for following a time series rather than time itself can collapse an experiment to a more manageable duration.

As in a chronosequence, a retrospective study can also provide valuable and often cost-effective information needed in the design of long-term empirical experiments. Such results can provide a far better idea on the magnitude of the changes to be expected, how the long-term experiment should be established, and the duration that it should run.

Weaknesses associated with this approach include:

1. Adequate documentation should be available concerning the initial conditions of the forest stand. In many retrospective studies a true control treatment will not exist.

2. The stand should not be affected by ancillary factors such as insect and disease problems or climatic factors such as wind damage and temperature extremes. Such factors confound the interpretation of the observations.

3. As only two points typically occur in the time series, a linear relation with time must be assumed. The true rate of change at any given point in time is not known, only the average rate over the entire time series. It is possible, in cases dealing with growth and yield studies, to reconstruct a yearly growth function through a stem analysis in which annual rings or leader growth rates have been measured.

4. Because of potential documentation and experimental design limitations associated with the initial stand conditions it can be difficult to provide an adequate statistical treatment of the results.

Examples of Retrospective Studies

Retrospective studies have played a very important role in long-term forestry research. In particular they have been used to examine the effect of major impacts such as burning or mechanical site preparation. They have less application for investigating more subtle impacts such as different intensities of forest harvesting.

Windrowing: As previously discussed, it is preferable to conduct a retrospective study on an existing trial design than in situations where information on initial conditions is lacking. Ballard [21] and later Dyck *et al.* [22] investigated the impacts of windrowing on growth of second-rotation radiata pine in Kaingaroa Forest, New Zealand, in an area windrowed following the clearfelling of a first-rotation radiata pine stand. Conveniently, only part of the compartment was windrowed and the remainder planted with no site preparation. Volume production in the windrowed areas at both ages 7 [21] and 17 [22] was reduced by approximately 40% as a consequence of slash and topsoil displacement, primarily a nutritional effect. Despite potential problems associated with this type of study, the authors are confident that results (Figure 2.8) provided a good indication of the effects of windrowing treatment on tree growth.

Mechanical Site Preparation and Burning: A randomized block experiment, established in 1977 on the Canterbury Plains, New Zealand, had two main objectives: a) to test the cost effectiveness of various site preparation methods, and b) to examine the effects of different site preparation methods on initial tree survival and growth to age 13 years [23]. Site preparation treatments were replicated three times and included: broadcast burn, windrow and burn, windrow, line blade, and slash retention. Weed competition was uniform at the start of the study because of chemical weed control. Initial growth measurements were made to age 3 years, after which the trial was temporarily abandoned. Broom (*Cytisus scoparius*) became established throughout the trial area by 1983, but competition was not uniform across treatments [22].

At tree age 10 years the study was revisited with a new objective, which was to determine the effect of the original site preparation operations on medium-term productivity. Results at age 10 years indicated that slash retention on these dry, nitrogen-limited sites can significantly improve tree growth relative to other site preparation treatments (Figure 2.9) [23]. This result is considered to be somewhat conservative because weed competition was greatest for the slash retention plots.

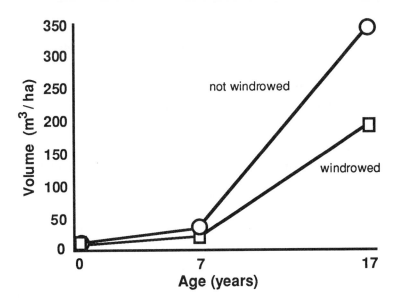

FIGURE 2.8 Retrospective analysis of the effect of windrowing on radiata pine volume production at ages 7 and 17 years [22].

Although the study was not ideal because the trial was abandoned for 7 years and then analyzed retrospectively, the results have been used to alter the way forests are now managed on the Canterbury Plains (J. Balneaves pers. comm.).

Burning: The opportunity to conduct a comprehensive retrospective study on the impacts of burning on long-term site productivity was provided by a slash burning study installed by W.G. Morris [24] between 1946 and 1952. The original objectives of the Morris study were to investigate the effects of burning old-growth slash on fire hazard and conifer regeneration using paired plots in a randomized block design. Miller *et al.* [25] reestablished plots (burned and unburned) at 44 locations, most of which were revegetated in Douglas-fir. The objectives of the later study were to determine the effect of burning on the seven plant associations represented on the sites and to construct a model capable of predicting effects of burning on volume growth. Determining the effects of burning on productivity was seriously confounded by weed competition, which varied

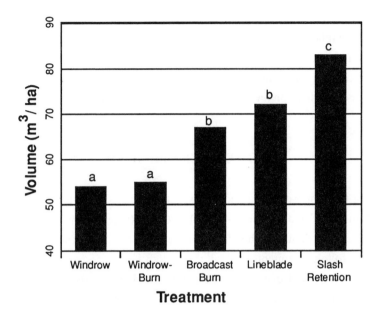

FIGURE 2.9 Site preparation effects on radiata pine volume production at age 10 years [23].

considerably across sites [25]. This study emphasized one of the major short-comings of retrospective studies which is the lack of control by the researcher over potentially confounding factors.

Harvesting: In a review of the strengths and weaknesses of retrospective re-search, Powers [26] critiqued the study of Helms *et al.* [27] in which he examined the effects of soil compaction on the growth of a ponderosa pine (*Pinus ponderosa*) plantation in California. Helms *et al.* used stem measurements to reproduce tree growth trends and regressed tree height against current soil bulk density. Soil compaction was reported to have reduced height growth in the first years after planting, and by age 15 years, trees growing in the soils of highest bulk density (ranging up to 1.27 g/m^3) were 13% shorter than trees in soils of lower bulk density (down to 0.68 g/m^3).

This study highlighted some of the problems associated with retrospective studies [26]. Incorrect deductions may have been made because present conditions may not have reflected the "causal" conditions of the past. For example, soil bulk density is likely to have decreased because of tree growth and other factors. Ideally, initial soil bulk density should have been measured at the time the trees were planted. Replication on other sites would have improved the study [26].

LONG-TERM EXPERIMENTS

In contrast to chronosequence and retrospective studies, long-term experiments start at "time zero" and have long-term objectives. Rather than taking a snap shot of the present and deducing the effects of the past, long-term trials start with the present and follow treatments through time.

It is now apparent that other than for developing hypotheses, retrospective analyses of harvesting/site preparation impacts are inadequate and well-designed long-term experiments are required to provide conclusive results. Few countries have invested in long-term experiments to investigate the impacts of harvesting practices on site productivity, possibly because it was considered that any significant declines in productivity would be obvious, particularly through retrospective analysis, but also because funding philosophies have favoured short-term studies.

Sweden has been an exception, as long-term studies were installed up to 60 years ago, although not always with appropriate statistical design or objectives directly relevant to long-term site productivity [28].

In countries such as Australia and New Zealand, where intensive plantation management is practised based on relatively short rotations, site productivity issues have attracted considerable attention since Keeves [29] reported large growth reductions in second-rotation radiata pine in southern Australia. The response in Victoria was to install field trials to establish the cause of "second-rotation decline" and to alter site preparation practices to conserve organic matter and nutrients [30]. In South Australia, the response was to apply fertilizer to improve growth using the "Maximum Growth Sequence" regime [31].

New Zealand has not experienced second-rotation decline on the Australian scale, but in the 1960s reductions in productivity caused by biomass removal alone were reported on some nutrient-deficient sites [5]. The response in this case was simply to apply fertilizer, which is common practice on many New Zealand soil types even at the start of the first rotation [32]. Although harvesting impacts were generally investigated retrospectively, the establishment of a series of long-term field experiments has recently been initiated.

Design of Long-term Experiments

Scientifically-designed experiments are the norm for short-term forestry research projects and it is logical that the same principles that are used to guide the design and implementation of short-term studies should also apply to long-term site productivity research. As for any scientific experiment there must be clearly defined hypotheses to test and the design must be statistically valid. Long-term experiments differ from shorter-term ones primarily in time frame. The implications of maintaining an experiment that remains relevant to science and management for a rotation length or more demands flexibility in design to enable new hypotheses to be tested if the original hypotheses become irrelevant.

Long-term experiments will invariably use tree growth as an indicator of productivity and site quality. Therefore, treatment plots must be of an adequate size and contain a sufficient number of trees to avoid problems associated with edge effect between treatments and to ensure that adequate numbers of trees are available for measurement at the end of the experiment. Details of design criteria for long-term experiments are provided in Chapter 9.

Long-term field trials provide the opportunity to evaluate hypotheses related to ecosystem processes in the context of longer-term study and supporting base information. A primary objective of site productivity research should be to improve prediction.

One of the most critical factors to the success of long-term experiments is commitment, from both scientists and funding institutions. This is particularly important during the establishment and early stages of the experiment when, if not controlled, extraneous factors may confound results. However, it is also critical to ensure that interim results from process studies and early growth trends are captured before such opportunities are lost.

Strengths and Weaknesses

Compared to retrospective analyses (i.e., retrospective and chronosequence studies) the strengths of long-term experiments are enormous. In particular, the scientist has the opportunity for complete control over treatments and statistical design, and over potentially confounding factors such as weed competition. Although long-term, most experiments will produce valuable results from a very early stage, particularly if they include process studies.

The major weaknesses of long-term experiments are the long time frame and the high costs involved. By comparison, retrospective studies can provide immediate answers at a small fraction of the cost of a long-term study. However, if the long-term experiment has been designed correctly and confounding factors have either been controlled or fully accounted for during the experiment, then the scientist should have confidence in the results. A similar degree of confidence is not possible with retrospective studies.

Long-term experiments also require "vision" and a major commitment from scientists and institutions along with flexibility in design to incorporate unexpected shifts in research emphasis. Without this, long-term trials may still produce useful results, but are unlikely to yield to their full potential. There are numerous examples of long-term studies where the commitment has left with the researcher upon retirement.

Examples of Long-term Experiments

Long-term Nutrition Experiments: The Regional Forest Nutritional Research Program (RFNRP) was initiated by the University of Washington and sponsored by the forest industry and agencies of the Pacific Northwest to assess the forest nutritional problems of this region. As a part of this program the growth response of Douglas-fir and hemlock to N fertilization has been systematically followed for over 20 years. This long-term forest fertilization response study includes within its experimental design six major Douglas-fir and two hemlock provinces and 277 installations (total of 1905 plots). A part of this long-term study was designed to establish appropriate levels of fertilization as well as the frequency of application. At each installation 3 paired plots were established and treated with 0, 220, and 440 kg N/ha as urea. At year 8, the paired plots were split with one half refertilized at the rate of 220 kg N/ha. At years 12 and 16 these plots were again refertilized with 220 kg N/ha. At year 20 the plots were remeasured but not refertilized [33].

Subsequent to this initial series of plots, additional installations have been made to determine the effect of site quality and spacing on the duration and extent of the growth response. Results from this program have been invaluable in establishing forest fertilization guidelines in the Pacific Northwest.

Second-rotation Decline - Matched Sites: Concern over the demonstrated potential for second-rotation (2R) decline triggered the establishment of "1R/2R" experiments in the state of Victoria, Australia. The trials were designed to compare second-rotation growth of radiata pine to first-rotation growth on the same site, and/or for matching first-rotation growth on "native" forest sites [4, 30]. Comparison on the "same" sites relied on stem sectional analysis of 1R trees to reconstruct growth patterns. Comparison between 2R and ex-native 1R was considered necessary to avoid potential problems that might have been caused by differences in climate for 1R and 2R stands on the same site. Results from these studies (see Chapter 3) indicated that 2R decline was avoidable provided that nutrients were conserved on poorer sites.

Biomass Removal: Sweden has a history of long-term experiments, some dating back to the 1920s [28]. However, most of the long-term experiments were not designed specifically to examine the impacts of harvesting practices on long-term site productivity, but, because of flexible design, could serve dual purposes. Results from these earlier studies have shown that for some sites, biomass and nutrient removal associated with whole-tree thinning resulted in up to 20% reductions in volume production after 50 years [6].

More recently, well-replicated studies have been installed to investigate the potential impacts of intensive harvesting and whole-tree thinning operations on site productivity. These studies generally consist of two treatments (e.g., stem-only *v*. whole-tree removal) and four replications. Because nitrogen availability is considered to be potentially limiting to growth on many of these sites, nitrogen mineralization studies have been included in the experimental design (H. Lundkvist, pers. comm.) Results from these studies are discussed in Chapter 3.

Biomass Removal Plus Amelioration: To provide information on the sustainability of production forestry on major New Zealand soil types, a series of "intensive harvesting" studies was initiated in 1986. As of 1991, experiments had been installed on six major soil types. The experiments were designed to test two primary hypotheses: a) that radiata pine productivity was negatively correlated with increasing biomass removal at harvest; and b) that the reason for this was nutritional. The objective of the series was not only to provide information on the impact of current practices on productivity, but also to determine if productivity could be improved by altering silvicultural practices.

Other than the core treatments, design was flexible so that additional questions could be addressed where appropriate. For example, subplots were used to investigate compaction effects at three of the trial sites. Process studies, including nitrogen and phosphorus mineralization and uptake, leaf area production, photosynthetic efficiency, and water use were conducted at some sites as appropriate. One study site also incorporated a genetic overlay of various radiata pine families to investigate the interaction of genetics with site factors (M. Skinner, pers. comm.). Treatments were replicated either three or four times, depending on expected variability, and plots split so that one half of each main plot received fertilizer. This was done to separate nutritional from moisture effects. Complete weed control was maintained during the experiment as this component was considered critical to the success of the trial.

Each experimental site was designed to run for one rotation of radiata pine (30 years). Further design details are available in Dyck *et al.* [34] and Smith *et al.* [35].

Biomass Removal Plus Species Replacement: Similar in concept to the second-rotation pine experiment described above, Compton and Cole [36] examined second-rotation decline of Douglas-fir following three levels of harvest removal: stem-only, whole-tree, and whole-tree plus forest floor. This study was established in 1979 on recently harvested low site quality (24 m) and high site quality (34 m) stands of 55-year-old Douglas-fir. Duplicate 20-m^2 treatment plots were established at each site for all three levels of harvest removal. The sites have been systematically remeasured every two years with 10-year summary results reported in Compton and Cole [36]. After 4 years there appeared to be a systematic trend of growth reduction for all three treatments at both the high and low site quality areas. This decrease in production appeared to be directly related to the amount of biomass (and thus nitrogen) removed. To determine if this decrease in productivity was caused by the loss of nitrogen, one set of the duplicate plots were fertilized at year 5 (1984) with 200 kg N/ha applied as urea. Some clear trends have already emerged. Harvest removal can have a detrimental effect on second-rotation Douglas-fir production with the greatest effect occurring at those sites receiving the largest loss in biomass and nitrogen. As noted in the New Zealand study, this loss in productivity was corrected with the appropriate replacement of the deficient nutrient.

SUMMARY

Chronosequence and retrospective research strategies have similar strengths and weaknesses (Table 2.1) which are in direct contrast to those for long-term experiments. In particular, the considerably shorter time required to achieve results, together with lower cost, and lower level of commitment required, makes the former strategies appear very attractive. However, heavily stacked against these strengths are very large disadvantages including lack of control over initial and subsequent conditions, inability in most cases to apply inferential statistics, and generally much lower credibility of results.

Whereas chronosequence and retrospective approaches are very useful for developing hypotheses, and also useful, if carefully applied, for testing ecosystem model predictions, they are generally unsatisfactory for testing hypotheses related to long-term site productivity. This is mainly because the investigator has no control over conditions at the time "treatments" were applied, or over external factors, such as weed competition or insect attack, that may confound results.

The advantages of long-term experiments are very large, and in fact, this strategy is the only one capable of producing results with a high degree of credibility. However, the benefits of long-term experiments will only be realized if the researcher has an accurate vision as to future requirements, if the experiments are

correctly designed and implemented, and if there is a long-term commitment to
continue the studies through to a logical conclusion.

TABLE 2.1

Strengths and weaknesses of the three long-term site productivity research
strategies.

	Type of Long-term Research Strategy		
	Chronosequence	Retrospective	Long-term Experiment
Time frame	+	+	−
Cost	+/−	+	−
Commitment	+	+	−
Require "vision"	−	−	+
Initial control	−	−	+
Credibility	+/−	+/−	+
Opportunity for adequate statistical treatment	−	−	+
Repeatability	−	−	+
Confounding by external factors	−	−	+
Ease of locating sites	−	−	+
Usefulness for:			
a) testing models	+/−	+/−	+
b) developing hypotheses	+	+	+
c) testing hypotheses	−	−	+
Assist management	+	+	+
Opportunity for process studies	−	−	+

+   Advantageous
−   Inappropriate
+/−   Potentially either

LITERATURE CITED

1. Dyck, W.J. and Cole, D.W. Requirements for site productivity research. In: *Impact of Intensive Harvesting on Forest Site Productivity.* Proceedings, IEA/BE A3 Workshop, South Island, New Zealand, March 1989. (Eds.) W.J. Dyck and C.A. Mees. IEA/BE T6/A6 Report No. 2. Forest Research Institute, Rotorua, New Zealand, FRI Bulletin No. 159, 1990, pp. 159-170.

2. Johnson, D.W. The effects of harvesting intensity on nutrient depletion in forests. *In: IUFRO Symposium on Forest Site and Continuous Productivity.* (Ed.) R. Ballard and S.P. Gessel. USDA Forest Service, Pacific Northwest Range Experiment Station, Portland, OR, General Technical Report PN W-163, 1983, pp. 157-166.

3. Messina, M.G., Dyck, W.J., and Hunter, I.R. *The Nutritional Consequences of Forest Harvesting with Special Reference to the Exotic Forests in New Zealand.* IEA/FE Project CPC-10 Report No. 1, 1985.

4. Squire, R.O., Farrell, P.W., Flinn, D.W., and Aeberli, B.C. Productivity of first and second rotation stands of radiata pine on sandy soils. II. Height and volume growth at five years. *Australian Forestry*, 1985, **48**, 127-137.

5. Stone, E.L. and Will, G.W. Nitrogen deficiency of second-generation radiata pine in New Zealand. In: *Forest-Soil Relationships in North America.* (Ed.) C.T. Youngberg. Oregon State University Press, Corvallis, OR., 1965, pp. 117-139.

6. Lundkvist, H. Ecological effects of whole-tree harvesting - some results from Swedish experiments. In: *Predicting Consequences of Intensive Forest Harvesting on Long-term Productivity by Site Classification.* Proceedings, IEA/BE A3 Workshop, Georgetown, SC, October 1987. (Eds.) T.M. Williams and C.A. Gresham. IEA/BE A3 Report No. 6. Baruch Forest Science Institute of Clemson University, Georgetown, SC, USA, 1988, pp. 131-140

7. Cole, D.W. and Van Miegroet, H. Chronosequences: a technique to assess ecosystem dynamics. In: *Research Strategies for Long-term Site Productivity.* Proceedings, IEA/BE A3 Workshop, Seattle, WA, August 1988. (Eds.) W.J. Dyck and C.A. Mees. IEA/BE A3 Report No. 8. Forest Research Institute, Rotorua, New Zealand, FRI Bulletin 152, 1989, pp. 5-23.

8. Draper, N. and Smith, H. *Applied Regression Analysis, 2nd edition.* John Wiley and Sons, New York, 1981.

9. Crocker, R.L. and Major, J. Soil development in relationship to vegetation and surface age at Glacier Bay. *Journal of Ecology*, 1955, **43**, 427-448.

10. Christensen, N.L. and Peet, R.K. Convergence during secondary forest succession. *Journal of Ecology*, 1984, **72**, 25-36.

11. Heilman, P.E. Changes in distribution and availability on nitrogen with forest succession on north slopes in interior Alaska. *Ecology*, 1966, **47**, 825-834.

12. MacLean, D.A. and Wein, R.W. Changes in understorey vegetation with increasing stand age in New Brunswick forests: species composition, cover, biomass, and nutrients. *Canadian Journal of Botany*, 1977, **55**, 2818-2831.

13. Ruark, G.A. and Bockheim, J.G. Biomass, net primary production, and nutrient distribution for an age-sequence of *Populus tremuloides* ecosystems. *Canadian Journal of Forest Research*, 1988, **18**, 435-443.

14. Sprugel, D.G. Density, biomass, productivity, and nutrient cycling changes during stand development in wave-generated balsam fir forests. *Ecological Monographs*, 1984, **54**, 165-186.

15. Long, J.N., and Turner, J. Aboveground biomass of understorey in an age sequence of four Douglas-fir stands. *Journal of Applied Ecology*, 1975, **12**, 179-188.

16. Madgwick, H.A.I., Jackson, D.S., and Knight, P.J. Above-ground dry matter, energy and nutrient contents of trees in an age series of *Pinus radiata* plantations. *New Zealand Journal of Forestry Science*, 1977, **7**, 445-468.

17. Ovington, J.D. The circulation of minerals in plantations of *Pinus sylvestris* L. *Annals of Botany*, 1959, **23**, 229-239.

18. Switzer, G.L. and Nelson, L.E. Nutrient accumulation and cycling in loblolly pine (*Pinus taeda* L.) plantation ecosystems: the first twenty years. *Soil Science Society of America, Proceedings*, 1972, **36**, 143-147.

19. Forrest, W.G. and Ovington, J.D. Organic matter changes in an age series of *Pinus radiata* plantations. *Journal of Applied Ecology*, 1970, **7**, 177-186.

20. Turner, J. Nutrient cycling in an age sequence of western Washington Douglas-fir stands. *Annals of Botany*, 1981, **48**, 159-169.

21. Ballard, R. Effect of slash and soil removal on the productivity of second rotation radiata pine on pumice soil. *New Zealand Journal of Forestry Science*, 1978, **8(2)**, 248-258.

22. Dyck, W.J., Mees, C.A., and Comerford, N.B. Medium-term effects of

mechanical site preparation on radiata pine productivity in New Zealand - a retrospective approach. In: *Research Strategies for Long-term Site Productivity*. Proceedings, IEA/BE A3 Workshop, Seattle, WA, August 1988. (Eds.) W.J. Dyck and C.A. Mees. IEA/BE A3 Report No. 8. Forest Research Institute, Rotorua, New Zealand, FRI Bulletin 152, 1989, pp. 79-92.

23. Balneaves, J.M. Maintaining site productivity in second-rotation crops, Canterbury Plains, New Zealand. In: *Impact of Intensive Harvesting on Forest Site Productivity*. Proceedings, IEA/BE A3 Workshop, South Island, New Zealand, March 1989. (Eds.) W.J. Dyck and C.A. Mees. IEA/BE T6/A6 Report No. 2. Forest Research Institute, Rotorua, New Zealand, FRI Bulletin No. 159, 1990, pp. 73-83.

24. Morris, W.G. Influence of slash burning on regeneration, other plant cover, and fire hazard in the Douglas-fir region. USDA Forest Service Research Paper PNW-29, 1958.

25. Miller, R.E., Hazard, J.W., Bigley, R.E., and Max, T.A. Some results and design considerations from a long-term study of slash burning effects. In: *Research Strategies for Long-term Site Productivity*. Proceedings, IEA/BE A3 Workshop, Seattle WA, August 1988. (Eds.) W.J. Dyck and C.A. Mees. IEA/BE A3 Report No 8. Forest Research Institute, Rotorua, New Zealand, FRI Bulletin 152, 1989, pp. 63-78.

26. Powers, R.F. Retrospective studies in perspective: strengths and weaknesses. In: *Research Strategies for Long-term Site Productivity*. Proceedings, IEA/BE A3 Workshop, Seattle, WA, August 1988. (Eds.) W.J. Dyck and C.A. Mees. IEA/BE A3 Report No. 8. Forest Research Institute, Rotorua, New Zealand, FRI Bulletin 152, 1989, pp. 47-62.

27. Helms, J.A., Hipkin, C., and Alexander, E.B. Effects of soil compaction on height growth of a California ponderosa pine plantation. *Western Journal of Applied Forestry*, 1986, **1**, 104-108.

28. Andersson, F.O., and Lundkvist, H. Long-term Swedish experiments in forest management practices and site productivity. In: *Research Strategies for Long-term Site Productivity*. Proceedings, IEA/BE A3 Workshop, Seattle, WA, August 1988. (Eds.) W.J. Dyck and C.A. Mees. IEA/BE A3 Report No. 8. Forest Research Institute, Rotorua, New Zealand, FRI Bulletin 152, 1989, pp. 125-137.

29. Keeves, A. Some evidence of loss of productivity with successive rotations of *Pinus radiata* in the south-east of South Australia. *Australian Forestry*, 1966, **30**, 51-63.

30. Squire, R.O. Review of second rotation silviculture of *Pinus radiata* plantations in Southern Australia: establishment practice and expectations. *Australian Forestry*, 1983, **46**, 83-90.

31. Woods, R.V. Early silviculture for upgrading productivity on marginal *Pinus radiata* sites in the south-eastern region of South Australia. Woods and Forests Department, South Australia, Bulletin 24, 1976.

32. Will G.M. Nutrient deficiencies and fertiliser use in New Zealand exotic forests. New Zealand Forest Service, Forest Research Institute Bulletin No. 97, 1985.

33. Hazard, J.W. and Peterson, C.E. Objectives and analytical methods of the Regional Nutrition Research Project. Institute of Forest Resources, Contribution 53, University of Washington, Seattle, 1984.

34. Dyck, W.J., Hodgkiss, P.D., Oliver, G.R., and Mees, C.A. Harvesting sand-dune forests: Impacts on second-rotation productivity. In: *Long-term Field Trials to Assess Environmental Impacts of Harvesting*. Proceedings, IEA/BE T6/A6 Workshop, Florida, USA, February 1990. (Eds.) W.J. Dyck and C.A. Mees. IEA/BE T6/A6 Report No. 5. Forest Research Institute, Rotorua, New Zealand, FRI Bulletin No. 161, 1991, pp. 163-176.

35. Smith, C.T., Dyck, W.J., Beets, P.N., Hodgkiss, P.D., and Lowe, A.T. Nutrition and productivity of *Pinus radiata* following harvest disturbance and fertilization of coastal sand dunes. In: Proceedings, IEA/BE T6/A6 Workshop, Asa, Sweden, June 1991. *Forest Ecology and Management* (in press).

36. Compton, J.E. and Cole, D.W. Impact of harvest intensity on growth and nutrition of successive rotations of Douglas-fir. In: *Long-term Field Trials to Assess Environmental Impacts of Harvesting*. Proceedings, IEA/BE T6/A6 Workshop, Florida, USA, February 1990. (Eds.) W.J. Dyck and C.A. Mees. IEA/BE T6/A6 Report No. 5. Forest Research Institute, Rotorua, New Zealand, FRI Bulletin No. 161, 1991, pp. 151-161.

CHAPTER 3

EVIDENCE FOR LONG-TERM PRODUCTIVITY CHANGE AS PROVIDED BY FIELD TRIALS

L.A. MORRIS

Warnell School of Forest Resources, University of Georgia
Athens, GA 30602, USA

R.E. MILLER

USDA Forest Service, Pacific Northwest Research Station
3625 93Ave. SW, Olympia, WA 98512, USA

INTRODUCTION

In the past two decades, considerable emphasis has been placed on developing computer models that can be used to predict effects of management practices on long-term forest productivity [1, 2, 3, 4]. Although the value of such models as exploratory research tools is widely recognized, acceptance of their predicted changes in forest growth is limited. All models incorporate current understanding of forest growth, and produce results that are only as reliable as this understanding. Because our understanding of factors affecting forest growth is incomplete, results from carefully controlled field trials, rather than models, provide the more convincing evidence of productivity change.

In this chapter, we review the best available field evidence for changes in long-term site productivity associated with harvest and regeneration activities. This evidence includes comparisons of growth in successive rotations on the same plots, results from matched plots, and analyses of controlled treatment experiments. We begin by discussing assumptions required in analyses of long-term field trials and by offering minimum criteria that should be met before accepting results from field trials as reliable evidence for productivity change. We then evaluate evidence for productivity change associated with the following management activities: use of monocultures or restricted species, harvest utilization and nutrient removal, slash disposal, and ameliorative treatments including nutrients and tillage.

ASSUMPTIONS IN LONG-TERM FIELD TRIAL RESEARCH

Productivity Components

There is no generally agreed upon usage for the terms forest productivity and site productivity. While some authors treat these as interchangeable terms, other authors draw distinctions. *Forest productivity* is the more general term and is widely used by foresters, botanists, ecologists, and wildlife scientists. It refers to growth and maintenance of all or any part of the assemblage of plants and animals that exist in forests at scales that range from microplots to entire ecosystems. In contrast, *site productivity* generally has a more limited scale and usage. It is used by soil scientists and foresters when referring to the growth or capacity of a site to grow trees at the scale of individual forest stands.

Traditionally, productivity is expressed as a function of biotic factors (species, genotype) and abiotic factors (soil conditions, slope, precipitation). For the purpose of evaluating long-term site productivity, these factors are more conveniently divided into four components: plant potential, climate, soil capacity, and catastrophe that differ in degree to which they are controlled or assumed constant in long-term studies (Table 3.1). As defined in Chapter 2, "site quality" is determined by climate and soil properties (i.e., capacity) and is the major focus of long-term site productivity studies.

The first component, plant potential, includes major factors manipulated by forest management: tree species, genotype, stocking, and quality and quantity of competing vegetation. Although this component has a major influence on forest productivity as measured by timber or fiber yield, its influence relates more to utilization and allocation of site resources to the desired products than to long-term change in resources. Plant potential is not necessarily related to long-term site productivity. For instance, a decrease in the capacity of a site to produce biomass

TABLE 3.1

Factors affecting stand productivity, by groups subject to different levels of control in long-term studies. Factors associated with soil capacity are of prime importance for evaluating long-term changes in site productivity associated with harvest and associated practices.

GROUP	VARIABLES	EXPERIMENTAL CONTROL
Plant Potential	Species, genotype, stocking, age, plant competition	Minimize differences among "rotations" and treatments unless part of experimental treatments
Climate	Temperature, precipitation, humidity, growing season length, atmospheric transfers and pollution, sunlight	Differences removed by co-variance analyses for successive rotations; assumed not to interact with experimental treatments in field studies
Soil Capacity	Soil depth, rooting volume and restrictions, organic matter content, water-holding capacity, nutrient storage, nutrient mineralization, and availability	Experimental manipulation and measurement
Catastrophe	Hurricane damage, volcanic activity, extreme early/late freeze or drought, insect or disease epidemics not associated with treatments	None

can be masked by eliminating species that compete with crop trees. A major effort in productivity research is to control plant factors so that measurements of tree height, stand volume, or stand basal area will provide a reliable index of site productivity and change.

The second component includes climatic influences. For most field trials of long-term site productivity, climate is largely ignored in analyses because: 1) over the long term, climate is assumed to average out (i.e., years of good and poor growth contribute to observed growth); and 2) climate is not controlled or modified by management activities at a specific site. Although evidence for global climate change is increasing, comparisons among management treatments within a rotation remain valid, providing treatments do not interact with these broad-scale climatic conditions. In contrast, changing climatic conditions severely limit our ability to assess site productivity changes among successive rotations.

The third component, soil capacity, includes those factors that determine the ability of soil to provide support, water, and nutrients required by trees. This capacity is maintained through the activities of a broad range of soil biota that contribute to development of soil structure, decomposition of organic matter, and nutrient mineralization and transformation. This component is the main thrust of most long-term site productivity research. In theory, effects of management practices on soil can be assessed independently of tree growth measurements. In practice, however, such plant-independent evaluation of soil capacity is difficult and provides only circumstantial evidence of productivity change. No currently available soil extraction or bioassay can reliably index the complex soil chemistry and biological activity that affect mineralization and availability of nutrients, and conditions for root growth. For these reasons, soil capacity in forestry must ultimately be evaluated by measurement of tree growth.

The final component, catastrophe, plays an essential role in determining the structure and productivity of forest stands. Some catastrophic events like hurricanes and volcanic eruptions are beyond our control and can destroy field trials. Such disasters seldom affect field-trial results because the trial is discarded when such events occur. Other catastrophes, like insect and disease epidemic, may be related to, or induced by, forest practices; hence these should not be ignored or dismissed as random influences without investigation.

Acceptable Field-Trial Evidence for Productivity Change

Worldwide, numerous studies have been established to evaluate changes in site productivity resulting from forest management activities. Although such studies have provided useful insight about seedling establishment, stand growth, nutrient accumulation and cycling, and the role of insects and disease, they have generally failed to provide an answer to the central question: *How do specific management activities affect long-term site productivity?* To be acceptable evidence for a change in long-term site productivity, three conditions must be met. First, differences in tree growth must be attributable to differences in site conditions rather than to differences in resource allocation among target and non-target species or to differences in plant potential. Second, growth results must be available for a sufficient duration of time so that the influence of ephemeral differences in initial site conditions has diminished and so that the capacity of the site to support tree growth is stressed. Finally, adequate experimental control must exist. We shall address each condition in more detail.

Selecting Growth Measures: The degree to which a site can supply physical support, water, and nutrients to trees, and the extent to which tree roots can grow through the soil to utilize these resources, determines the capacity of a site for tree growth. Theoretically, site productivity can be evaluated by any measure that accurately reflects changes in these conditions. The merits of using net primary productivity (NPP), the total carbon fixed in organic matter minus respiration losses, as a measure of site productivity change has been discussed [5]. This measure clearly has value in ecological studies; however, from a practical standpoint, measurement of NPP is unworkable in long-term studies of forest management. Complete characterization of net primary productivity is difficult and subject to large measurement error, particularly where changes in species composition are involved. Hence, most long-term studies in forestry rely on growth of the target tree species for assessing changes in site productivity. Site index (or height growth) has been the most commonly used measure because height is readily measured and relatively less sensitive to stand conditions than other parameters. Stand basal area and bole volume growth or biomass accumulation provide more comprehensive measures of stand growth, but these can only be reliably interpreted where growth differences are not strongly confounded by differences in stand conditions.

Defining Long-Term Response: A second major consideration for evaluating results from field trials of long-term productivity involves the definition of "long-term." Many management practices have short-term impacts on stand or site conditions that are ephemeral and poorly related to long-term site productivity. Control of herbaceous competition, for example, often promotes rapid tree growth in young stands but has little influence on growth following crown closure (except, of course, in the case of nitrogen-fixing plants). When competition control does not result in subsequent accelerated erosion or nutrient loss, it largely affects tree growth through changes in resource allocation to trees, rather than through a change in total site resources or productive capacity. Such shifts in resource allocation will appear as an early increase in tree growth followed by a return to annual growth patterns that do not differ from growth patterns of untreated stands [6]. The original increase is maintained after crown closure, but

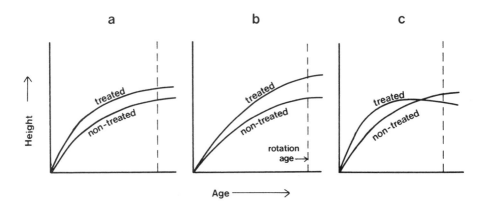

FIGURE 3.1 Possible patterns of dominant tree height growth following management treatment: (a) initial increase in resource allocation without site change; (b) increase in long-term productivity; and (c) transient increase followed by long-term decrease in site productivity (after [6]).

further growth improvements are not evident. Hence this pattern of growth response does not indicate a change in long-term site productivity (Figure 3.1(a)).

The growth pattern after herbaceous competition control exemplifies but one of three characteristic patterns of growth response to management. Figures 3.1(b) and (c) illustrate two other treatment-induced patterns in stand productivity. Figure 3.1(b) illustrates a difference in site productivity that is maintained; Figure 3.1(c) illustrates a transient change in productivity. The distinction between Figure 3.1(a) and Figure 3.1(b) is important. The growth pattern of Figure 3.1(b) indicates a change in site productivity that will most likely continue into future rotations. Although the growth pattern of Figure 1a does not necessarily rule out the possibility that a positive change in long-term site productivity may yet occur, it is more probable that the soil capacity is not fundamentally different than the control. The growth pattern in Figure 3.1(c) also represents a change in site productivity that will likely continue into future rotations. In this case, the pattern indicates a decrease in productivity.

Until the pattern of growth change is evident, one has little basis for predicting the consequences of management activities on long-term productivity. For slow-growing stands, a reliable pattern may require 20 or more years. For faster growing stands or short-rotation plantations, 10 years may be adequate to observe these changes and to predict their long-term consequences.

Experimental Control: According to Mead *et al.* [7], experimental designs for long-term studies require the following (see chapter 9 for further discussion on experimental design):
1. at least four replications in randomized or randomized complete block designs;
2. suitable pre-treatment site and crop information on individual experimental plots;
3. minimum measurement plot areas of 400 m² containing at least 12 remaining trees at the end of the experiment;
4. buffer areas of at least 10-m width surrounding each measurement plot that receive experimental treatments.

Few studies have been established which meet these experimental requirements and none have been carried long enough to provide unequivocal results. Thus, our current conclusions about long-term site productivity change are based on imperfect results and must be continually re-examined as new information becomes available. Only when results from numerous imperfect studies, and our understanding of forest growth processes lead to the same conclusions, can we ensure robust conclusions.

Research on possible declines in productivity resulting from removing species from a stand or replacing mixed-species stands with plantations of one tree species has been based largely upon results from plots matched with respect to management history and soil conditions or from randomly paired plots. In matched plots, treatment effects and initial site differences are easily confounded by unrecognized but important site factors or experimental bias. Such confounding is reduced in paired-plots where treatments are randomly assigned to experimental units.

Research on harvest utilization is more amenable to controlled experimentation. Comparisons among harvest utilization options within a rotation can be easily designed as valid replicated experiments (see Chapter 9). Unfortunately, much of the available information on harvest utilization comes from non-replicated studies on large, operationally harvested areas. This is clearly in response to nutrient mineralization and water quality concerns raised by early research on the Hubbard Brook watersheds [8]. Results from such large-scale studies are frequently confounded by differences in tree growth due to regeneration success, plant competition, and site variation.

Studies of site change after slash disposal or ameliorative treatments usually have the strongest experimental designs. Many of these experiments are well replicated and, in some, differences in stocking, plant competition, or insect damage have been minimized [9].

Regardless of study purpose, minimum requirements for an acceptable design include: 1) having appropriate control plots against which treated plots can be evaluated; and 2) being certain that all factors other than treatment have been minimized or measured in order to account for their contribution to observed differences in growth.

SPECIES CHANGE AND SINGLE SPECIES ROTATIONS

Plants and soils interact. Extension of roots through soil, uptake and return of various nutrients, contribution of organic matter, and the influence of vegetation on temperature and moisture distribution profoundly affect soils. Conversely, chemical and physical characteristics of the soil affect plant growth and species distribution.

Concern about the effects of species change on long-term site productivity, specifically the effect of conifer establishment, emerged from Europe more than 70 years ago. Between the late 1800's and early 1900's, conifer plantations were established on marginal crop and forest lands throughout Europe. Limited experience in plantation establishment, coupled with the degraded condition of the lands, resulted in regeneration failures and poor growth. By 1953, it was generally assumed that site productivity was degraded by conversion of pasture, farmland, or hardwood forests to conifers [10]. Evidence for such declines in site productivity was weak, however, because it was largely developed from observations of stands growing on different sites with different histories. These early reports were followed by reports of productivity decline in second-rotation plantings of *Pinus radiata* in Australia [11, 12] and New Zealand [13], and *Pinus patula* in Swaziland [14].

None of these studies of species conversion meet all of the previously discussed criteria for inferring long-term site productivity change. If only studies that have adequate growth records and plots matched for similar management history and site conditions are considered (Table 3.2), we cannot conclude that conversion to conifer plantations leads to long-term productivity decline. Growth of second- and third-generation conifer plantations is as likely to increase as to decrease, relative to that of the first generation. For instance, Evans [15] described results from second-rotation *Pinus patula* in Swaziland where growth on 129 plots near rotation age (13-15 years) was compared with growth of the first rotation on the same plots (Table 3.3). Overall, the second rotation averaged slightly less volume production than the first; a substantial productivity decline was largely confined to one forest block. In the other four blocks, productivity in the second rotation equalled productivity in the first rotation. This study is particularly noteworthy in that earlier evaluations of the same plots by this author suggested a general productivity decline had occurred [14]. It illustrates the benefits of extended observations – and the problems of short-term data. Squire *et al.* [16] evaluated productivity of second-rotation radiata pine in Australia using carefully matched plots and stem analysis to compare growth in the current rotation with that in previous rotations. These authors found that on both low- and high-quality sites, growth through age 5 was greater during the second rotation. As in most such studies, the reasons for the improved growth were speculative, but appeared to be due to improved planting and establishment procedures.

Issues of species change and second-rotation decline are not easily disentangled from associated questions about harvest utilization, slash disposal, and soil tillage to be discussed later in this chapter. This dilemma is illustrated by two recent studies that combine results from designed field trials with retrospective comparisons of previous rotation growth on the same site [17, 18].

Cellier *et al.* [17] assessed factors affecting height growth of second-rotation *Pinus radiata* plantations in South Australia in the same area as the afore-

TABLE 3.2

Summary of selected investigations of second-rotation decline in conifer plantations

Location	Species	Experimental Design	Date Begun[a]	Last Eval.n Age (yr)	Productivity Variables	Results (in comparison to first rotation)	Comments	Source
Australia (Victoria)	*Pinus radiata*	Matched plots, successive rotations	1975	5	Dominant tree height, stand volume	Second rotation dominant height: +63 to +84% on poor sites, +26 to +44% on good sites; volume >100% better in second rotation		[16]
Swaziland	*Pinus patula*	Matched plots	1964	10-14	Mean tree height and volume, stand volume	Volume of second rotation averaged 7% lower, ranged from -20% to +16%	Second rotation reduction attributed to winter drought	[14, 15]
Australia	*Pinus radiata*	Successive rotation (retrospective)		16	Basal area	Marked declines in site class across a range of site qualities and stand conditions		[11]
New Zealand	*Pinus radiata*	Matched plots	1965	12-21	Height, basal area, stand volume	Declines in second-rotation productivity occurred variably, but most common on ridge positions; second-rotation growth generally better than first-rotation growth in lower slope positions.	Most declines exhibited growth trends of Figure 1a.	[13]

[a] If applicable

TABLE 3.3

Comparisons of rotation-aged *Pinus patula* height and volume for first and second rotations
in Swaziland (source: [15]).

Forest Block		14-year rotation			13-year rotation	
	Plots	First	Second	Plots	First	Second
	no.	---- Height (m) ----		no.	---- Height (m) ----	
A	20	18.7	16.8***	4	18.5	16.4
B	11	17.4	17.3	9	16.4	16.5
C	10	17.8	18.7	8	17.2	18.0*
D & E	12	18.0	17.8	11	17.4	16.4*
		— Volume (m³ ha⁻¹) —			— Volume (m³ ha⁻¹) —	
A	20	301	239***	4	293	264
B	11	287	286	9	268	275
C	10	301	316	8	270	314
D & E	12	306	297	11	294	271*

*, ***   indicate significant differences between first and second rotations for $p = .10$ and .01,
respectively.

mentioned studies by Keeves [11] and Bednall [12]. These investigators established *P. radiata* plantations after harvesting matched first- and second-rotation stands. They compared factorial combinations of site-preparation treatments and nutrient additions. Weed competition on all plots was minimized by herbicides. At age 7, height of the current rotation exceeded height of the previous rotation at the same age. Although the optimal treatments on both first- and second-rotation sites resulted in similar growth, the treatments that produced this optimal growth differed between first- and second-rotation sites. Specifically, height growth was increased on second-rotation, but not on first-rotation sites, by treatments that included nitrogen additions. These results suggest that removal of nitrogen in two rotations reduced site productivity on these sites.

Allen *et al.* [18] reported results from a replicated study established after harvest of part of a *Pinus taeda* plantation. Growth of the previous plantation was determined by stem analysis and compared with current rotation growth on plots established after either harvest of merchantable pine stems (pine stem-only) or complete above-ground harvest of pines and competing hardwoods followed by either chop-burn or shear-pile and disk-harrowing. Each of these four treatments were split into two subplots where competition was either minimized through repeated herbicide applications or allowed to grow without herbicides.

Like Cellier *et al.* [17], these investigators found that height growth through 8 years of the current (second) rotation was greater than 8-year height growth of the first rotation for all treatments evaluated. Although growth among the four harvest and site preparation treatments differed, competition control had the greatest influence on short-term growth. Harvest utilization had no detectable effect on 8-year height growth. In plots where competition was minimized by repeated herbicide application, pine growth was only slightly greater in the more disturbed shear-pile and harrowed site-prepared areas than in chop-burn areas. In plots where natural competition was not controlled, differences between site preparation treatments were larger because site preparation provided some control of vegetation.

Despite evidence to the contrary, some authors claim as fact that multiple forest rotations with conifers decrease site productivity. For instance, Sheppard [19] writes: "A decline in productivity between successive crops of spruce in Europe was reported around the turn of the century and similar reports have come forward since from a number of places for other species in other circumstances. Therefore, we must acknowledge that there is ... a danger in growing tree crops in monoculture."

We conclude that current evidence does not indicate such declines occur or, if they do occur, are probably not related simply to establishment of conifer species.

In contrast to these results, evidence from field trials clearly shows that addition of nitrogen-fixing tree species to stands on N-deficient sites increases site productivity. Hence, removing such beneficial species could reduce long-term productivity.

In western Washington, USA, four plots (0.08 ha each) in a 48-year-old *Pseudotsuga menziesii* plantation were matched topographically beside four plots in a strip of the same plantation interplanted with *Alnus rubra* [20]. Initial tree density at planting averaged 1,700 Douglas-fir and 3,100 red alder per hectare. By age 48, the four plots in the pure stand averaged 970 Douglas-fir and those in the mixed stand averaged 660 Douglas-fir and 730 alder. The largest 100 Douglas-fir per hectare averaged 20% taller in the mixed stand and were larger in diameter, despite the greater stand density. Douglas-fir volume averaged 217 m^3ha^{-1} in the mixed stand versus 203 m^3ha^{-1} in the pure. The additional alder volume in the mixed stand averaged 175 m^3ha^{-1}.

The growth-enhancing benefits of N_2-fixing alder at this site are not surprising because 30 years after interplanting, soil in the mixed stand averaged 33% more nitrogen (1,000 kg ha^{-1}; [21]). Moreover, 15-year response in a N-fertilization trial in a nearby portion of the same Douglas-fir plantation averaged 88% (116 m^3ha^{-1} after a single application of 314 kg N ha^{-1}; [22]). The large, extended response to N fertilizer at this location is unusual and attributed to improved nutrition and to secondary effects of increased growth on stand structure. On more naturally-productive sites, however, growth-promoting benefits of alder in mixed stands are less apparent, perhaps explained by greater amounts of soil nitrogen at naturally productive sites and the early competitive advantage of this species [23].

Growth-enhancing effects of a tropical N_2-fixing tree species, *Albizia falcataria* (L) Fosb., in an equal-aged mixture with *Eucalyptus saligna* Sm were quantified in Hawaii [24]. At 48 months, *Eucalyptus* trees in the mixed plantings that contained 34% or more *Albizia* were equal to or larger than those in pure stands receiving four fertilizer applications, each equivalent to 40 kg N, 18 kg P, and 33 kg K per hectare [24]. In contrast to these results on the west coast of Hawaii Island, *Albizia* grew poorly and provided no growth enhancement in a companion test in a much drier location of the island [88].

On N-deficient sites, both N_2-fixing species and nitrogen fertilizers are potential means to enhance tree growth, hence site productivity. Use of N_2-fixing plants, however, is more complicated than use of fertilizers for several reasons: 1) the N_2-fixing plants must survive and thrive in sufficient numbers and duration; 2) some N_2-fixing tree species may result in associated trees being overtopped or damaged – this problem can be avoided with early spacing control; and 3) as summarized by Binkley [23], larger quantities of nitrate and organic N are leached from pure red alder stands or alder mixtures than from pure Douglas-fir. This can increase leaching losses of nutrient cations and lower soil pH. In summary, successful use of N_2-fixing species to enhance stand and site productivity requires careful planning and execution.

HARVEST UTILIZATION AND NUTRIENT REMOVAL

Nutritional Basis for Harvesting Impacts

Concerns about species-induced reductions in site productivity are not supported by existing evidence, yet concerns about the influence of forest harvest on long-term productivity should not be summarily dismissed. Direct removal and indirect loss of essential plant nutrients will occur in forests managed for wood and fiber production. Without nutrient replacement by natural sources or by fertilization, productivity could decline. Current questions about impacts of harvest utilization on site productivity center on the relative balance between nutrient inputs and exports, and the capacity of soil to supply nutrients to a growing forest.

Excellent literature reviews on nutrient distribution and cycling in major forest biomes are available. General reviews of nutrient cycling in the boreal [25], temperate [26], and tropical [27] forest biomes were published in the proceedings of the Fifth North American Forest Soils Conference. Literature reviews about specific forest ecosystems are also available; for example, about the distribution and cycling of nutrients in deciduous forests of Belgium [28], and in tropical rain forests of Central America [29].

Distribution of nutrients in representative forest stands within the boreal, temperate, and tropical forest biomes are presented in Table 3.4. Within each biome, variability among stand types is large and tends to mask differences in total nutrient content among the biomes. Clearly, however, comparatively larger quantities of N, P, K, Ca, and Mg are stored in the vegetation of tropical forests and in the forest floor of boreal forests.

Depletion of nutrients under various harvesting regimes has been documented or can be calculated using data from ecosystem studies in most commercially important forest types. Most nutrient depletion is associated with direct biomass removal; loss of nutrients through accelerated erosion, leaching, or gaseous loss is relatively small. Comprehensive reviews of removal rates in short-rotation plantations [37], medium-age forests [38], and long-rotation systems [39] were completed for a symposium on the impacts of intensive harvesting on forest nutrient cycling. General summaries [40, 41] and summaries for specific regions or forest ecosystems [42, 43, 44] are also available. The general consensus of these

TABLE 3.4

Nutrient content and distribution in major forest biomes.

Cover Type	Vegetation					Forest floor					Mineral soil					Source
	N	P	K	Ca	Mg	N	P	K	Ca	Mg	N	P	K	Ca	Mg	
								kg/ha								
BOREAL																
Picea mariana	167	42	84	277	27	1214	213	382	102	430	296	468	16802	2215	2605	[30] a
Picea-Abies	387	52	159	413	36	1465	100	1052	253	154	559	114	9383	766	1113	[30] b
Pinus banksiana	171	15	85	114	19	328	43	524	319	116	3729	6100	165354	98708	53839	[311] c
Pinus banksiana	346	29	146	294	52	544	40	37	254	33	5554	495	500	1727	289	[32] d
TEMPERATE																
Mixed Quercus	406	32	245	868	81	33	1	15	74	5	4480	920	26800	132900	6460	[28] e
Fagus spp.	285	39	187	152	43	180	11	20	51	14	6640	42	254	365	249	[33] f
Pinus elliottii-P. palustris	174	13	51	148	44	256	11	11	134	28	2698	19	81	406	82	[34] g
Pseudotsuga menziesii	326	67	227	342	-	175	26	32	137	-	2809	3878	234	741	-	[35] h
TROPICAL																
Celtis-triplochiton	1530	103	702	2140	318	514	34	204	530	72	4592	13	650	2576	3969	[36] i
Cavanillesa-ceiba	ND	241	4598	4702	437	-	-	-	-	-	-	33	508	18582	2830	[29] j

a Overstory vegetation only, soil depth to 35 cm (approx. rooting depth), soil cations determined by HF/HClO$_4$ digestion.

b Overstory vegetation only, soil depth to 26 cm (approx. rooting depth), soil cations determined by HF/HClO$_4$ digestion.

c Soil depth to 100 cm, soil cations determined by HF/HClO$_4$ digestion.

d Soil depth to 100 cm, soil P determined by Bray No. 1, soil cations by NH$_4$OAc extraction.

e Forest floor includes leaf litter only, soil to 40 cm depth, totals by unspecified digestion.

f Soil depth to 70 cm, P, and cations extracted, using H$_2$SO$_4$/HClO$_4$/HNO$_3$ digestion.

g Soil depth to base of B$_h$ horizon (approx. 80 cm), soil P and cations determined by H$_2$SO$_4$/HCl extraction.

h Soil depth to 60 cm, P by H$_2$O$_2$/H$_2$SO$_4$ digestion, cations by NH$_4$O Ac extraction.

i Forest floor includes "litter" and "trash", soil depth to 30 cm, soil P by Truog, cations by extraction in NH$_4$O Ac.

j Soil depth to 30 cm, soil P and cations by H$_2$SO$_4$/HClO$_4$/HNO$_3$ digestion.

reviewers is that conventional stem-only harvests of mid-age stands will have little impact on tree nutrition in future rotations. Atmospheric inputs of N and P (and probably S) occur in sufficient quantities to approximately balance harvest removals, and soil reserves of K, Ca, and Mg are generally considered sufficient for numerous rotations, even in the absence of atmospheric or weathering inputs. A report to the contrary [32] is based on estimated mineral reserves of shallow soils as determined by weak extraction techniques.

By comparison, large nutrient removals occur in more intensive above-ground harvests of short-rotation forestry. In an early review, Kimmins [40] estimated that a change from conventional, merchantable stem-only harvest to complete above-ground tree harvest of various northern conifer types (hemlock-cedar, pine, spruce, spruce-fir, hemlock-fir) and cottonwood, would increase removals of N by 86-288%, P by 54-367%, K by 14-236% and Ca by 15-179%. Wells *et al.* [45] estimated that removals of P, K, Ca, and Mg would double for 16-year-old *Pinus taeda* stands in the southeastern U.S., using a complete above-ground harvest instead of a merchantable-stem harvest. Even greater removals would occur under intensive management schemes such as proposed by Koch [46]. Koch proposed a harvesting scheme for *P. taeda* which would incorporate multiple thinnings in a 35-year rotation. At the end of the rotation, crop trees would be harvested with their taproot and crown intact. Utilizing Koch's estimates of projected biomass removals, and available data on nutrient concentration of *P. taeda* tissue, we calculated total nutrient removals under such a system (Table 3.5). The 22, 3, 14, 16, and 4 kg/ha, respectively, of N, P, K, Ca, and Mg that would be removed on an annual basis are about 35% greater than removals in two 16-year rotations with complete above-ground harvest [45]. Results from these earlier analyses have been confirmed in a more recent review [41]. It seems clear that increased biomass utilization and/or shorter rotations at some locations will create an imbalance in nutrient input-export that could lead to productivity declines.

Direct Evidence of Productivity Decline with Increased Biomass Removal
Although nutrient-balance analyses provide indirect evidence for projecting declines in long-term site productivity in managed forests, direct evidence of such declines is rare. Most cases where decreased site productivity has been associated

with increased utilization of the previous stand occur on low-quality sites in cool climates (Table 3.6) where the consequences of N removals are greatest. Lundkvist [47], after reviewing results of matched-plot experiments of whole-tree harvesting in *Pinus sylvestris* and *Picea abies*, concluded that, for most sites, growth of subsequent stands was greatest when harvest slash was left in place rather than removed. In general, differences in height growth among utilization regimes were greatest during the first 10 years after planting; mean tree height in plots planted after harvest with slash left in place ranged from 30% to 40% greater than height in plots where slash was removed. This early height-growth advantage, however, diminished thereafter. By age 16 on the best quality sites, a 3% reduction in height growth was associated with lower biomass removals, reversing the earlier trend of improved growth. Compton and Cole [48] reported similar results for Douglas-fir growth after complete above-ground biomass removal that included forest floor removal, complete-tree harvest (stem, branches and foliage), or stem-only removal. On a low-quality and a medium-quality site, 10-year height growth was lowest where biomass removal was greatest. On both sites, application of N fertilizer removed or reduced growth differences.

It should be noted that subordinate vegetation was not controlled in the studies evaluated by either Lundkvist or Cole; thus, differences in vegetation among treatments could contribute to the observed tree growth patterns. Hence, the effect of harvest treatments on soil properties and tree growth could not be separated from the effects of competing vegetation.

Remaining field trials provide weak evidence for productivity declines after increased harvest utilization. For the longest term study of its kind in the USA, Williams *et al.* [49] described results of harvest utilization in *P. banksiana* in Minnesota. Two harvest-utilization levels were investigated: complete tree (above-ground removal of stem, branch, and foliage) or tree length (removal of merchantable portions of the stem). The treatments were applied at two study sites with similar soils but were not replicated within a site. At one site, 14 years following planting, average tree height was 11% less on the complete-tree harvested plot than on the tree-length harvested plot. Conversely, on the second site at the same age, height was 21% greater in the complete-tree harvested plot than on the tree-length harvested plot. In the previously described study by Allen *et al.* [18], average 8-year height growth of *P. taeda* following stem-only harvest versus complete above-ground harvest of pine and hardwoods did not differ. This occurred despite large differences in biomass and nutrient removals between the two harvest treatments [50]. Furthermore, despite differences in growth among site preparation and herbicide treatments within each harvest intensity, there was no evidence of interaction between growth response to harvest and these treatments. It must be noted that these latter results are for young stands that do not entirely fulfil our previously stated requirements for acceptable field evidence. The possibility that response may yet follow the pattern of Figure 3.1(c) can not be dismissed.

SLASH DISPOSAL

Slash Disposal Impacts

As is harvest utilization, slash disposal is also important in evaluating management impacts on long-term productivity. Mechanized removal of slash into piles or windrows has the greatest potential for negative site impacts because it can displace large quantities of organic matter, soil, and associated nutrients on much of the site [50, 54, 55, 56]. In *Pinus elliotii - P. palustris* forests in the southern United States, Morris *et al.* [56] reported that amounts of N, P, and K displaced into windrows during piling with a blade exceeded removals from merchantable stem harvests by six-fold. The displaced nutrients represented more than 10% of the total soil nutrient content to a 1-m depth. Similar results have been reported for New Zealand [54] and the Piedmont of the southern U.S. [50], and, hence, can probably be generalized to many sites prepared for planting by these mechanical techniques (Table 3.7).

Burning is another common means for slash disposal. Nitrogen loss from these fires can be as large as displacement during windrowing. In boreal forests where debris and forest floor accumulations are large, more than 500 kg N/ha can be lost during slash burning [61, 62]. Losses from other forest types, however, are typically lower and may be similar to N removal during merchantable stem harvest [63]. Slash burning is likely to cause further reductions of N stored in vegetation and soil components.

TABLE 3.5

Estimated nutrient removals in a 35-year rotation of intensively managed loblolly pine (*Pinus taeda*).

Component	Harvested biomass[a] – kg/ha –	Nutrient concentration[b] %					Nutrient removal kg/ha				
		N	P	K	Ca	Mg	N	P	K	Ca	Mg
Corridor thinning (age 4)											
Whole tree	24,000	.165	.020	.106	.120	.030	40	5	26	29	7
Intermediate thinnings											
Bolewood, bark, and taproot	170,700	.092	.012	.071	.090	.023	157	20	121	154	39
Tops, branches, and foliage	47,700	.455	.051	.244	.240	.054	217	24	116	114	26
Final harvest											
Lateral roots	27,300	.176	.047	.168	.143	.060	48	13	46	39	16
Bolewood, bark, and taproot	152,900	.092	.012	.071	.090	.023	141	18	109	138	35
Tops, branches, and foliage	32,900	.455	.051	.244	.240	.054	150	17	80	79	18
Total removed							753	92	498	553	141

a Source: Koch [46].

b Computed as a weighted average using the data of Wells and Jorgensen [45]; taproot concentrations were assumed to be equal to bolewood concentrations.

TABLE 3.6

Influence of harvest utilization (biomass removal) on productivity of subsequent stands.

Location	Species	Experimental Design[a]	Date Begun[b]	Age at last eval. (yr)	Productivity Variables	Results	Comments	Source
USA - North Carolina	*Pinus taeda*	Harvest was main plot of split-plot in RCB (3 reps)	1981	6	Height, diameter, volume index; total accumulated biomass (above ground)	No differences in subsequent pine growth associated with harvest utilization	For plots with and without competition control	[18]
USA - Minnesota	*Pinus banksiana*	Non-replicated within site, 2 sites	1969, 1971	14, 17	Height, diameter, and stand basal area	Results were reversed on two sites; following complete-tree vs. stem-only harvest, pine height: -11% on one site, +19%[a] on second site	Differences in regeneration success and soil disturbance confound results	[49]
USA - Washington	*Pseudo-tsuga menziesii*	CRD on 2 sites (no reps)	1979	8	Total height, height growth	Most complete biomass removal resulted in 27% decrease on low quality sites, 25% increase on high quality site	N-fertilization eliminated treatment differences	[52, 53]
Sweden	*Picea abies Pinus sylvestris*	Summary of numerous RCB experiments	var.	21-25	Height growth, volume	Growth less in studies where biomass removal greatest; differences between treatments smallest on high quality sites and diminish with stand age		[47]

[a] CRD = completely randomized; RCB = randomized complete block
[b] Estimated from Figure 2, [49].

Apart from their effects on nutrient budgets, slash disposal and site preparation have numerous direct and indirect influences on soil physical and chemical properties. Use of heavy equipment can lead to soil compaction and disturbance which have been associated with reduced growth in some studies (e.g., [64, 65, 66, 67]). Compaction will be discussed as a separate topic in a subsequent section.

Growth Following Mechanical Slash Disposal
Virtually all available data from carefully designed field experiments show that survival and early growth of planted seedlings is greatest on sites where biomass removals were greatest and where organic and mineral soil horizons were most severely disturbed during mechanical slash disposal (Table 3.7). In the short-term, mechanical slash disposal often enhances plantation growth because of: 1) greater availability of nutrients, particularly N; 2) greatly reduced plant competition; and on some sites, 3) amelioration of soil compaction during associated tillage operations. Nevertheless, tree response to these treatments in older controlled experiments indicates a shift toward reduced growth as the forest matures (pattern of Figure 3.1(c)). Such results are consistent with retrospective site-preparation research that shows growth reductions associated with nutrient removal and soil disturbance do not appear until after crown-closure. For instance, after comparing a series of stands differing in years since site preparation in the Piedmont of southeastern USA, Burger and Kluender [68] concluded that growth of stands aged 5 years or less was improved by site preparation that included piling or severe soil disturbance; in older stands, however, decreased tree height and stand volume were associated with these treatments. Evaluation of the oldest of the available results presented in Table 3.7 seems to confirm this relationship. For instance, in the Piedmont of the southeast United States, early growth of *Pinus taeda* was greater on sites prepared by shear-pile and disk harrowing than on sites prepared by chopping followed by a light burn, even when differences in vegetation competition were minimized by repeated herbicide applications [18]. Examination of annual height growth curves (Figure 3.2) for this study indicates that most of the height growth advantage on shear-pile and harrowed plots occurred before age 6 and suggests that height growth on chop and burn plots may be greater after age 6. A retrospective study on a similar Piedmont site [69] found that

TABLE 3.7

Influence of slash disposal and soil disturbance on growth of subsequent stands.

Operations evalued	Species	Location	Experimental design	Date init.	Age at last eval. (yr)	Productivity Variables	Results	Comments	Source
Windrowing and burning	P. taeda	USA, North Carolina	subplots in main plot comparing harvest utilization (3 reps)	1981	8	Dominant tree height and total accumulation of biomass	Pine growth greatest in plots prepared by shear-pile/disking - least in chop/burn plots, biomass trends reversed; productivity greater in current rotation	No detectable influence of harvest utilization	[18]
Windrowing, bedding	P. radiata	Australia, Victoria	RCB (4 reps)	1972	8.5	Mean tree height and diameter, stand volume	Growth increased by bedding, decreased by windrowing		[57]
Windrowing, burning	P. radiata	New Zealand	CRD (3 reps)	1978	8.5	Mean tree height, diameter and stem volume*	No significant difference in areas with heavy weed competition. Weight: -8%, diameter: -7% or inter windrowed areas when only areas with low competition compared		[58]
Windrowing (root rake), bedding, P-fertilization	P. taeda	USA, South Carolina	RCB (2 reps, 3 sites)	1968-1973	5	Mean tree height	Height increased by fertilizer and bedding, decreased by root raking		[59]
Scalping[b], slash disposal, tillage, burning, disk harrowing	P. elliottii	USA, Florida, Georgia	Split plot (2 reps, 5 sites)	1958	17	Mean tree height and dbh, merchantable volume	Burning following by disk harrowing increased growth; burning followed by scalping decreased growth	Spacing and soil effects also evaluated	[60]

Operations evaluated	Species	Location	Experimental design	Date init.	Age at last eval. (yr)	Productivity Variables	Results	Comments	Source
Burning	Mixed conifers	USA, Washington and Oregon	Matched plots (non-replicated) at 44 locations	1946-51	40	SI, stand volume, trees > 4.1 cm dbh	No change in average long-term productivity associated with slash burning	Burning favored Douglas-fir & hardwood species; no vegetation control; or planting	[71]
Windrowing, V-blade, disk harrowing	P. taeda	USA, Georgia and South Carolina	CRD (12 reps)c	1981	6	Mean tree height and diameter, stand volume and total biomass	Windrowing followed by disk harrowing was the best treatment on 50% of the sites through age 3 but early growth advantages diminished by age 6	Growth response was site dependent, each of 7 treatments resulted in greatest growth on at least 1 site	[90]
Windrowing, bedding, burning, and P-fertilization	P. taeda	USA, Southeastern states	Split-plot (4 reps) (numerous locations)	1978-1981	8	Mean tree height, stand volume	Variable response to slash disposal, sustained response associated with tillage and fertilization	Site preparation and fertilization differed among sites, results grouped by category	[18]
Windrowing, burning	P. radiata	New Zealand	CRD (3 reps)	1978	8.5	Mean tree height, diameter and stem volume	No significant difference in areas with heavy weed competition. Weight: -8%, diameter - 7% or inter-windrowed areas when only areas with low competition compared		[58]

a RCB = randomized complete block, CRD = completely randomized.
b Intentional removal of surface litter and mineral soil.
c 12 sites, no replication within site.

standing volume at age 31 of *P. taeda* planted on a site that had been windrowed was 267 m³/ha compared to 345 m³/ha on an adjacent site that had been burned prior to planting. At index age 25, height of dominant and co-dominant *P. taeda* planted in interwindrow areas of the windrowed site was 3.3 m less than height of trees near windrows or planted on the adjacent burned site (a 14% reduction). Stem analyses of felled trees indicated heights of trees planted in burned-only and windrowed sites were similar until about age 7. After age 7, height growth was less in windrowed areas than in the burned-only areas.

One reason for the early growth advantage often associated with windrowing is improved competition control. In the previously described study by Allen *et al.* [18], only on chop-and-burn plots where competition differences were minimized by herbicide application did height growth equal or exceed growth on windrowed plots after age 6. On non-herbicided plots, height growth remained greater in windrowed plots through age 8. Parallel results have been reported by Dyck *et al.* [58], (Table 3.8). At locations where heavy weed competition existed, trees in non-windrowed or burned areas grew at similar rates through age 9. At locations where weed competition was low, growth was lower in windrowed plots.

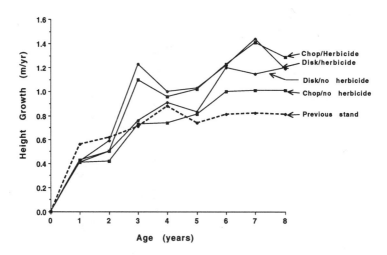

FIGURE 3.2: Annual height growth of dominant trees during the previous rotation and first 8 years of current rotation following four site preparation-weed control treatments in the North Carolina Piedmont (after [18]).

TABLE 3.8
Growth of 8.5-year-old *Pinus radiata* in windrowed and burned plots versus in non-prepared plots, all with different levels (open, partial, heavy) of competing vegetation (source: [58]).

	Height* (m)	Diameter* (cm)	Stem Volume* (m³/tree)
Windrowed and burned:			
Open	8.8[a]	14.6[a]	.066[a]
Partial	8.6[ab]	13.8[ab]	.057[ab]
Heavy	8.0[b]	12.3[b]	.043[b]
Non-prepared:			
Open	9.6[a]	15.7[a]	.082[a]
Partial	9.2[b]	13.9[b]	.061[b]
Heavy	8.7[b]	12.5[b]	.049[b]

* Dissimilar superscripts within a column indicate significant differences among treatments (t-test, a=.01).

We conclude from available field trials and retrospective research that long-term productivity is frequently deleteriously affected by mechanical slash removal such as windrowing but not by chopping or other operations that leave the majority of slash in place.

Productivity Change Following Slash Burning
Most studies have not found consistent differences in growth on areas burned after harvest versus those that were not burned [70, 71]. Moreover, early differences in tree growth between burned and nonburned plots are frequently not substantiated by later measurements [70]. As with results from most mechanical slash removal studies, results from slash burning studies are normally confounded by differences in vegetation regrowth and seedling establishment. Slash burning can reduce total soil nutrients while increasing their availability. Slash burning also increases vegetation competition in some places yet decreases it in others. It is difficult to isolate these effects without a vegetation-control treatment. This is true of older studies such as the Morris plots [72] evaluated by Miller and Bigley [71]. These matched plots were established in 1946 through 1951 to measure slash burning effects on fire hazard and natural regeneration. The plot-pairs were not replicated at any of the 44 locations in western Washington and Oregon, USA. Some plot pairs regenerated naturally; other pairs were planted. Burning had

several effects on subsequent stands. By reducing numbers of advance regeneration, fire shifted species composition away from shade-tolerant conifers. Slash burning enhanced regeneration of *P. menziesii*, primarily because more mineral soil was exposed for seeding. When site index of *P. menziesii* or volume production of all species were used as measures of site productivity, there was no evidence to generalize a productivity decline. Slashburning apparently decreased stand productivity at some locations, but increased it at others. This research is of particular interest because it represents the longest and most widely distributed data base of its kind. Nevertheless, the results are ambiguous because of the several confounding factors and design problems described above.

SOIL PHYSICAL CONDITIONS

Apart from effects on nutrient budgets, harvesting and other forest management activities can have direct and indirect effects on soil factors affecting site productivity. In particular, use of heavy equipment to remove logs can lead to soil compaction, particularly on the 10-25% of the site in skid trails, and to reduced tree growth in some locations [64, 65, 66], but not all [73]. Where soil compaction occurs, recovery can occur quickly as reported by Thorud and Frissel [74] following harvesting in Minnesota, or can require more than 50 years as reported by Perry [75] in North Carolina, and Grecean and Sands [76] in Australia. The risk to long-term site productivity depends upon the speed of recovery. Risks are low where compaction is limited to the surface soil and/or the ameliorating effects of freezing and thawing, wetting and drying, fauna activity, and root growth are large. Deep compaction, particularly when associated with degradation of soil structure (puddling) in warm climates with low shrink-swell soils, poses a significant long-term risk.

AMELIORATIVE TREATMENTS

Reviews of research on long-term site productivity tend to focus on harvest and associated management activities considered to have potentially negative impacts.

Several harvest- and regeneration-associated practices, however, have the potential to improve productivity throughout the rotation and, perhaps, through multiple rotations.

P Fertilization at Planting

Phosphorus fertilization may be the best example of a management practice that improves long-term site productivity (Figure 3.1(b)). On P-deficient sites, fertilization at planting has improved growth throughout a full rotation [77] and into a second rotation. For instance, Gentle *et al.* [78] report *P. radiata* growth response to fertilization with two forms of phosphorus at age 16 in both the first rotation and second rotation (Table 3.9). During the first rotation, height and volume accumulation curves diverged throughout the rotation indicative of a change in site productivity (Figure 3.1(b)). Information presented in Table 3.9 suggests this trend has continued in the second rotation.

While P fertilization can provide long-term growth responses, this is not always the case (e.g., [79]). As Ballard [89] points out, the general persistence of response is due to the large amounts of applied P (50-100 kg/ha) in comparison to removals in forest harvest (5-70 kg/ha) and the sustained availability provided by this amount of phosphorus.

TABLE 3.9

Response of *Pinus radiata* to a single application of phosphorus, in the rotation fertilizer was applied and in the second rotation following fertilization (Source: [78]).

	Dominant tree height* (m)	Stem volume[a]* (m³/ha)
Initial Rotation		
No fertilization	13.2[a]	38[a]
Rock phosphate	16.9[b]	118[b]
Super phosphate	17.4[b]	128[b]
Second Rotation		
No fertilization	13.7[a]	52[a]
Rock phosphate	16.2[b]	135[b]
Super phosphate	16.3[b]	145[b]

*   Within a rotation, dissimilar superscripts indicate significant differences (a=.05).
[a]   To a 10 cm (i.b.) diameter

Liming (Ca additions)

In contrast to widespread use of lime in agriculture, addition of Ca and/or Mg to forests is relatively rare. Although some forests in Germany were limed as early as 1860, liming of forests has never been widespread. Wiedemann ([80] - p. 279) cites several examples of growth increases of 20 to 25% during two or more decades after quick lime (CaO) or limestone ($CaCO_3$) was added to established stands of pole-sized trees with heavy accumulation of forest floor (raw humus). Baule and Fricker ([81] - p. 56) concluded: "In the forest, liming therefore in most cases is conducted less with the object of providing the trees with a supply of calcium but pre-eminently of improving the physical, chemical and biological properties of the soil. It is actually possible to make acid soils or those with a low calcium content capable of carrying broad-leaved trees by liming to a sufficient depth." In the last two decades, interest in liming has been stimulated by recognition of the threat to long-term site productivity posed by acid precipitation.

Andersson and Lundkvist [82] recently summarized results from liming experiments completed since 1912 in Sweden, Finland, and Germany. These authors concluded that long-term increases in site productivity are generally associated with liming; however, response to liming often follows a pattern similar to, but the reverse of, the pattern illustrated in Figure 3.1(c). In most cases, growth rate of limed trees first decreased in comparison to the unlimed control and, only after about 40 years, then increased in comparison to the unlimed control. The authors speculated that this growth pattern may occur because lime additions stimulated microbial activity that immobilized N normally available for tree uptake. They also suggested that micronutrient deficiencies may have been induced by excessive liming and that positive growth responses only occurred after natural acidification compensated for lime-induced deficiencies. Regardless of the cause, chlorosis and growth retardation can be a short-term consequence of lime addition on many forest sites.

Comparable long-term results are not generally available outside Europe. Although liming has often been included in programs designed to evaluate forest fertilization, few programs have carried liming trials for extended periods of time because initial results were not encouraging. For instance, Ca additions were included as part of the original fertilization trials in slash pine in the southeastern

U.S. but were not included in subsequent research because Ca additions failed to increase tree growth [83].

Tillage

Other ameliorative practices that may have long-term benefits are bedding, disk harrowing, or ripping which can improve physical conditions restricting root growth. While the short-term benefits of such treatments are clear, long-term benefits are not. Many responses to treatment seem to fall into responses illustrated in Figure 3.1(a). Trees planted after these treatments achieve an early growth advantage which can be maintained throughout the rotation, but the advantage or response does not increase; thus, this is not a long-term change in site productivity. Such growth responses have been reported for *P. elliottii* and *P. taeda* in the United States following bedding [6], bedding and disk harrowing [84, 85, 86], and following bedding and subsoiling for *P. radiata* in New Zealand [87].

The pattern of response to such tillage treatments may relate to the reasons for the observed initial response. Responses may be temporary where benefits of tillage are largely due to: 1) decreased vegetative competition; 2) accelerated occupation of rooting volume without an increase in absolute rooting volume; 3) short-term increases in N mineralization; and 4) beneficial changes in thermal regimes or water infiltration and distribution. Responses appear to be long-term where amelioration of soil physical conditions which restricted total rooting volume were largely responsible for initially poor growth. Thus, bedding of sandy sites does not appear to have long-term benefits because much of the initial benefit is associated with ephemeral increases in nitrogen mineralization and competition control. Responses to bedding may be long-term, however, on fine-textured soils where bed configuration is maintained throughout the rotation and surface drainage improved.

CONCLUSIONS

Our understanding of factors affecting forest growth is incomplete; therefore, results from carefully controlled field trials, rather than models, provide more convincing evidence of productivity changes caused by harvesting practices. To be acceptable evidence of change in long-term site productivity, three conditions must be met. First, differences in tree growth must be attributable to differences in site conditions, rather than to differences in resource allocation among target and non-target species or to differences in plant species or genotype. Second, growth results must be available for a sufficient time so that the influence of ephemeral differences in initial site conditions has diminished and so that capacity of the site to support tree growth is stressed. Finally, adequate experimental control must exist. Shifts in resource allocation will appear as an early increase in tree growth. This increase is maintained after crown closure, but further growth improvements are not evident. Hence this pattern of growth response does not indicate a change in long-term site productivity. Few studies meet adequate experimental requirements and none have been carried long enough to provide long-term results. Regardless of study purpose, minimum requirements for acceptance are: 1) appropriate control plots against which treatment differences can be evaluated; and 2) factors other than treatments have been minimized or measured to account for their influence on observed differences in growth.

Issues of species change and second-rotation decline are not easily disentangled from associated questions on harvest utilization, slash disposal, and soil tillage. Despite lack of evidence to support the concept that replacing mixed vegetation with conifers decreases site productivity, some authors continue to accept species-induced declines in productivity as fact. We conclude that current evidence does not indicate such declines occur or, if they do occur, are probably not related simply to establishment of conifers and their continued management. In contrast, evidence from field trials clearly shows that addition of nitrogen-fixing tree species to stands on N-deficient sites increases site productivity. Hence removing such beneficial species could reduce long-term productivity.

Concerns about forest harvest on long-term productivity should not be summarily dismissed. Direct removal and indirect loss of essential plant nutrients will occur in forests managed for wood and fiber production. Without replacement by natural sources or by fertilization, productivity could decline. Although nutrient-balance analyses provide indirect evidence for projecting declines in long-term site productivity in managed forests, direct evidence of tree growth rarely shows such declines. Most field trials provide weak evidence for productivity declines after increased harvest utilization because of poor study design and because a number of rotations that may be required to induce productivity decline. Apart from their effects on nutrient budgets, slash disposal and site preparation have numerous direct and indirect influences on soil physical and chemical properties. Virtually all available data from carefully designed field experiments show that survival and early growth of planted seedlings is greatest on sites where biomass removals were greatest and where organic and mineral soil horizons were most severely disturbed during mechanical slash disposal. One reason for the early growth advantage often associated with windrowing is improved competition control. We conclude from available field trials and retrospective research that long-term productivity is frequently deleteriously affected by mechanical slash removal. In contrast, most studies have not found consistent differences in growth on areas burned after harvest versus those that were not burned. As with results from most mechanical slash removal, effects on soil properties are normally confounded by differences in vegetative regrowth and seedling establishment.

Native soil productivity can be improved. On P-deficient sites, fertilization at planting can improve growth throughout a full rotation and into a second rotation. Other ameliorative practices that can have long-term benefits are bedding, disk harrowing, or ripping which can improve physical conditions restricting root growth. While the short-term benefits of such treatments to tree growth are clear, long-term benefits are less evident.

The cumulative effects of repeated harvests and stand regeneration practices remain unquantified. Effectively designed and implemented field trials over a range of soil and climatic conditions are needed to provide reliable quantification. Data from such trials are needed to provide independent validation of computer models and our understanding of factors affecting tree and stand performance.

LITERATURE CITED

1. Pastor, J. and Post, W.M. Development of a linked forest productivity-soil process model. Oak Ridge National Laboratory, Environmental Sciences Division Publication No. 2455, 1985, 162 p.

2.   Kimmins, J.P. Community organization : methods of study and predictions of the productivity and yield of forest ecosystems. *Canadian Journal of Botany*, 1988, **66**, 2654-2672.

3.   Reed, D.D., Holmes, M.J., Jones, E.A., Liechty, H.O., and Mroz, G.D. An ecological growth model for four northern hardwood species in upper Michigan. In: *Process Modeling of Forest Growth Responses to Environmental Stress*. (Eds.) R.K. Dixon, R.S. Meldahl, G.A. Ruark, and W.G. Warren. Timber Press, Portland, 1990, pp. 288-293.

4.   Weinstein, D.A. and Beloin, R. Evaluating effects of pollutants on integrated tree processes: a model of carbon, water, and nutrient balances. In: *Process Modeling of Forest Growth Responses to Environmental Stress*. (Eds.) R.K. Dixon, R.S. Meldahl, G.A. Ruark, and W.G. Warren. Timber Press, Portland, USA, 1990, pp. 313-323.

5.   Wisiol, K. and Hesketh, J.D. (Eds.). *Plant Growth Modeling for Resource Management*. CRC Press, Boca Raton, FL USA, 1987.

6.   Hughes, J.H., Campbell, R.C., Duzan, H.W., and Dudley, C.S. Site index adjustments for intensive forest management treatments at North Carolina. Weyerhaeuser Forest Research Technical Report 042-1404/79/24, 1979.

7.   Mead, D.J., Whyte, A.G.D., Woollons, R.C., and Beets, P.N. Designing long-term experiments to study harvesting impacts. In: *Long-term Field Trials to Assess Environmental Impacts of Harvesting*. Proceedings, IEA/BE T6/A6 Workshop, Feb. 1990, Florida, USA. (Eds.) W.J. Dyck and C.A. Mees. New Zealand Forest Research Institute Bulletin 161, 1991, pp. 107-124.

8.   Likens, G.E., Bormann, F.H., Johnson, N.M., Fisher, D.W., and Pierce, R.S. Effects of forest cutting and herbicide treatment on nutrient budgets in the Hubbard Brook watershed-ecosystem. *Ecological Monographs*, 1970, **40**, 23-47.

9.   Dyck, W.J. and Mees, C.A. (Eds.). *Long-Term Field Trials to Assess Environmental Impacts of Harvesting*. Proceedings, IEA/BE T6/A6 Workshop, February 1990, Florida, USA. New Zealand Forest Research Institute Bulletin 161, 1991.

10.   Ovington, J.D. Studies of the development of woodland conditions under different trees. *Journal of Ecology*, 1953, **111**, 12-34.

11.   Keeves, A. Some evidence of loss of productivity with successive rotations of *Pinus radiata* in the south-east of South Australia. *Australian Forestry*, 1966, **30**, 51-63.

12. Bednall, B.H. The problem of lower volumes associated with second rotations in *Pinus radiata* plantations in South Australia. Woods and Forest Dept, South Australia, Bulletin 17, 1968.

13. Whyte, A.G.D. Productivity of first and second crops of *Pinus radiata* on the Moutere soils of Nelson. *New Zealand Journal of Forestry*, 1973, **18**, 87-103.

14. Evans, J. Two rotations of *Pinus patula* in the Usutu forest, Swaziland. *Commonwealth Forestry Review*, 1975, **54**, 64-81.

15. Evans, J. A further report on second rotation productivity in the Usutu Forest, Swaziland - results of the 1977 reassessment. *Commonwealth Forestry Review*, 1978, **57**, 253-261.

16. Squire, R.O., Farrell, P.W., Flinn, D.W., and Aeberli, B.C. Productivity of first and second rotation stands of radiata pine on sandy soils. *Australian Forestry*, 1985, **48**, 127-137.

17. Cellier, K.M., Boardman, R., Boomsma, D.B., and Zed, P.G. Response of *Pinus radiata* D. Don to various silvicultural on adjacent first-and second-notation sites near Tantanoola, South Australia. I. establishment and growth up to age 7 years. *Australian Forest Research*, 1985, **15**, 431-447.

18. Allen, H.L., Morris, L.A., and Wentworth, T.R. Productivity comparisons between successive loblolly pine rotations in the North Carolina Piedmont. In: *Long-term Field Trials to Assess Environmental Impacts of Harvesting*. Proceedings, IEA/BE T6/A6 Workshop, Feb. 1990, Florida, USA. (Eds.) W.J. Dyck and C.A. Mees. New Zealand Forest Research Institute Bulletin 161, 1991, pp. 125-136.

19. Sheppard, K.R. *Plantation Silviculture*. Martinus Nijhoff, Dordrecht, 1986.

20. Miller, R.E. and Murray, M.D. The effects of red alder on growth of Douglas-fir. In: *Utilization and management of alder*. (Comp.) D.G. Briggs, D.S. DeBell, and W.A. Atkinson. USDA Forest Service General Technical Report PNW-70. Pacific Northwest Forest and Range Experiment Station, Portland, Oregon, 1978, pp. 283-306.

21. Tarrant, R.F. and Miller, R.E. Accumulation of organic matter and soil nitrogen beneath a plantation of red alder and Douglas-fir. *Soil Science Society of America Journal*, 1963, **27**, 231-234.

22. Miller, R.E. and Tarrant, R.F. Long-term growth response of Douglas-fir to ammonium-nitrate fertilizer. *Forest Science*, 1983, **29**, 127-137.

23. Binkley, D. Mixtures of nitrogen$_2$-fixing and non-nitrogen$_2$-fixing tree species. In: *The Ecology of Mixed-species stands of trees*. (Eds.) M.G.R. Cannell *et al.* Special Publication No. 11, British Ecological Society Oxford, Blackwell Scientific Publications, London, 1992, pp. 99-123.

24. DeBell, D.S., Whitesell, C.D., and Schubert, T.H. Using N$_2$-fixing *Albizia* to increase growth of *Eucalyptus* plantations in Hawaii. *Forest Science*, 1989, **35**, 64-75.

25. Krause, H.H., Weetman, G.F., and Arp, P.A. Nutrient cycling in boreal forest ecosystems of North America. In: *Forest Soils and Land Use*. Proceedings, Fifth North American Forest Soils Conference, Fort Collins, CO, Aug. 1978. (Ed.) C.T. Youngberg. Colorado State Univ. Press, Fort Collins, 1979, pp. 287-319.

26. Ralston, C.W. Mineral cycling in temperate forest ecosystems. In: *Forest Soils and Land Use*, Proceedings, Fifth North American Forest Soils Conference, Fort Collins, Colorado, August 1978. (Ed.) C.T. Youngberg. Colorado State Univ. Press, Fort Collins, 1979, pp. 320-340.

27. Cole, D.W. and Johnson, D.W. Mineral cycling in tropical forests. In: *Forest Soils and Land Use*. Proceedings, Fifth North American Forest Soils Conference, Fort Collins, Colo., August 1978. (Ed.) C.T. Youngberg. Colorado State University Press, Fort Collins, 1979, pp. 341-356.

28. Duvigneaud, P. and Denaeyer-DeSmet, S. Biological cycling of minerals in temperate deciduous forests. In: *Analysis of Temperate Forest Ecosystems*. (Ed.) D.E. Reichle. Springer-Verlag, New York, 1970, pp. 199-225.

29. Golley, F.B., McGinnis, J.T., Clements, R.G., Child, G.I., and Duever, M.J. *Mineral Cycling in a Tropical Moist Forest Ecosystem*. University of Georgia Press, Athens, 1975, 248 p.

30. Weetman, G.F. and Webber, B. The influence of wood harvesting on the nutrient status of two spruce stands. *Canadian Journal of Forest Research*, 1972, **2**, 351-369.

31. Foster, N.W. and Morrison, I.K. Distribution and cycling of nutrients in a natural *Pinus banksiana* ecosystem. *Ecology*, 1976, **57**, 110-120.

32. Green, D.C. and Grigal, D.F. Nutrient accumulations in jack pine stands on deep and shallow soils over bedrock. *Forest Science*, 1980, **26**, 325-333.

33. Ovington, J.D. Quantitative ecology and the woodland ecosystem concept. *Advances in Ecological Research*, 1962, **1**, 103-192.

34. Burger, J.A. The effects of harvest and site preparation on the nutrient budget of an intensively managed southern pine forest. Ph.D. Dissertation, University of Florida. Univ. Microfilms, Ann Arbor, Mich. (Diss. Abstr. 40:4047B), 1979, 184 p.

35. Cole, D.W., Gessel, S.P., and Dice, S.F. Distribution and cycling of nitrogen, phosphorus, potassium, and calcium in a second growth Douglas-fir ecosystem. In: *Proceedings, Symposium on Primary Productivity and Mineral Cycling in Natural Ecosystems.* (Ed.) H.E. Young. University of Maine Press, Orono, 1967, pp. 197-232.

36. Greenland, D.J., and Kowal, J.M.L. Nutrient content of the moist tropical forests of Ghana. *Plant and Soil*, 1960, **12**, 154-173.

37. Hansen, E.H. and Baker, J.B. Biomass and nutrient removal in short rotation intensively cultured plantations. In: *Proceedings, Symposium on Impact of Intensive Harvesting on Forest Nutrient Cycling*, Syracuse, NY, 13-16 August 1979. Northeastern Forest Experiment Station, Broomall, PN, 1979, pp. 130-151.

38. Morrison, I.K. and Foster, N.W. Biomass and element removal by complete-tree harvesting of medium rotation stands. In: *Proceedings, Symposium on Impact of Intensive Harvesting on Forest Nutrient Cycling.* Syracuse, NY, August 1979. USDA NE Forest Experiment Station, Broomall, PN, 1979, pp. 98-110.

39. Marian, G.M. Biomass and nutrient removal in long rotation stands. In: *Proceedings, Symposium on Impact of Intensive Harvesting on Forest Nutrient Cycling*, Syracuse, NY, 13-16 August 1979. USDA Northeastern Forest Experiment Station, Broomall, PN, 1979, pp. 98-110.

40. Kimmins, J.P. Evaluation of the consequences for future tree productivity of the loss of nutrients in whole-tree harvesting. *Forest Ecology and Management*, 1977, **1**, 169-183.

41. Mann, L.K. and others (numerous). Effects of whole-tree and stem-only clearcutting on post harvest hydrologic losses, nutrient capital and regrowth. *Forest Science*, 1988, **34**, 412-428.

42. Silkworth, D.R. and Grigal, D.F. Determining and evaluating nutrient losses following whole-tree harvesting of aspen. *Soil Science Society of America Journal*, 1982, **46**, 626-631.

43. Johnson, D.W. and Todd, D.E. Nutrient export by leaching and whole-tree harvesting in a loblolly pine and mixed oak forest. *Plant and Soil*, 1987, **102**, 99-109.

44. Miller, R.E., Hazard, J.W., Bigley, R.E., and Max, T.A. Some results and design considerations from a long-term study of slash burning effects. In: *Research Strategies for Long-term Site Productivity*. Proceedings, IEA/BE A3 Workshop, Seattle, WA, August 1988. (Eds.) W.J. Dyck and C.A. Mees. IEA/BE A3 Report 8. New Zealand Forest Research Institute Bulletin 152, 1989, pp. 63-78.

45. Wells, C.G., Jorgensen, J.R., and Burnette, C.E. Biomass and mineral elements in a thinned loblolly pine plantation at age 16. USDA Forest Service Research Paper SE-126, 1975, 10 p.

46. Koch, P. A concept for southern pine plantations operation in the year 2020. In: *Forest Plantations - the Shape of the Future*. Proceedings, Weyerhaeuser Science Symposium No. 1, Tacoma, WA, 1980, pp. 69-85.

47. Lundkvist, H. Ecological effects of whole tree harvesting - some results from Swedish field experiments. In: *Predicting consequences of Intensive Forest Harvesting on Long-term Productivity by site classification*. (Eds.) T.M. Williams and C.A. Gresham. IEA/BE Project A3, Report 6. Baruch Forest Science Institute, Clemson University, Georgetown, SC, USA, 1988, pp. 131-140.

48. Compton, J.E. and Cole, D.W. Impact of harvest intensity of growth and nutrition of successive rotations of Douglas-fir. In: *Long-term Field Trials to Assess Environmental Impacts of Harvesting*. Proceedings, IEA/BE T6/A6 Workshop, Feb. 1990, Florida, USA. (Eds.) W.J. Dyck and C.A. Mees. New Zealand Forest Research Institute Bulletin 161, 1991, pp. 151-161.

49. Williams, T.M., Alm, A.A., and Mace, A.C. Jr. Fifteen-year jack pine growth on full-tree harvesting experiments in Northeastern Minnesota. In: *Research Strategies for Long-term Site Productivity*. Proceedings, IEA/BE A3 Workshop, Seattle, WA, August 1988. (Eds.) W.J. Dyck and C.A. Mees. IEA/BE A3 Report 8. New Zealand Forest Research Institute Bulletin 152, 1989, pp. 111-121.

50. Tew, D.T., Morris, L.A., Allen, H.L., and Wells, C.G. Estimates of nutrient removal, displacement and loss resulting from harvest and site preparation of a *Pinus taeda* plantation in the Piedmont of North Carolina. *Forest Ecology and Management*, 1986, **15**, 257-267.

52. Cole, D.W. Impact of whole-tree harvesting and residue removal on productivity and nutrient loss from selected soils of the Pacific Northwest. College of Forest Resources, University of Washington, Report AR-10, Seattle, WA, 1988, 12 p.

53. Bigger, C.M. and Cole, D.W. Effects of harvest intensity on nutrient losses and future productivity in high and low productivity red alder and Douglas-fir stands. In: *Proceedings, Symposium on Forest Site and Continuous Productivity*. USDA Forest Service General Technical Report PNW-163, 1983, pp. 167-187.

54. Webber, B. Potential increases in nutrient requirements of *Pinus radiata* under intensified management. *New Zealand Journal of Forestry Science*, 1978, **8**, 146-159.

55. Ballard, R. Effect of slash and soil removal on the productivity of second rotation radiata pine on a pumice soil. *New Zealand Journal of Forestry Science*, 1978, **8**, 248-258.

56. Morris, L.A., Pritchett, W.L., and Swindel, B.F. Displacement of nutrients into windrows during site preparation of a pine flatwoods forest. *Soil Science Society of America Journal*, 1983, **17**, 591-594.

57. Turvey, N.D. and Cameron, J.N. Site preparation for a second rotation of radiata pine: soil foliage chemistry and effect on tree growth. *Australian Forest Research*, 1986, **16**, :9-19.

58. Dyck, W.J., Mees, C.A., and Comerford, N.B. Medium-term effects of mechanical site preparation on radiata pine productivity in New Zealand - a retrospective approach. In: *Research Strategies for Long-term Site Productivity*. Proceedings, IEA/BE A3 Workshop, Seattle, WA, August 1988. (Eds.) W.J. Dyck and C.A. Mees. IEA/BE A3 Report 8. New Zealand Forest Research Institute Bulletin 152, 1989, pp. 79-92.

59. Wilhite, L.P. and McKee, W.H. Jr. Site preparation and phosphorus application alter early growth of loblolly pine. *Southern Journal of Applied Forestry*, 1985, **9**, 103-109.

60. Sarigumba, T.I. and Anderson, G.A. Response of slash pine to different spacings and site-preparation treatments. *Southern Journal of Applied Forestry*, 1979, **3**, 91-94.

61. Kimmins, J.P. and Feller, M.C. Effect of clearcutting and broadcast slash-burning on nutrient budgets, streamwater chemistry and productivity in

Western Canada. In: Proceedings, Div. 1, XVI IUFRO World Congress, 1976, pp. 186-197.

62. Zavitkovski, J. and Newton, M. Ecological importance of snowbush *Ceonothus velutinus* in the Oregon Cascades. *Ecology*, 1968, **49**, 1134-1145.

63. Wells, C.G. and Morris, L.A. Maintenance and improvement of soil productivity. p. 306-318. In: *Proceedings, Symposium on the Loblolly Pine Ecosystem*, December 1982. N.C. State University, Raleigh, NC, USA, 1983, pp. 306-318.

64. Hatchell, G.E., Ralston, G.W., and Foil, R.R. Soil disturbances in logging. *Journal of Forestry*, 1970, **68**, 772-775.

65. Helms, J.A. and Hipkin, C. Effects of soil compaction on tree volume in a California ponderosa pine plantation. *Western Journal of Applied Forestry*, 1986, **1**, 121-124.

66. Froehlich, H.A., Miles, D.W.R., and Robbins, R.W. Growth of young *Pinus ponderosa* and *Pinus contorta* on compacted soils in central Washington. *Forest Ecology and Management*, 1986, **15**, 285-294.

67. Clayton, J.L., Kellogg, G., and Forrester, N. Soil disturbance–tree growth relations in central Idaho clearcuts. USDA Forest Service Research Note INT-372, 1987, 6 p.

68. Burger, J.A. and Kluender, R.A. Site preparation - Piedmont. In: *Symposium on the Loblolly Pine Ecosystem (East Region)*. (Eds.) R.C. Kellison and S.A. Gingricle. USDA Forest Service and North Carolina State University, Raleigh, N.C., USA, 1982, pp. 58-74.

69. Fox, T.R., Morris, L.A., and Maimone, R.A. Windrowing reduces growth in a loblolly pine plantation in the North Carolina Piedmont. In: Proceedings, Fifth Biennial Southern Silvicultural Research Conference, 1-3 Nov. 1988, Memphis, TN, USA. (Comp.) J.H. Miller. USDA Forest Service General Technical Report 50-74, 1989, pp. 133-139.

70. Curran, M.P. and Ballard, T.M. Some slashburning effects on soil and trees in British Columbia. In: *Sustained Productivity of Forest Soils*, Proceedings of the 7th North American Forest Soils Conference. (Eds.) S.P. Gessel *et al.* University of British Columbia, Faculty of Forestry Publication, Vancouver, B.C., Canada, 1990, pp. 355-361.

71. Miller, R.E. and Bigley, R.E. Effects of burning Douglas-fir logging slash on stand development and site productivity. In: *Sustained Productivity of Forest Soils*, Proceedings of the 7th North American Forest Soils Conference. (Eds.) S.P. Gessel *et al.* University of British Columbia, Faculty of Forestry Publication, Vancouver, BC, Canada, 1990, pp. 362-376.

72. Morris, W.G. Influence of slash burning on regeneration, other plant cover and fire hazard in the Douglas-fir region. USDA Forest Service Research Paper 29. Pacific Northwest Forest and Range Experiment Station, Portland, OR, USA, 1958.

73. Miller, R.E., Stein, W.I., Heninger, R.L., Scott, W., Little, S., and Goheen, D. Maintaining and improving site productivity in the Douglas-fir region. In: *Maintaining the Long-Term Productivity of the Pacific Northwest Forest Ecosystems*. (Eds.) D.A. Perry *et al.* Proceedings, Symposium, 1987. Timber Press, Portland, Oregon USA, 1987, pp. 98-136.

74. Thorud, D.B. and Frissel, S.S. Time changes in soil density following compaction under an oak forest. Minnesota Forestry Research Note 257, School of Forestry, University of Minnesota, St. Paul, MN, 1976.

75. Perry, T.O. Soil compaction and loblolly pine tree growth. *Tree Planters' Notes*, 1964, **67**, 9.

76. Grecean, E.L. and Sands, R. Compaction of forest soils: a review. *Australian Journal of Soil Research*, 1980, **18**, 163-188.

77. Pritchett, W.L. and Comerford, N.B. Long-term response to phosphorus fertilization on selected southeastern Coastal Plain soils. *Soil Science Society of America Journal*, 1982, **46**, 640-644.

78. Gentle, S.W., Humphreys, F.R., and Lambert, M.J. An examination of a *Pinus radiata* fertilizer trial fifteen years after treatment. *Forest Science*, 1986, **32**, 822-829.

79. Mead, D.J. and Gadgil, R.L. Fertilizer use in established radiata pine stands in New Zealand. *New Zealand Journal of Forestry Science*, 1978, **8**, 105-134.

80. Wiedemann, E. Ertragskundliche und waldbauliche Grundlagen der Forstwirtschaft. Sauerländer's Verlag. Frankfurt am Main. 3rd edition, 1950, 346 p.

81. Baule, H. and Fricker, C. *The Fertilizer Treatment of Forest Trees.* BLV Verlag Syeselkchaft, Munich, 1970.

82. Andersson, F.O. and Lundkvist, H. Long-term Swedish field experiments in forest management practices and site productivity. In: *Research Strategies for Long-term Site Productivity.* Proceedings, IEA/BE A3 Workshop, Seattle, WA, Aug. 1988. (Eds.) W.J. Dyck and C.A. Mees. IEA/BE A3 Report 8. New Zealand Forest Research Institute Bulletin 152, 1989, pp. 125-137.

83. Pritchett, W.L. and Smith, W.H. Forest Fertilization in the U.S. Southeast. In: *Forest Soils and Forest Land Management*, Proceedings, Fourth North American Forest Soils Conference, Laval University, August 1973, Laval University Press, Quebec, 1975, pp. 467-476.

84. Haywood, J.D. Planted pines do not respond to bedding on an Acadia-Beauregard-Kolen silt loam site. USDA Southern Forest Experiment Station Research Note 50-259, 1980.

85. Tiarks, A.E. Effect of site preparation and fertilization on slash pine growing on a good site. In: Proceedings, Second Biennial Southern Silviculture Research Conference, November 1982, Atlanta, USA. USDA Forest Service General Technical Report SE-24, 1983, pp. 34-39.

86. McKee, W.H. Jr., and Wilhite, L.P. Loblolly pine response to bedding and fertilization varies by drainage class on lower Atlantic Coastal Plain sites. *Southern Journal of Applied Forestry*, 1986, **10**, 16-21.

87. Mason, E.G., Cullen, A.W.J., and Rijkse, W.C. Growth of two *Pinus radiata* stock types on ripped and ripped/bedded plots at Karioi forest. *New Zealand Journal of Forestry Science*, 1988, **18**, 287-296.

CHAPTER 4

IMPACTS OF HARVESTING AND ASSOCIATED PRACTICES ON OFF-SITE ENVIRONMENTAL QUALITY

D.G. NEARY

USDA Forest Service, Rocky Mountain Station,
Flagstaff, AZ 86001, USA

J.W. HORNBECK

USDA Forest Service, NE Forest Experiment Station
Durham, NH 03824, USA

INTRODUCTION

Since the inception of forestry as a profession and science, foresters have pondered over and debated the effects of tree harvesting on site productivity. That particular topic is the major focus of this volume. However, in the past four decades, the harvesting of trees for production of wood products or biomass fuels has grown increasingly more controversial for other reasons. Debates over the environmental and ecological consequences of harvesting in forest ecosystems have intensified despite research investments to identify and quantify these impacts. Initially much of the controversy centered around easily-viewed and conceptualized on-site phenomena such as aesthetics, wildlife habitat, and erosion. Beginning in the 1970's, awareness of the potential off-site effects of forest harvesting and associated practices became more apparent. This led to examination of landscape-scale environmental impacts and the development of the concept of cumulative effects [1]. Now the debate encompasses off-site effects which occur over tremendous spatial scales ranging from local to global.

In many countries, laws have been enacted to address concerns about environmental quality arising from a variety of land management activities. In the United States, numerous national laws affect both the methods and intensity of forest harvesting and site preparation on public and private lands. Because of public reaction to clearcut logging, the U.S. Forest Service has, among other actions, limited the use, frequency, and size of harvesting units. In states like California, legislation has been enacted to regulate forestry management practices which can impact environmental quality. Concerns over the environmental impacts of pesticides have led to a ban on herbicide use in Sweden's forests without a truly scientific analysis. In Canada, the number of individual herbicides which can be used is severely limited, and the province of Saskatchewan has suspended use of herbicides in forests altogether.

The primary on-site and off-site characteristics of the environment which can be affected as a result of harvesting forest lands include air quality, surface water quantity and quality, groundwater quality and quantity, terrestrial and aquatic wildlife habitat, and biological diversity. The environmental changes which result from harvesting forest lands can have positive as well as negative impacts on other components of the ecosystem. After decades of research on the environmental effects of forest harvesting, much is known about the impacts on the plant-soil-water-atmosphere continuum on-site [2, 3]. Most environmental concerns now relate to landscape-scale impacts and the cumulative effects of all land management disturbances (forestry, agriculture, urban, industrial, etc.) over time (past, present, and future). The issue of global climate change and how forest clearing and harvesting affects atmospheric emissions and greenhouse gases has elevated these once local and regional concerns to national and global issues.

Landscape-scale physical, chemical, and biological impacts of forest harvesting and associated disturbances have been under study at many sites for the past 20 to 60 years [3, 4, 5]. Most commonly, watersheds have been used world-wide as the basis for conducting ecosystem-level research on both undisturbed and disturbed forest ecosystems [6]. Much of the "off-site" environmental impact information discussed in this chapter was obtained by watershed-scale scientific methodology.

The objective of this chapter is to synthesize the impacts of forest harvesting and associated site preparation practices on off-site environmental quality. The term "off-site" here refers to landscape units outside of actual harvested areas but within reasonably contiguous ecosystems. It includes non-harvested riparian zones within cutting units as well as adjacent units of the landscape and first- to fourth-order watersheds. Groundwater systems beneath the biologically-active soil zone of harvested stands are also considered "off-site" for the purposes of the discussion in this chapter. Regional, national, or global environmental impacts may be mentioned, but are beyond the scope of this analysis. The only "on-site" environmental impacts which will be examined are ones that relate to biological diversity and wildlife habitat. On-site disturbances associated with forest harvesting may affect the diversity and wildlife habitat values of surrounding forest ecosystems.

AIR QUALITY

Air quality is not really affected by forest harvesting *per se* but by associated site preparation practices. Burning and herbicide applications are the chief activities which can result in "off-site" air quality problems. Fires can generate varying amounts of smoke, organic by-products of pyrolysis, and oxides of carbon (C), sulphur (S), and nitrogen (N) depending on fire intensity, fuel type, fuel loadings, and climatic conditions. When used in association with herbicides (i.e., as "brown and burn" vegetation-control methods), fires have a limited potential under certain conditions to result in air-transport of these pesticides. However the primary mechanism of "off-site" herbicide transport in air is aerosol drift during application operations.

Smoke
Fire is used throughout the world's forested regions to reduce organic debris left after forest harvesting or clearing which might interfere with future land management. The "off-site" effects of smoke that can impinge upon environmental quality include visual deterioration, volatilization of C, N, and S as gases, and transport of natural and synthetic organic compounds.

Visual Hazards: The amount of smoke generated by prescribed slash-reduction fires is dependent on fuel load, moisture, and climatic conditions. Carefully prescribed burns are usually timed to minimize smoke production. Excessive smoke can reduce the scenic qualities of forests as well as create localized vision hazards on transportation corridors. In some areas, legal restrictions or narrow prescription conditions have been imposed on burning in forest lands to maintain air quality.

C, N, and S Oxides: Both prescribed fires after forest harvesting and wild-fires produce volatilization losses of oxides of C, N, S, potassium (K), etc. Again, the degree of volatilization is dependent on burning conditions. The main off-site impact is on greenhouse gases and ultimately atmospheric processes. This topic is analyzed in more detail by Schlesinger [7].

As an example, Woodwell et al. [8] estimated that of the annual carbon flux from forests to the atmosphere of 1.8 to 4.7 x 10^9 Mg, 23-35% originates from the harvesting of forests (3 to 15 x 10^6 ha/yr). Clearing for agriculture and grazing constituted the largest source of forest-stored carbon. Of the estimated annual forest-to-atmosphere C flux, the majority came from tropical forest clearing [9]. Harvesting and clearing in intensively managed temperate forests at the rate common in the early 1980's did not significantly contribute to the release of carbon compounds into the atmosphere (<2%).

Organic Chemical Transport: Forest fires can affect air quality off-site by the emission of natural polynuclear aromatic hydrocarbons (PAH) as well as suspended particulate matter [10]. Estimates of PAH production in the United States range from 10-130 Mg/yr. Some of these PAH's are known carcinogens (e.g., benzo[a]pyrene) as well as allergens.

With the advent of "brown and burn" forest weed control methods, public concerns grew in the mid-1980s about the "off-site" transport of herbicide residues in smoke. Also, utilization of herbicide-killed hardwoods for firewood raised questions about the relationships between "off-site" exposure in wood smoke and human health and safety. McMahon *et al.* [11] found that cool (<500 °C) and smouldering fires can volatilize varying amounts of pesticide (11-92%) from burning wood. Hot fires (>600 °C) result in complete thermal decomposition of forestry-use pesticides. However, even with smouldering fires, pesticides such as dicamba, 2,4-D, picloram, triclopyr, dichlorprop, lindane, and chlorpyrifos did not exceed air quality standards in closed environments, and exposure to residues did not exceed recommended human health no-effect levels [12, 13].

McMahon and Bush [14] reported on forest worker exposures to imazapyr, triclopyr, hexazinone, and picloram during site preparation burns in the southern United States 30 to 169 days after herbicide application. Worker exposure depended on fire size, but, even in high density smoke zones, herbicide exposure for all four herbicides was well below permissible exposure limits. Thus, herbicide exposure to the general public down wind would be orders of magnitude less and not constitute a health hazard.

Pesticide Movement in Air

Herbicides are the main class of pesticides used in association with forest harvesting. They are frequently used in intensively managed forests to retard weed growth after cutting. Herbicides can move off-site by either drift during application or later volatilization [15]. The amount of spray or vapour drift is dependent on a complex of factors including herbicide chemistry and physical properties (e.g., vapour pressure, physical form, solubility), formulation (solid or liquid), application system (air, ground, broadcast, spot, injection, manual, etc.), distribution system (nozzle types, pressures, etc.), spray droplet diameter distribution, and meteorological conditions. Most documented cases of herbicide transport in air masses and subsequent deposition in dryfall or rainwater involve agricultural pesticides [15]. Pesticide seasonal deposition in dryfall/wetfall is generally 10^3- to 10^4-fold less than application rates. Because of different use patterns and application systems, air transport of forestry herbicides is not considered a significant environmental quality problem.

SURFACE WATER QUANTITY AND QUALITY

Quantity

Water Yield: The effects of harvesting on water yield from forested watershed studies throughout the world have been well documented [16, 17]. The latter

TABLE 4.1 Effect of forest harvesting on water yield the first year after cutting.

Forest Type; Location	Precip. (mm)	Cut (%)	Increased Yield (%)	Reference
Pinyon-Juniper; Utah, USA	457	100	0	[18]
Aspen-Conifer; Colorado, USA	536	100	22	[19]
Ponderosa Pine; Arizona, USA	570	100	63	[20]
Oak Woodland; California, USA	635	99	77	[21]
Chaparral*; Arizona, USA	660	90	111	[22]
Hardwood-Conifer; Japan	1153	100	71	[23]
Coastal Redwoods; California, USA	1200	67	34	[24]
N. Hardwoods; New Hampshire, USA	1219	100	48	[25]
Hardwoods-Pine; Georgia, USA	1219	100	54	[26]
Slash Pine; Florida, USA	1450	56	126	[27]
Eucalyptus; Australia	1520	100	106	[28]
Mixed Hardwoods; West Virginia, USA	1524	85	22	[29]
Mixed Hardwoods; North Carolina, USA	2070	100	16-53	[30]
Douglas-fir; Oregon, USA	2286	100	32	[31]
Beech-Podocarp; New Zealand	2600	100	43	[32]

*   Herbicided, but not harvested in the traditional manner.

review is quite extensive as it summarizes results from 93 individual watershed experiments. Selected catchments from that review, as well as more recently published results, representing a range of ecosystems and hydrologic regimes, are listed in Table 4.1.

For the most part, water yields increase when mature forests are harvested. The only exception to this rule occurs in regions where fog is abundant and scavenging of moisture out of the air contributes a significant part of water yield [33]. The magnitude of measured water yield increases the first year after cutting can vary greatly at one location and between locations depending on climate, precipitation, geology, soils, watershed aspect, tree species, and the proportion of forest vegetation harvested [34]. Since the measured increases in water yield after forest harvesting are due primarily to reductions in the transpiration component of evapotranspiration (ET), yield increases have been found to be greater in ecosystems with high ET. Streamflow increases produced by harvesting decline as both woody and herbaceous vegetation regrows. This recovery period can range from a few years [27] to several decades [34].

Timing of Yield Changes: The time of year when water yield changes occur after forest harvesting is determined by vegetation and climate. Processes such as evapotranspiration, snow deposition and melt, and fog scavenging, as well as the timing and intensity of precipitation, control the timing and direction of yield changes.

In areas where sizeable snowpacks occur and conifers predominate, most of the yield increases due to cutting will occur as snowmelt runoff. Harvesting can have two important effects on snowmelt runoff. There may be an increase because more snow reaches and accumulates in the snowpack. Also, the timing of melt may be advanced in the absence of shading by a full forest canopy. In high altitude spruce-fir forests of Colorado, nearly all the annual flow increase of 90 mm associated with a 39% strip cutting resulted from increased trapping of snowpack and came during the May-June snowmelt runoff [35]. Snowmelt runoff was advanced 4 to 8 days as a result of felling a northern hardwood overstory [25].

Where snow is less of a factor or hardwoods predominate, water yield increases resulting from harvesting are more likely to occur during growing season months. For example, nearly all of the annual water yield increase of 310 mm due to clearfelling of a northern hardwood forest in New Hampshire occurred in the low-flow months of June through September [25]. Similar results have been observed in the Southern Appalachians [34], and Australia [28]. In most instances, water yield increases resulting from forest harvesting occur at a time of the year when they are beneficial in terms of augmenting downstream water supplies or maintaining habitat for aquatic organisms.

Peak Flows: Reduction of transpiration and interception losses produces wetter soils with less opportunity for storing rainfall. Consequently, harvests which remove most of the vegetation can increase peak flows, especially where regrowth is slow or covers only part of the cleared area, or where snowmelt contributions are sizeable [36]. The magnitude of peak flow increases depends upon the amount and timing of precipitation. Where precipitation is abundant, there is little opportunity for large differences in water storage between cut and uncut forests, and thus only small changes in peak flows are likely to occur [37]. Harvesting in drier forest zones, such as pinyon pine-juniper or chaparral stands in the southwestern United States, can cause significantly greater increases in peak flows [22].

Increased peak flows can be undesirable when they produce flooding, additional sediment transport, or channel scouring. The magnitude of these adverse impacts will depend primarily on the size of harvested area and the magnitude of peak flows. For most carefully managed forests, only small portions of forested headwaters are cutover at any given time. Increased peak flows or advances in snowmelt runoff quickly lose identity as they join flow from larger, uncut areas. Thus, impacts on flood potential are localized and probably do not extend beyond one-to-two stream orders downstream. Extra precautions may be necessary in the capacity design of culverts, bridges, and road drainage systems downstream of cutting areas.

Large-scale forest clearing, as currently being practised in parts of Brazil, Indonesia, Thailand, and elsewhere in the tropics, can produce substantial peak flow increases. The corresponding flooding, erosion, and channel geometry changes are proving to be environmentally catastrophic.

Quality

Forest harvesting is designated as a nonpoint source of water pollution. Although it is commonplace to protect water quality by closely regulating logging and site preparation, water quality continues to be a major concern regarding off-site impacts of harvesting. The broad areas of concern regarding water quality are sediment, temperature, and chemical content. While most of the attention has focused on surface waters, the quality of groundwater is a growing international concern.

Sediment: Sedimentation, or the erosion and transport of rocks, mineral soil, and organic debris to streams, has long been the most obvious and important concern regarding water quality. Sediment yields from major river systems range from 1,200 kg/ha/yr (Columbia; USA) to 140,000 kg/ha/yr (Huang Ho; China), and reflect the climate, hydrology, geology, soils, vegetation, physiographic regions, and land use history of each basin [38]. Natural rates of sediment yield from smaller, forested watersheds are normally low (<100 kg/ha/yr) but can vary tremendously (up to 5 orders of magnitude; [39]). In streams emanating from forested watersheds, water quality is even more important since these streams have been typically used for water supplies. In addition, these streams are important as habitat and refugia for aquatic biota.

Except during catastrophic mass wasting events, floods, or where bedrock is naturally highly erosive, sediment is usually not an important problem in undisturbed forest ecosystems. Debris avalanches can cause major sediment problems in harvested forests of the Pacific Rim and other steeplands (Table 4.2). These episodic, spectacular events can account for much of the erosion from harvested stands, and seriously affect other forest resources and values such as water quality, fish habitat, engineering structures, buildings, recreation areas, reservoir capacity, etc. The loss of soil strength on steep slopes due to root decay 4 to 8 years after cutting is usually the mechanism predisposing slopes to avalanching [40]. Forest harvesting increases both the erosion rate and frequency of debris avalanches, but not necessarily the average size [41]. Road construction aggravates all three debris avalanche parameters. In summary, harvesting alone can increase natural rates of erosion produced by debris avalanches by a factor of four, but roading increases the rate to about 120 times that of undisturbed steepland forests.

In the eastern United States, natural erosion rates from forested watersheds are usually low (<100 kg/ha/yr; [46]). However, the disturbances that accompany forest harvesting and site preparation, especially road construction, can cause sediment yields to increase. In some physiographic regions with highly erosive soils, sediment yields after cutting and site preparation have increased temporarily by as much as 278 fold up to the 9,000 to 14,000 kg/ha range ([47]; Table 4.3).

Similar patterns in natural forest and disturbed-site sediment yields have been measured elsewhere (Tables 4.4 and 4.5). Again, it is evident that disturbances which create large areas of bare soil, aggravated by high rainfall, erosive soils, and steep terrain, produce the most sediment yield. Unstable geologic formations can also be a major contributing factor [42]. Except for some situations, erosion losses

TABLE 4.2
Effect of forest harvesting, road construction, and other forest disturbances
on sediment yield by debris avalanches.

Location	Treatment	Yield ($m^3/km^3/yr$)	Reference
Siuslaw Forest, Oregon, USA	Uncut	28	[41]
	Clearcut	111	
	Roads	3500	
Andrews Forest, Oregon, USA	Uncut	36	[42]
	Clearcut	132	
	Roads	1770	
Olympic Mountains, Washington, USA	Uncut	72	[41]
	Roads	11800	
Coast Mountains, B.C., Canada	Uncut	11	[43]
	Clearcut	24	
	Roads	282	
Notown, Westland, New Zealand	Uncut	100	[44]
	Clearcut	2200	
San Dimas Forest, California, USA	Uncut	7	[45]
	Burned	1907	

from harvest-disturbed forested lands usually do not approach those of agriculture (5,000 to 13,000 kg/ha/yr; [67]). They also do not persist from the same landscape units, as do sediment losses from agricultural land uses, if normal forest regeneration occurs.

Sedimentation and turbidity have received considerable research attention. Operational guidelines in the form of Best Management Practices have been developed in different regions and countries to protect water resources from sedimentation during and after logging operations [68, 52]. Most of these guidelines relate to constructing and maintaining logging roads, skid trails, and landings, which are the primary sources for 90% of the sediment generated by harvesting [69]. The underlying principles of these guidelines are to minimize disturbances in the vicinity of stream channels, reduce the erosive power of water on bare road surfaces, and to protect the normally high infiltration capacity of forest soils.

Sediment is a constant off-site environmental quality concern since re-entrainment after initial deposition in ephemeral or perennial stream channels can

TABLE 4.3
Effect of forest harvesting and site preparation on sediment yield,
eastern United States.

Location	Treatment	Yield (kg/ha)	Reference
Hubbard Brook, New Hampshire, USA	Uncut	42	[5]
	Clearcut	365	
Moonshine Creek, Georgia, USA	Uncut	67	[48]
	Herbicide, Cut	170	
Clemson Forest, South Carolina, USA	Uncut	20	[49]
	Cut/Burn	151	
Piedmont, North Carolina, USA	Uncut	35	[50]
	Cut/Blade	9730	
Gulf Coast, Mississippi, USA	Uncut	620	[51]
	Cut/Bed	14250	
Bradford Forest, Florida, USA	Uncut	3	[52]
	Cut/Windrow	36	
Ouachita Mountains, Arkansas, USA	Uncut	71	[53]
	Cut/Shear	535	
	Cut/Chemical	251	
Ouachita Mountains, Arkansas, USA	Uncut	12	[54]
	Select Cut	36	
	Clearcut	237	
Terre Noir Creek, Arkansas, USA	Uncut	4	[55]
	Select Cut	13	
	Clearcut	264	
Angelina N.F., Texas, USA	Uncut	36	[56]
	Cut/Chop	170	
	Cut/Shear	306	

keep sediment moving downstream for long time periods and distances. Turbidity and sedimentation can harm aquatic communities by altering habitat, and affecting respiration and photosynthesis [70]. Turbid water can affect human use by making it unacceptable for drinking or recreation. Nutrients, heavy metals, and organic compounds adsorbed onto sediments pose long-term threats to lakes and impoundments. While deposited sediments can be a long-term sink for chemicals,

TABLE 4.4
Effect of forest harvesting and site preparation on sediment yield,
western United States.

Location	Treatment	Yield (kg/ha)	Reference
Mogollon Rim, Arizona, USA	Uncut	8	[57]
	Clearcut	11	
	Cut/Skid	56	
	Cut/Road	89	
Entiat Forest, Washington, USA	Uncut	28	[58]
	Wildfire	2353	
Johnson Gulch, Montana, USA	Uncut	18	[59]
	Clearcut	137	
Coast Range, Oregon, USA	Uncut	530	[60]
	Cut/Road	1460	

TABLE 4.5
Effect of forest harvesting and site preparation on sediment yield;
Europe, South America, Asia, New Zealand.

Location	Treatment	Yield (kg/ha/yr)	Reference
Britain, U.K.	Undisturbed	30	[61]
	For. Drain	123	
Wales, U.K.	Undisturbed	37	[62]
	For. Drain	90	
Oxapampa, Peru	Uncut	121	[63]
	Cut/Pasture	542	
Hong Kong, China	Uncut	2000	[64]
	Partial Cut	67000	
	Clearcut	97000	
Koolau, Hawaii, USA	Uncut	536	[65]
	Cut/Ag.	2090	
Tawhai Forest, New Zealand	Uncut	429	[66]
	Clearcut*	611	
	Clearcut[†]	3432	

*   Cable logged and burned, no riparian buffer
†   Skidder logged and burned, 20-m riparian buffer.

desorption or biologic incorporation of organic and inorganic chemicals adsorbed to suspended or bottom sediments can produce long-term eutrophication or toxicity problems [71].

Dissolved Inorganic Ions: Undisturbed forests usually have tight cycles for major cations and anions, resulting in low concentrations in streams. Harvesting interrupts uptake by vegetation and speeds mineral weathering, nitrification, and decomposition. These processes in turn increase the concentration of inorganic ions in soil solution and leaching to streams via interflow.

TABLE 4.6
Effect of forest harvesting on maximum NO_3-N concentrations
in streamflow.

Location	Forest Type	Maximum NO_3-N (mg/L)	Reference
Hubbard Brook, New Hampshire, USA	Northern hardwoods	6.1	[5]
Hubbard Brook, New Hampshire, USA	Northern hardwoods	17.8*	[72]
Fernow Forest, West Virginia, USA	Mixed hardwoods	1.4	[73]
Coweeta Lab, North Carolina, USA	Converted to grass	0.7	[74]
Coweeta Lab, North Carolina, USA	Oak-Hickory hardwoods	0.2	[74]
Andrews Forest, Oregon, USA	Douglas-fir	0.6	[75]
Alsea Basin, Oregon, USA	Douglas-fir	2.1	[76]
Narrows Mountains, N.B., Canada	N. hardwoods, conifers	1.6	[77]
Haney, B.C., Canada	Western hemlock	0.5	[78]
Okanagan, B.C., Canada	Spruce-fir	0.4	[79]
Tawhai Forest, New Zealand	Beech-podocarp	0.2	[80]
Tawhai Forest, New Zealand	Beech-podocarp	0.4†	ibid.

*   Treated with Bromacil to stop regrowth

†   Post-burn.

For example, clearcutting of a northern hardwood stand in New Hampshire caused nitrate-nitrogen (NO_3-N) concentrations to increase from <0.7 mg/L to a maximum of 6.1 mg/L in the second year after harvest (Table 4.6). Harvesting a nearby watershed in a series of strips over a 4-year period created a peak NO_3-N concentration of 2.0 mg/L [5]. When regrowth of vegetation on a harvested watershed was prevented by herbicide applications, the NO_3-N peak soared to 17.8 mg/L [72]. This caused a considerable controversy since the data were interpreted as being typical of all clearcut harvests. However, a number of studies since then have indicated that the high NO_3-N loss at Hubbard Brook was an anomaly. By contrast in the Southern Appalachian Mountains, maximum NO_3-N concentrations after harvesting never exceeded 0.7 mg/L due to quick regrowth of the reestablished forest [74]. Elsewhere in North America and the world, peak concentrations of NO_3-N in streamflow after harvesting tend to remain less than 2.1 mg/L.

Most inorganic ions found in forest streams are not in toxic concentrations, and standards are not particularly rigorous regarding water supplies. For example, the NO_3-N increases discussed above, while below the recommended standard of 10 mg/L NO_3-N for drinking water, are high enough to stimulate objectionable increases in algal growth in oligotrophic streams. Increases in stream cations and anions are normally short-lived (usually <3 years duration), and are quickly diluted when streams from uncut areas merge to form higher-order basins.

A number of reviews of the water quality effect of dissolved nutrients released after forest clearcutting have concluded that there is no significant adverse impact [74, 81, 82]. This is due to the short-term nature of harvesting disturbances and subsequent regrowth of vegetation. Table 4.7 indicates the magnitude of nutrient responses to forest harvesting in outputs in streamflow. While distinct increases in nutrient outputs from harvested forest watersheds do occur, forest harvesting by itself does not result in a significant off-site export of nutrients or serious threats to downstream water quality. Site preparation treatments in preparation to planting usually produce the biggest effects on stream quality.

Temperature: Forest vegetation shades stream channels from solar radiation, promoting stream temperatures that are cooler and less variable than for unshaded sites (Brown, [84]). Increases in temperature resulting from cutting trees next to stream channels affect physical, chemical, and biological processes. For example, a rise in stream temperature decreases the concentration of dissolved oxygen in water while at the same time causing oxygen demand of fish to be increased. A rise in stream temperature also increases the solubility of other chemicals, and may influence the effects of pollutants on aquatic life.

Changes in stream temperatures within harvested sites can usually be prevented by leaving a buffer strip of uncut trees along perennial channels [68]. In cases where riparian buffer strips are not feasible, Brown [84] developed models for estimating changes in stream temperature due to clearcutting, and for routing the warmed water downstream. The model can then be used to make decisions as to whether or not clearcut harvesting will be acceptable in terms of both on- and off-site impacts to stream temperature. As with water yield, the cumulative effects of changes in stream temperature due to forest harvesting can be quickly masked as streams from cut areas merge with those from larger uncut areas. However, it should be recognized that rises in stream temperature on some reaches could produce adverse effects on sensitive populations such as salmonid fishes.

Dissolved Gases: Dissolved oxygen (DO) is important in maintaining the character, productivity, and species diversity of forest streams. Concentrations of DO measured in aquatic ecosystems differ considerably depending on climate, time of year, stream geomorphology, riparian conditions, and altitude. Adequate levels vary by species but generally need to be above 5 mg/L for warm water. Levels are significantly affected by temperature, biological oxygen demand (BOD), and degree of natural aeration. Forest harvesting can adversely affect DO concentrations, particularly if large amounts of fine debris are allowed to fall into and remain in streams [85]. Coarse woody debris does not affect BOD like fine organic material and is necessary to maintain habitat and stabilize low-order stream channels. Removal of riparian vegetation during harvesting can decrease DO by raising stream temperatures. In some colder, light-limited streams there are tradeoffs between temperature effects on DO and reduced primary productivity due to low solar radiation.

Dissolved Organic Carbon: In forest streams, dissolved organic carbon

TABLE 4.7

Nutrient losses in streamflow after forest harvesting and site preparation.

Location	Treatment	NO$_3$-N	NH$_4$-N	PO$_4$-P (kg/ha/yr)	Ca	K	Na	Mg	Reference
Hubbard Brook, New Hampshire, USA	Uncut	2.8	0.2	***	12.2	2.4	8.8	3.4	[72]
	Clearcut, herbicide	146.9	0.5	***	93.1	36.5	19.0	18.6	
Fernow Forest, West Virginia, USA	Uncut	0.6	1.1	0.1	5.4	3.6	2.9	2.4	[73]
	Clearcut	3.0	1.7	0.2	6.5	4.9	4.6	3.4	
Marcell Forest, Minnesota, USA	Uncut	0.3	0.9	0.1	6.9	4.4	1.8	3.0	[74]
	Clearcut	0.5	1.6	0.2	7.9	4.4	2.1	2.9	
Clemson Forest, South Carolina, USA	Uncut	<0.1	<0.1	<0.1	0.7	1.0	0.9	0.5	[49]
	Burn, clearcut	0.1	<0.1	<0.1	1.4	2.5	1.4	1.0	
Moonshine Creek, Georgia, USA	Uncut	0.1	<0.1	<0.1	0.6	3.5	1.3	0.5	[48]
	Herbicided, cut	4.9	0.8	0.1	6.0	12.9	3.4	4.2	
Coweeta Lab, North Carolina, USA	Uncut	<0.1	<0.1	<0.1	5.4	4.7	11.4	3.0	[74]
	Clearcut	0.4	<0.1	0.1	10.6	6.1	11.2	4.5	
Bradford Forest, Florida, USA	Uncut	<0.1	0.1	<0.1	0.4	0.1	***	0.6	[52]
	Cut, disk, bed, burn	<0.1	<0.1	<0.1	0.5	0.2	***	0.4	
Andrews Forest, Oregon, USA	Uncut	0.3	<0.1	0.2	26.3	3.1	18.8	7.5	[83]
	Clearcut, burned	0.7	1.5	0.6	81.4	6.0	35.5	26.2	
Haney, B.C., Canada	Uncut	0.4	0.1	***	20.3	2.4	12.9	4.6	[78]
	Clearcut	2.4	<0.1	***	24.7	3.9	14.7	4.5	
	Clearcut, burned	1.7	<0.1	***	53.0	4.4	20.5	6.3	
Tawhai Forest, New Zealand	Uncut	0.5	***	0.1	6.3	3.2	28.2	3.3	[66]
	Clearcut, cable, burn	10.4	***	0.4	44.8	35.4	58.7	22.0	
	Clearcut, skid, burn	6.0	***	0.4	24.6	23.8	56.9	12.0	

(DOC) plays an important role in the aquatic food chain [86]. Fungi and bacteria that utilize DOC as a carbon source are important to the food webs of higher trophic levels. Relatively recent research indicates that harvesting results in both depressed and elevated levels of DOC depending on the degree and type of disturbance to the forest. Moore [87] reported elevated DOC (1.2 to 2.4 greater) in New Zealand streams 8 to 10 years after logging disturbance. He attributed the elevated levels to a combination of inherent catchment differences, vegetation changes, microclimate, hydrologic changes after cutting, and the quantities of logging debris left in channels. Long-term implications to aquatic ecology are still being studied.

Particulate Organics: Particulate organic matter in forest streams is an important food source for filter-feeding aquatic invertebrates. These species are critical parts of the food web for aquatic vertebrates such as amphibians and fish. Thus changes in the abundance of particulate organics can have an impact on the ecology of forest streams. Forest harvesting initially increases the amount of coarse particulate organic matter in streamflow [85]. Aquatic insects play a key role in shredding coarse organic material into finer particles that are utilized by filter feeders. Ultimately, fine organic particulates decline due to reduced inputs of leaf litter after harvesting and reductions in the numbers of woody debris dams that tend to retain particulate organic material in the stream channels [88, 89].

Coarse Woody Debris: Large pieces of woody debris were once considered completely deleterious to water quality in forest streams because of BOD concerns. It is now recognized that woody debris plays an important ecological/ geomorphological role in stabilizing stream channels, storing sediment, holding coarse and fine particulate organic matter within upper stream reaches, creating pool-riffle channel sections, and providing complex habitat for fish [89, 90, 91, 92].

Pesticide Residues: In countries conducting intensive forest harvesting and management, pesticides are sometimes used during reforestation that follows forest harvesting. Although herbicides, insecticides, and fungicides are used for reforestation objectives, herbicides account for the largest usage. A large number of herbicides are registered by various countries for all types of vegetation management, but less than a dozen account for the majority of the silvicultural usage, both in frequency and total amounts applied. These herbicides are: 2,4-D , 2,4-DP, dicamba, fosamine, glyphosate, hexazinone, imazapyr, picloram, sulfometuron methyl, tebuthiuron, and triclopyr. Other forestry herbicides used less commonly include atrazine, bifenox, DCPA, diphenamid, napropamide, oxyfluorofen, sethoxydim, simazine and terbuthylazine.

Michael and Neary [93] discuss recent herbicide fate studies in southern United States forest ecosystems in some detail. Most of the research in this field in the past decade has been conducted in forests of the "South" because of intensive and extensive herbicide use. Much of the previous work in forestry herbicide fate was done by Norris [94] in the Pacific Northwest of the United States. Some herbicide fate studies have been conducted more recently in Canada and Australia. Michael and Neary [93] listed some 24 studies, and examined sampling matrices and streamflow peak concentrations in some detail (Table 4.8). The following discussion briefly highlights several aspects of the behavior of herbicides in forested watersheds using hexazinone, the most studied of these chemicals.

Hexazinone residue fate and transport in forested watersheds has been better documented in the southern United States than anywhere else for the most commonly used forestry herbicides. Miller and Bace [102] reported high concentrations (up to 2,400 mg/m^3) from direct fall of hexazinone pellets into a perennial stream. In an aerial application in Tennessee, pellets were applied to <20% of a large watershed but hexazinone was never detected in streamflow during a 7-month period following application [100]. In a detailed study in the upper Piedmont of Georgia, first-order watersheds were broadcast-treated with hexazinone pellets at a rate of 1.68 kg/ha) [101]. Residues peaked in the first of 26 storms (442 mg/m^3) and declined steadily thereafter. Loss of hexazinone averaged 0.53% with 2 storms accounting for nearly 60% of the off-site transport. Subsurface movement in baseflow lasted for less than 2 weeks and produced a short-term pulse with a peak of 24 mg/m^3. In contrast, hexazinone was applied to a watershed in Arkansas as a liquid spot application with somewhat different results [98]. Baseflow continued to carry low levels of hexazinone (<14 mg/m^3)

TABLE 4.8

Effect of buffer strips on forestry pesticide concentrations
in streamflow (from [95]).

Pesticide	Application Rate (kg/ha*)	Application System	Buffer (m)	Pesticide Peak (mg/m³)	Reference
Azinphosmethyl	3.4	Air spray	0	1540	[96]
Carbofuran	19.0	Ground pellet	0	7820	[96]
Chlopyralid	2.5	Air spray	20	17	[97]
Hexazinone	2.0	Ground spotgun	1	9	[98]
Hexazinone	1.4	Ground spotgun	15	16	[99]
Hexazinone	1.7	Air pellet	20	0	[100]
Hexazinone	1.7	Ground pellet	0	442	[101]
Hexazinone	1.8	Air pellet	0	2400	[102]
Imazapyr	2.2	Air spray	0	680	[103]
Imazapyr	2.2	Air spray	15	0	[103]
Picloram	5.0	Ground pellet	140	10	[104]
Picloram	5.6	Air pellet	0	442	[105]
Sulfometuron methyl	0.4	Ground spray	5	7	[106]
Sulfometuron methyl	0.4	Aerial spray	15	44	[107]
Triclopyr	11.2	Air spray	0	80	[108]
Triclopyr	2.0	Ground spray	5	2	[109]
Triclopyr	3.8	Aerial spray	0	350	[110]

*   Active ingredient rate.

for over a year. The total amount of herbicide transported out of the watershed
was 2-3% of that applied, four to six times that reported by Neary *et al.* [101].

One aspect of the research on herbicide movement being conducted in the
South is the determination of dynamics of transport in stormflow and baseflow.
There is a general understanding that most residue transport occurs in the first three
to six storms [101, 106]. However questions still exist as to when to sample
relative to storm hydrographs. Storm event duration and intensity, distance of

residues from stream channels, routing, and mechanism of transport affect the timing of pesticide residue peaks. The general conclusions of these studies has been that forestry herbicides rarely occur in streamflow at concentrations high enough to affect even the most sensitive aquatic plants or invertebrates, much less fish.

Buffer strips, or zones of undisturbed vegetation alongside riparian areas and other surface waters, are frequently employed as "Best Management Practices" to reduce the impact of pesticides on aquatic ecosystems [111]. The efficacy of buffer strips in mitigating pesticide, or other nonpoint source pollutant transport into wetlands or riparian zones is quite varied due to the many factors that can affect pesticide transport. In virtually all of the environmental fate studies summarized by Comerford *et al.* [111], no attempt was made to investigate the effects and functions of differing buffer strip sizes. Where buffer strips were used, other criteria determined the buffer strip size.

Pesticide chemistry, application rate, distribution method, buffer size, and weather conditions are very important in determining how well buffer strips work. In all cases listed in Table 8 where resulting streamflow concentrations were high (>100 mg/m^3), no buffer strips were used or the buffer was violated during the pesticide application. Generally speaking, buffer strips of 15 m or larger are effective in minimizing pesticide residue contamination of streamflow (Table 4.8). The user of buffer strips can keep pesticide residue concentrations within water quality standards. However, they are not absolute and one as large as 140 m did not keep residues out of a perennial stream [104].

GROUNDWATER

Quantity

The increases in water yield produced by forest harvesting and measured at a watershed scale are the result of increased flow through the soil and geologic matrices (see previous discussion on water yield). Surficial groundwater is mainly affected by cutting due to the alteration of the transpiration component of evapotranspiration. In some areas of the world, regional confined and unconfined aquifers are recharged in forested areas. The Coastal Plain pine forests of the

Southeastern United States are a good example. The net effect of forest harvesting is to increase groundwater recharge, normally a very positive result [112]. However, high water tables that commonly occur in low relief, poorly drained soils after harvesting can create short-term access and long-term regeneration and productivity problems [113].

Quality

Anions/Cations: Any disturbance to a forest ecosystem can alter the equilibrium in biogeochemical cycling and ultimately, via diffusion or macropore flow leaching pathways, produce changes in the quality of groundwater. An excellent discussion of this topic is provided by Schlesinger [7]. Most concerns regarding groundwater quality after forest harvesting deal with nitrate nitrogen (NO^3-N) where the drinking water standard is 10 µg/L.

Vitousek and Melillo [114] examined the patterns and processes of nitrate losses from disturbed forests throughout the world. They reported NO^3-N concentrations the first year after disturbance ranging from 25.0 to 0.01 µg/L. The only instances (3) where NO^3-N levels exceeded the 10 µg/L standard involved additions of herbicides or use of other methods to inhibit resprouting vegetation. In the absence of regrowing plants to take up nitrogen, newly mineralized nitrogen was free to leach from the ecosystems into surface waters or groundwater. The results from 27 different studies discussed by Vitousek and Melillo [114] clearly indicated that forest harvesting does not adversely affect groundwater quality.

Pesticides: Pesticide contamination of groundwater has become a priority environmental issue in the past few years because of growing incidents of residues being detected in well samples. In most rural areas residences are dependent upon groundwater for a water supply. Also, significant forested areas utilize surficial and regional groundwater aquifers for major municipal water sources. For example, in the southern region of the United States, 98-100% of the rural population relies on groundwater while 14-89% of the public water supply population does. Thus it is important to address the issue of potential groundwater pollution from operational use of forestry pesticides.

In general terms, forestry use of pesticides associated with timber harvesting and stand reestablishment poses a low pollution risk to groundwater because of its use pattern. For instance, herbicide use in forestry is only 10% of agricultural usage and likely to occur only once or twice in rotations of 25 to 75 years. Application rates are generally low (<2 kg/ha) and animal toxicities are low. Some of the silvicultural herbicides can affect non-target plants at low concentrations (< 20 mg/m^3) and could affect the quality of water for irrigation purposes. Within large watersheds where extensive groundwater recharge occurs, intensive use of silvicultural herbicides would only affect <5% of the area in any one year. The greatest potential hazard to groundwater comes from stored concentrates, not operational application of diluted mixtures.

Regional, confined groundwater aquifers are not likely to be affected by silvicultural herbicides [115]. Surface, unconfined aquifers in the immediate vicinity of herbicide application zones have the most potential for short-term contamination. It is these aquifers that are directly exposed to leaching of residues from the root zone. This discussion will focus on the effect of silvicultural herbicides on these surface aquifers.

Research conducted in the southern United States over the past 15 years has documented the occurrence of short-term, low levels (<35 mg/m^3) of herbicides in surface groundwater or their complete absence after operational applications [98, 101, 104, 106, 109, 116]. Some herbicides such as sulfometuron methyl and metsulfuron methyl rapidly degrade in acidic surface groundwater. The only serious contamination of surface groundwater associated with forestry use in the region was due to the spillage of concentrated chemical in a pesticide storage area. Operational use of forestry herbicides for site preparation has not caused any groundwater problems.

HABITAT FRAGMENTATION AND BIOLOGICAL DIVERSITY

Habitat Fragmentation

One important aspect of the off-site impacts of forest harvesting is the cumulative effect of many individual cutting units at a landsape scale. Natural (wildfire, insect outbreaks, wind storms, etc.) and man-made (harvesting, controlled burns, site preparation, and conversion to urban or agricultural uses) disturbances produce changes in the structure, characteristics, and composition of the physical environment of a forest that affect biota in positive and negative ways. The result, when viewed at a landscape scale, is a fragmented forest [117]. Species that tolerate a wide range of habitats and are very mobile are better adapted to coping with disturbances to the forest environment. Problems arise with those species that have narrow habitat tolerances, limited or no short-term mobility, or large habitat area requirements.

Foresters have long known about the advantages of manipulating forest vegetation to enhance wildlife populations such as white tailed deer (*Odoecoilus virginiana*) [118]. However, consideration of fragmentation as an adverse impact of forest harvesting did not come to the forefront until the issue of man-induced species extinctions and biological diversity was really addressed in developed countries. In the United States that took the form of the Endangered Species Act of 1973.

Biodiversity

An aspect of the landscape-scale impacts of forest harvesting that is linked to forest fragmentation is biodiversity [119]. This is a topic of concern for both individual forest harvesting units as well as larger landscapes. While forest managers are concerned about producing fiber and other natural resources at a reasonable cost to meet societal demands, conservation biologists are concerned about maintaining the genetic diversity of the fauna and flora contained in forests. Striking a balance between the two will require some difficult societal and political decisions on both the national and global scales. A complete discussion of this topic as it relates to forestry is beyond the scope of this chapter.

Adequate long-term data for analyzing the positive and negative impacts of forest harvesting on plant and animal diversity are not very abundant since this type of forestry research is still fairly recent. However, several points are illustrated by the following.

Plant Diversity: Some controversy exists over what constitutes a measure or index of diversity. Swindel *et al.* [120] and Conde *et al.* [121, 122] used rarity as an intrinsic measure of plant diversity. They examined the effects of intensive site preparation (burning, bedding, windrowing, and discing) following harvesting in Coastal Plain pine flatwoods. It was determined that plant species diversity increased for 5 years after harvesting, mechanical site preparation, and planting of slash pine (*Pinus elliottii* var. *elliottii*). Rare plant species, which increased the diversity indices, are adapted to disturbance and flourished after the dominance of slash pine was interrupted by harvesting. Preliminary research conducted by Neary *et al.* [123] indicated, as might be expected, that herbicide use in site preparation can reduce diversity because of the plant species-selective properties of some herbicides. However, they concluded that there are insufficient long-term data to adequately analyze the plant diversity impacts of normal herbicide use. The matrix of individual herbicides, times of application, combinations and rates of herbicides, and different plant species and ecosystems are too great to allow adequate evaluation.

Threatened and Endangered Species: In some parts of the world, legal requirements relating to threatened, endangered, and sensitive (TES) species are affecting forest harvesting. Most notable are two high profile TES species, the red cockaded woodpecker in the southern United States and the northern spotted owl in the Pacific Northwest region. Harvesting restrictions have been imposed on federal land managers in order to preserve or improve habitat for both species. Listing of other species as TES in the near future may place harvesting restrictions on additional forest lands. In many instances, these decisions have been made with very incomplete knowledge. One of the problems with decisions based on a few, well-publicised TES species is that an ecosystem approach is not used. Then, other less well-known TES species are ignored. In the case of the red cockaded woodpecker, forest management decisions made to favor the bird in sandhills ecosystems may adversely affect other rare animals and plants adapted to disturbance.

Fish, amphibians, and aquatic plants are the TES species at most risk from off-site impacts of harvesting and associated practices. Salmonid species such as salmon and native trout are particularly sensitive to changes in the aquatic environment (e.g. sediment, temperature, dissolved gases, etc.). Considerable research is currently in progress to understand the linkages between forest management and aquatic and riparian TES species.

CUMULATIVE EFFECTS AND IMPACT ASSESSMENT

Background

Research on the off-site environmental impacts of forest harvesting and its associated silvicultural practices has traditionally used approaches like small-scale lab studies, field plots, runoff plots, and watersheds. As indicated in the literature reviewed in this chapter, there are considerable bodies of knowledge about individual environmental impacts. Some of these effects are positive and some are negative. They also occur at different periods over a range of scales including the entire forest landscape. However, a major question on the cumulative effects of all these impacts on downstream waters, adjacent biota, and environmental quality remains largely unanswered and certainly poorly understood [1].

The need for cumulative effects analyses stems from both scientific concern and legal requirements. A major problem with conducting cumulative effects analyses is that very little of the enormous scientific data base on the environment as a whole, much less forest ecosystems, involves studies established in such a manner as to obtain cumulative effects data. For example, of 162 articles in the Journal of Environmental Quality between 1972 and 1989 dealing with pesticide fate, only 10% had information directly applicable to cumulative effects analysis. Process-based watershed-scale models offer a means of extrapolating small-scale information to that of large watersheds. However, their limitations need to be recognized since many of these models are designed for sublandscape scales and have inadequate validation data bases.

Types

A cumulative effects approach to evaluating the environmental effects of forest harvesting or any other type of anthropogenic disturbance differs from conventional approaches in that the spatial and temporal boundaries are greatly expanded [124]. In addition, cumulative effects analyses consider the broader concept of processes within ecosystems as well as interactions with other ecosystems components. Also, cumulative effects can result in positive as well as negative impacts.

There are five general categories for analyzing cumulative effects [125]. They are: 1) time-crowded; 2) space-crowded; 3) indirect; 4) synergistic; and 5) nibbling. For the purpose of this discussion we will use water quality as an example of the environmental characteristic of concern. Time-crowded effects are those that occur sufficiently close in time on the same unit of landscape that water quality does not recover during the interval. Space-crowded effects are those that happen sufficiently close together on adjacent landscape units that the impacts on water quality do not dissipate. Indirect effects are those that are not directly related to a particular characteristic of water quality or are delayed in time or space. Synergistic effects result from interactions of two or more water quality effects that produce results that are quantitatively and qualitatively different from either alone. Nibbling effects are such that each of a series of events or conditions do not produce a significant water quality effect alone, but together they do so in a stepwise manner.

As one example of an indirect cumulative effect related to the discussion here, Neary *et al.* [48] reported on sediment, anion, and cation exports from small forest watersheds after application of the herbicide hexazinone for site preparation. The use of a herbicide for site preparation had a positive indirect effect on water quality since it greatly reduced sediment yield on a site prone to erosion because of past, abusive land uses.

CONCLUSIONS

This chapter has briefly touched on a number of aspects of a complex subject, the off-site environmental effects of forest harvesting. Numerous references to the comprehensive literature on the impacts of harvesting on the forest environment provide a basis for more in-depth inquiry.

A common thread that runs through all the research on this subject is that the environmental effects of harvesting tend to be small to moderate and are short-term, especially when compared to other land uses. This is due to the infrequent nature of forest harvesting and the generally good stewardship practised by forestry professionals. The environmental impacts of forest harvesting are, and will continue to be, controversial. However, the ecological information that has been generated on this subject provides a scientific basis for guiding the decisions of forest land managers.

LITERATURE CITED

1. Sidle, R.C. and Hornbeck, J.W. Cumulative effects: A broader approach to water quality research. *Journal of Soil and Water Conservation*, 1991, **46**, 268-271.

2. Brooks, K.N., Ffolliot, P., Gregerson, H.M., and Thames, J.L. *Hydrology and the Management of Watersheds*. Iowa State University Press, Ames, 1991, 392 p.

3. Swank, W.T. and Crossley, D.A. (Eds.). *Forest Hydrology and Ecology at Coweeta*. Springer-Verlag, New York, NY, 469 p.

4. Troendle, C.A. and King, R.M. The effect of timber harvest on the Fool Creek watershed, 30 years later. *Water Resources Research*, 1985, **21**, 1915-1922.

5. Hornbeck, J.W., Martin, C.W., Pierce, R.S., Bormann, F.H., Likens, G.E., and Eaton, J.S. The northern hardwood forest ecosystem: 10 years of recovery from clearcutting. Research Paper NE-596. Broomall, PA. USDA Forest Service, Northeastern Forest Experiment Station, 1987, 30 p.

6. Franklin, J.F. Chapter 30: Past and future of ecosystem research - Contribution of dedicated experimental sites. In: *Forest Hydrology and Ecology at Coweeta*, . (Eds.) W.T. Swank and D.A. Crossley. Springer-Verlag, New York, NY, 1988, pp. 415-424.

7. Schlesinger, W.H. *Biogeochemistry: An Analysis of Global Change*. Academic Press, San Diego, CA, 1991, 443 p.

8. Woodwell, G.M., Hobbie, J.E., Houghton, R.A., Melillo, J.M., Moore, B., Peterson, B.J., and Shaver, G.R. Global deforestation: contribution to atmospheric carbon dioxide. *Science*, 1983, **222**, 1081-1085.

9. Detwiler, R.P. and Hall, A.S. Tropical forests and the global carbon cycle. *Science*, 1988, **239**, 41-47.

10. McMahon, C.K. and Tsoukalas, S.N. Polynuclear aromatic hydrocarbons

in forest fire smoke. In: *Carcinogenesis, Volume 3, Polynuclear Aromatic Hydrocarbons*. Raven Press, New York, NY., 1978, pp. 61-73.

11.   McMahon, C.K., Clements, H.B., Bush, P.B., Neary, D.G., and Taylor, J.W. Pesticides released from burning treated wood. In: Proceedings, 8th Conference on Fire and Forest Meteorology, Detroit, MI, 1985, pp. 145-154.

12.   Bush, P.B., Neary, D.G., McMahon, C.K., and Taylor, J.W. Suitability of hardwoods treated with phenoxy and pyridine herbicides for use as firewood. *Bulletin of Environmental Contamination and Toxicology*, 1987, **16**, 333-341.

13.   Bush, P.B., Taylor, J.W., McMahon, C.K., and Neary, D.G. Residues of lindane and chlorpyrifos in firewood and woodsmoke. *Journal of Entomological Science*, 1987, **22**, 131-139.

14.   McMahon, C.K. and Bush, P.B. Forest worker exposure to airborne herbicide residues in smoke from prescribed fires in the southern United States. *American Industrial Hygiene Association Journal*, 1992, **53**, 265-272.

15.   Grover, R. Nature, transport, and fate of airborne residues. In: *Environmental Chemistry of Herbicides, Volume II*. (Eds.) R. Grover and A.J. Cessna. CRC Press, Boca Raton, FL, 1991, pp. 90-117.

16.   Hibbert, A.L. Forest treatment effects on water yield. In: *International Symposium on Forest Hydrology*. (Eds.) W.E. Sopper and H.W. Lull. Pergamon Press, New York, 1967, pp. 527-543.

17.   Bosch, J.M. and Hewlett, J.D. A review of catchment experiments to determine the effect of vegetation changes on water yield and evapotranspiration. *Journal of Hydrology*, 1982, **55**, 3-23.

18.   Brown, H.E. Evaluating watershed management alternatives. Proc. Am. Soc. Civ. Eng. J. Irrig. Drain. Div., 1971, **97(IR1)**, 93-108.

19.   Bates, C.G. and Henry, A.J. Forest and streamflow experiment at Wagon Wheel Gap, Colorado. USDA Weather Bureau, Monthly Weather Review, Supplement No. 30, 1928, 79 p.

20.   Brown, H.E., Baker, M.B., Rogers, J.L., Clary, W.P., Kovner, J.L., Larson, F.R., Avery, C.C., and Campbell, R.E. Opportunities for increasing water yields and other multiple use values on ponderosa pine forest lands. Rocky Mountain Forest and Range Experiment Station, Fort Collins, Colorado, USDA Forest Service Research Paper RM-129, 1974, 36 p.

21. Lewis, D.C. Annual hydrologic response to watershed conversion from oak woodland to annual grassland. *Water Resources Research*, 1968, **4**, 59-72.

22. Hibbert, A.L. Increase in streamflow after converting chaparral to grass. *Water Resources Research*, 1971, **7**, 71-80.

23. Nakano, H. Effect on streamflow of forest cutting and change in regrowth on cutover area. Reprint, Bull. U.S. Gov. Forest Experiment Station, No. 240, 1971, 249 p.

24. Keppler, E.T. and Ziemer, R.R. Logging effects on streamflow: water yield and summer low flows at Caspar Creek in northwestern California. Water Resources Research, 1990, **26**, 1669-1679.

25. Hornbeck, J.W., Pierce, R.S., and Federer, C.A. Streamflow changes after forest clearcutting in New England. *Water Resources Research*, 1970, **6**, 1124-1132.

26. Hewlett, J.D. Forest water quality: an experiment in harvesting and regenerating Piedmont Forest. Research Paper, School of Forest Resources, Athens, Ga., 1979, 22 p.

27. Swindel, B.F., Lassiter, C.J., and Riekerk, H. Effects of clearcutting and site preparation on water yields from slash pine forests. *Forest Ecology and Management*, 1982, **4**, 101-113.

28. Bren, L.J. and Papworth, M. Early water yield effects of conversion of slopes of a eucalypt forest catchment to radiata pine plantation. *Water Resources Research*, 1991, **27**, 2421-2428.

29. Patric, J.W. and Reinhart, K.G. Hydrologic effects of deforesting two mountain watersheds in West Virginia. *Water Resources Research*, 1971, **7**, 1182-1188.

30. Swift, L.W. Jr. and Swank, W.T. Long term responses of streamflow following clearcutting and regrowth. In: *Symposium on Influence of Man on the Hydrological Regime with Special Reference to Representative and Experimental Basins*, Helsinki, Finland, IAHS-AISH Publication No. 130, 1980, 245-256.

31. Rothacher, J. Increases in water yield following clear cut logging in the Pacific Northwest. *Water Resources Research*, 1970, **6**, 653-658.

32. Pearce, A.J., O'Loughlin, C.L., and Rowe, L.K. Effects of clearfelling and

slash burning on water yield and storm hydrographs in evergreen mixed forests, western New Zealand. In: Symposium on Influence of Man on the Hydrological Regime with Special Reference to Representative and Experimental Basins, Helsinki, Finland, IAHS-AISH Publication No. 130, 1980, pp. 119-127.

33. Harr, R.D. Effects of timber harvest on streamflow in the rain-dominated portion of the Pacific Northwest. Proceedings of a Workshop on Scheduling Timber Harvest For Hydrological Concerns, Portland, Oregon, 1979, 45 p.

34. Swank, W.T., Swift, L.W. Jr., and Douglass, J.E. Chapter 22: Streamflow changes associated with forest cutting. species conversions, and natural disturbances. In: *Forest Hydrology and Ecology at Coweeta.* (Eds.) W.T. Swank and D.A. Crossley. Springer-Verlag, New York, NY, 1988, pp. 297-312.

35. Leaf, C.F. Watershed management in the Rocky Mountains subalpine zone: The status of our knowledge. Research Paper RM-137. Fort Collins, CO, USDA Forest Service, Rocky Mountain Forest and Range Experiment Station, 1975, 31 p.

36. Anderson, H.W., Hoover, M.D., and Reinhart, K.G. Forests and water: effects of forest management on floods, sedimentation, and water supply. General Technical Report PSW-18/1976. Berkeley, CA, USDA Forest Service, Pacific Southwest Forest and Range Experiment Station, 1976, 115 p.

37. Hornbeck, J.W. Storm flow from hardwood forested and cleared watersheds in New Hampshire. *Water Resources Research*, 1973, **9**, 346-354.

38. Brown, L., Pavich, M.J., Hickman. R.E., Klein, J., and Middleton, R. Earth Surface Processes and Landforms, 1988, **13**, 441-457.

39. O'Loughlin, C.L. and Ziemer, R.R. The importance of root strength and deterioration rates upon edaphic stability in steepland forests. In: Proceedings, IUFRO Workshop. P.I.07-00 Ecology of Subalpine Zones. (Ed.) R.H. Waring. Corvallis, Oregon, 1982, pp. 70-78.

40. Ziemer, R.R. Roots and the stability of forested slopes. In: *Proceedings of the Symposium, Erosion and Sediment Transport in Pacific Rim Steeplands*, Christchurch, New Zealand. (Eds.) T.R.H. Davies and A.J. Pearce. Publication 132, International Association of Hydrological Science, 1981, pp. 343-361.

41. Swanson, F.J., Swanson, M.M., and Woods, C. Analysis of debris-avalanche

erosion in steep forested lands: an example from Mapleton, Oregon. In: *Proceedings of the Symposium on Erosion and Sediment Transport in Pacific Rim Steeplands*, Christchurch, New Zealand. (Eds.) T.R.H. Davies and A.J. Pearce. Publication 132, International Association of Hydrological Science, 1981, pp. 65-75.

42. Swanson, F.J. and Dyrness, C.T. Impact of clearcutting and road construction on soil erosion by landslides in the western Cascade Range, Oregon. *Geology*, 1975, **1**, 393-396.

43. O'Loughlin, C.L. An investigation of the stability of the steepland forest soils in the Coast Mountains, southwest British Columbia. Ph.D. Thesis, University of British Columbia, Vancouver, British Columbia, Canada, 1972, 147 p.

44. O'Loughlin, C.L. and Pearce, A.J. Influence of Cenozoic geology on mass movement and sediment yield response to forest removal. *Bulletin of the International Association of Engineering Geology*, 1976, **14**, 41-46.

45. Wells, W.G. II. Some effect of brushfires on erosion processes in Coastal Southern California. In: *Proceedings of Symposium, Erosion and Sediment Transport in Pacific Rim Steeplands*, Christchurch, New Zealand. (Eds.) T.R.H. Davies and A.J. Pearce. Publication 132, International Association of Hydrological Science, 1981, pp. 305-342.

46. Maxwell, J.R. and Neary, D.G. Vegetation management effects on Sediment yields. In: Proceedings, Fifth Interagency sediment Conference, Volume 2. (Eds.) Shou-Shou, T. and K. Yung-Huang. Fed. Energy Reg. Comm., Washington, DC., 1991, pp. 12-55 tp 12-63.

47. Riekerk, H., Neary, D.G., and Swank, W.T. Magnitude of upland silvicultural nonpoint source pollution in the South. In: Proceedings of a Symposium: *The Forested Wetlands of the Southern United States*. (Eds.) D.D. Hook and R. Lea. Orlando, FL. General Technical Report SE-50, Asheville, NC, USDA Forest Service, Southeastern Forest Experiment Station, 1989, pp. 8-18.

48. Neary, D.G., Bush, P.B., and Grant, M.A. Water quality of ephemeral forest streams after site preparation with the herbicide hexazinone. *Forest Ecology and Management*, 1986, **14**, 23-40.

49. Van Lear, D.H., Douglass, J.E., Fox, S.K., and Augspurger, M.K. Sediment and nutrient export in runoff from burned and harvested pine watersheds in the South Carolina Piedmont. *Journal of Environmental Quality*, 1985, **14**, 169-174.

50. Douglass, J.E. and Godwin, R.C. Runoff and soil erosion from site preparation practices. In: *U.S. Forestry and Water Quality: What Course in the 80's?* Richmond, Virginia, Water Pollution Control Federation, Washington, D.C., 1980, pp. 51-73.

51. Beasley, R.S. Intensive site preparation and sediment loss on steep watersheds in the Gulf Coastal Plain. *Soil Science Society of America Journal*, 1979, **43**, 412-417.

52. Riekerk, H. Impacts of silviculture on flatwoods runoff, water quality, and nutrient budgets. *Water Resources Bulletin*, 1983, **19**, 73-79.

53. Beasley, R.S., Granillo, A.B., and Zillmer, V. Sediment losses from forest management: mechanical vs. chemical site preparation after cutting. *Journal of Environmental Quality*, 1986, **15**, 413-416.

54. Miller, E.L., Beasley, R.S., and Lawson, E.R. Forest harvest and site preparation effects on erosion and sedimentation in the Ouachita Mountains. *Journal of Environmental Quality*, 1988, **17**, 219-225.

55. Beasley, R.S. and Granillo, A.B. Sediment and water yields from managed forests on flat Coastal Plain sites. *Water Resources Research*, 1988, **24**, 361-366.

56. Blackburn, W.H., Knight, R.W., Wood, J.C., and Pearson, H.A. Stormflow and sediment loss from intensively managed forest watersheds in east Texas. *Water Resources Bulletin*, 1990, **26**, 465-477.

57. Heede, B.H. Overland flow and sediment delivery five years after timber harvest in a mixed conifer forest. *Journal of Hydrology*, 1987, **91**, 205-216.

58. Helvey, J.D. Effects of a north central Washington wildfire on runoff and sediment production. *Water Resources Bulletin*, 1980, **16**, 627-634.

59. Anderson, R. and Potts, D.F. Suspended sediment and turbidity following road construction and logging in western Montana. *Water Resources Research*, 1987, **23**, 681-690.

60. Beschta, R.L. Long-term patterns of sediment production following road construction and logging in the Oregon Coast Range. *Water Resources Research*, 1978, **6**, 1011-1016.

61. Robinson, M. and Blyth, K. The effect of forestry drainage operations on upland sediment yields: a case study. *Earth Surface Processes and Landforms*, 1982, **7**, 85-90.

62. Francis, I.S. and Taylor, J.A. The effect of forestry drainage operations on upland sediment yields: a study of two peat-covered catchments. *Earth Surface Processes and Landforms*, 1989, **14**, 73-83.

63. Plamondon, A.P., Ruiz, R.A., Morales, C.F., and Gonzalez, M.C. Influence of protection forests on soil and water conservation (Oxapampa, Peru). *Forest Ecology and Management*, 1991, **38**, 227-238.

64. Lal, R. Soil erosion from tropical arable lands and its control. *Advances in Agronomy*, 1984, **27**, 183-248.

65. Doty, R.D., Wood, H.B., and Merriam, R.A. Suspended sediment production from forested watersheds on Oahu, Hawaii. *Water Resources Bulletin*, 1981, **17**, 399-405.

66. O'Loughlin, C.L., Rowe, L.K., and Pearce, A.J. Sediment yield and water quality responses to clearfelling of evergreen mixed forests in western New Zealand. In: Proceedings of the Helsinki Symposium, *The Influence of Man on the Hydrological Regime with Special Reference to Representative Basins*. Publication 130, International Association of Hydrological Science, Gentbrugge, Belgium, 1980, pp. 285-292.

67. Larsen, V.E., Pierce, F.J., and Dowdy, R.H. The threat of soil erosion to long-term crop production. *Science*, 1983, **219**, 458-465.

68. Hornbeck, J.W., Corbett, E.S., Duffy, P.D., and Lynch, J.W. Forest hydrology and watershed management. In: *Forestry Handbook*. (Ed.) K.F. Wenger. John Wiley and Sons, New York, 1984, pp. 637-677.

69. Swift, L.W. Jr. Chapter 23: Forest access roads: Design, maintenance, and soil loss. In: *Forest Hydrology and Ecology at Coweeta*. (Eds.) W.T. Swank and D.A. Crossley. Springer-Verlag, New York, NY, 1988, pp. 313-324.

70. Phillips, R.W. Effects of sediment on the gravel environment and fish production. In: *Forest Land Use and Stream Environment*. Oregon State University, Corvallis, 1971, pp. 64-74.

71. Olsen, L.A. Effects of contaminated sediment on fish and wildlife: a review and annotated bibliography. National Coastal Ecosystem Team, U.S. Fish and Wildlife Service, U.S. Department of the Interior, 1984, 103 p.

72. Pierce, R.S., Hornbeck, J.W., Likens, G.E., and Bormann, F.H. Effect of elimination of vegetation on stream water quality and quantity. *International Association of Hydrological Science*, 1970, **96**, 311-328.

73. Aubertin, G.M. and Patric, J.H. Water quality after clearcutting a small watershed in West Virginia. *Journal of Environmental Quality*, 1974, **3**, 243-249.

74. Swank, W.T. Chapter 25: Stream chemistry responses to disturbances. In: *Forest Hydrology and Ecology at Coweeta*. (Eds.) W.T. Swank and D.A. Crossley. Springer-Verlag, New York, NY, 1988, pp. 339-357.

75. Fredricksen, R.L., Moore, D.G., and Norris, L.A. The impact of timber harvest, fertilization, and herbicide treatments on streamwater quality in western Oregon and Washington. In: *Forest Soils and Forest Land Management*. (Eds.) B. Bernier and C.H. Winget. Laval University Press, Quebec, Canada, 1975, pp. 283-314.

76. Brown, G.W., Gahler, A.R., and Marston, R.B. Nutrient losses after clearcut logging and slash burning in the Oregon Coast Range. *Water Resources Research*, 1973, **9**, 1450-1453.

77. Krause, H.H. Nitrate formation and movement before and after clearcutting of a monitored watershed in central New Brunswick, Canada. *Canadian Journal of Forestry Research*, 1982, **12**, 922-930.

78. Feller, M.C. and Kimmins, J.P. Effects of clearcutting and slash burning on streamwater chemistry and watershed nutrient budgets in southwestern British Columbia. *Water Resources Research*, 1984, **20**, 29-40.

79. Hetherington, E.D. Dennis Creek: A look at water quality following logging in the Okanagan Basin. Environ. Can. For. Serv. 1976, 28 p.

80. Neary, D.G., Pearce, A.J., O'Loughlin, C.L., and Rowe, L.K. Management impacts on nutrient fluxes in beech-podocarp hardwood forests. *New Zealand Journal of Ecology*, 1978, **1**, 19-26.

81. Brown, G.W. Logging and water quality in the Pacific Northwest. In: Proceedings of a National Symposium on Watersheds in Transition, American Water Research Association, Urbana, IL, 1972, pp. 330-334.

82. Neary, D.G. Impact of timber harvesting on nutrient losses in stream flow. *New Zealand Journal of Forestry*, 1977, **22(1)**, 53-63.

83. Fredricksen, R.L. Comparative chemical water quality - natural and disturbed streams following logging and slash burning. In: A Symposium, Forest Land Uses and Stream Environment, 19-21 October 1970. Oregon State University, Corvallis, Oregon, 1971, pp. 125-137.

84. Brown, G.W. *Forestry and Water Quality*. Oregon State University Book Stores, Inc., Corvallis, 1980, 124 p.

85. Ponce, S.L. The biochemical oxygen demand of finely divided logging debris in stream water. *Water Resources Research*, 1974, **10**, 983-988.

86. Meyer, J.L., Tate, C.M., Edwards, R.T., and Crocker, M.T. Chapter 20: The trophic significance of dissolved organic carbon in streams. In: *Forest Hydrology and Ecology at Coweeta*. (Eds.) W.T. Swank and D.A. Crossley. Springer-Verlag, New York, NY, 1988, pp. 269-278.

87. Moore, T.R. Dynamics of dissolved organic carbon in forested and disturbed catchments, Westland, New Zealand. 1. Maimai. *Water Resources Research*, 1989, **25**, 1321-1330.

88. Wallace, J.B. Chapter 19: Aquatic invertebrate research. In: *Forest Hydrology and Ecology at Coweeta*. (Eds.) W.T. Swank and D.A. Crossley. Springer-Verlag, New York, NY, 1988, pp. 257-268.

89. Webster, J.R., Benfield, E.F., Golladay, S.W., Kazmierczak, R.F. Jr., Perry, W.B., and Peters, G.T. Chapter 21: Effects of watershed disturbance on stream seston characteristics. In: *Forest Hydrology and Ecology at Coweeta*. (Eds.) W.T. Swank and D.A. Crossley. Springer-Verlag, New York, NY, 1988, pp. 279-296.

90. Bilby, R.E. Role of organic debris dams in regulating export of dissolved and particulate matter from a forested watershed. *Ecology*, 1981, **62**, 1234-1243.

91. Bissen, P.A., Bilby, R.E., Bryant, M.D., Dolloff, C.A., Grette, G.B., House, R.A., Murphy, M.L., Koski, K.V., and Sedell, J.R. Chapter 5: Large woody debris in forested streams in the Pacific Northwest: past, present, and future. In: *Streamside Management: Forestry and Fisheries Interactions*. (Eds.) E. Salo and T. Cundy. University of Washington, College of Forest Resources, Contribution No. 57, Seattle, WA, 1987, pp. 143-190.

92. Mosley, M.P. The influence of organic debris on channel morphology and bedload transport in a New Zealand forest stream. *Earth Surface Processes Landforms*, 1981, **6**, 571-579.

93. Michael. J.L. and Neary, D.G. Herbicide dissipation studies in forest ecosystems. *Environmental Toxicology and Chemistry*, 1993.

94. Norris, L.A. The movement, persistence, and fate of phenoxy herbicides and TCDD in the forest. *Residue Reviews*, 1981, **80**, 65-135.

95. USDA Forest Service. Final Environmental Impact Statement, Vegetation Management in the Piedmont and Coastal Plain. Southern Region Management Bulletin R8-MB-23, Atlanta, GA, 1989, 1248 pp.

96. Bush, P.B., Neary, D.G., Taylor, J.W., and Nutter, W.L. Effects of pesticide use in a pine seed orchard on pesticide levels in fish. *Water Resources Bulletin*, 1986, **22**, 817-827.

97. Leitch, C. and Fagg, P. Chlorpyralid herbicide residues in streamwater after aerial spraying of a *Pinus radiata plantation. New Zealand Journal of Forestry Science*, 1985, **15**, 195-206.

98. Bouchard, D.C., Lavy, J.L., and Lawson, E.R. Mobility and persistence of hexazinone in a forested watershed. *Journal of Environmental Quality*, 1985, **14**, 229-233.

99. Lavy, T.L., Mattice, J.D., and Kochenderfer, J.N. Hexazinone persistence and mobility in a steep forested watershed. *Journal of Environmental Quality*, 1989, **18**, 507-514.

100. Neary, D.G. Monitoring herbicide residues in springflow after an operational application of hexazinone. *Southern Journal of Applied Forestry*, 1983, **7**, 217-223.

101. Neary, D.G., Bush, P.B., and Douglass, J.E. Offsite movement of hexazinone in stormflow and baseflow from forest watersheds. *Weed Science*, 1983, **31**, 543-551.

102. Miller, J.H. and Bace, A.C. Streamwater contamination after aerial application of pelletized herbicide. USDA Forest Service, Southern Forest Experiment Station, Research Note SO-255, 1980, 4 p.

103. Michael, J.L. Final Report: Fate of ARSENAL in forest watersheds after aerial application for forest weed control. USDA Forest Service, Southern Forest Experiment Station, Auburn, AL, 1986.

104. Neary, D.G., Bush, P.B., Douglass, J.E., and Todd, R.L. Picloram movement in an Appalachian hardwood forest watershed. *Journal of Environmental Quality*, 1985, **14**, 585-592.

105. Michael, J.L., Neary, D.G., and Wells, M.J.M. Picloram movement in soil solution and streamflow from a coastal plain forest. *Journal of Environmental Quality*, 1989, **18**, 89-95.

106. Neary, D.G. and Michael, J.L. Effect of sulfometuron methyl on ground water and stream quality in coastal plain forest watersheds. *Water Resources Bulletin*, 1989, **25**, 617-623.

107. Michael. J.L. and Neary, D.G. Final Report: Movement of sulfometuron methyl in forest watersheds after aerial application of OUST for herbaceous weed control. USDA Forest Service, Southern Forest Experiment Station, Auburn, AL, 1987, 39p.

108. McKellar, R.L., Schubert, O.E., Byrd, B.C., Stevens, L.P., and Norton, E.J. Aerial application of GARLON 3a to a West Virginia watershed. *Down To Earth*, 1982, **38(2)**, 15-19.

109. Bush, P.B., Neary, D.G., and Taylor, J.W. Effect of triclopyr amine and ester formulations on groundwater and surface runoff water quality in the coastal plain. *Proceedings Southern Weed Science Society*, 1988, **39**, 262-270.

110. Thompson, D.G, Staznik, B., Fontaine, D.D., Mackay, T., Oliver, G.R., and Troth, J. Fate of triclopyr ester (RELEASE) in a boreal forest stream. *Environmental Toxicology and Chemistry*, 1991, **10**, 619-632.

111. Comerford, N.B., Neary, D.G., and Mansell, R.S. The effectiveness of buffer strips for ameliorating offsite transport of sediment, nutrients, and pesticides from silvicultural operations. National Council of the Paper Industry for Air and Stream Improvement, Technical Bulletin No. 631, New York, 1992, 48 p.

112. Riekerk, H. Influence of silvicultural practices on the hydrology of pine flatwoods. *Water Resources Research*, 1989, **25**, 713-719.

113. Terry, T.A. and Hughes, J.H. Effects of intensive management on loblolly pine (*Pinus taeda* L.) growth on poorly drained soils of the Atlantic Coastal Plain. In: *Forest Soils and Forest Land Management*. (Eds.) B. Bernier and C.H. Winget. Laval University Press, Quebec, Canada, 1975, pp. 351-377.

114. Vitousek, P.M. and Melillo, J.M. Nitrate losses from disturbed forests: patterns and processes. *Forest Science*, 1979, **25**, 605-619.

115. Neary, D.G. Fate of pesticides in Florida's forests: An overview of potential impacts on water quality. *Soil and Crop Science Society of Florida, Proceedings*, 1985, **44**, 18-23.

116. Bush, P.B., Michael, J.L., Neary, D.G., and Miller, K.V. Effect of hexazinone on groundwater quality in the Coastal Plain. *Proceedings Southern Weed Science Society*, 1990, **43**, 184-194.

117. Harris, L.D. *The Fragmented Forest*. The University of Chicago Press, Chicago, IL, 1984, 211 p.

118. Leopold, A. *A Sand County Almanac*. Oxford University Press, New York, NY, 1949, 228 p.

119. Wilson, E.O. and Peter, F.M. (Eds.). *Biodiversity: Proceedings of the National Forum on Biodiversity*, Washington, D.C., September 21-24, 1986. National Academy Press, Washington, D.C., 1988.

120. Swindel, B.F., Conde, L.F., and Smith, J.E. Index-free diversity orderings: Concept, measurement, and observed response to clearcutting and site preparation. *Forest Ecology and Management*, 1987, **20**, 195-208.

121. Conde, L.F., Swindel, B.F., and Smith, J.E. Plant species cover, frequency, and biomass: Early responses to clearcutting, chopping and bedding in *Pinus elliottii* flatwoods. *Forest Ecology and Management*, 1983, **6**, 307-317.

122. Conde, L.F., Swindel, B.F., and Smith, J.E. Plant species cover, frequency, and biomass: Early responses to clearcutting, burning, windrowing, discing, and bedding in *Pinus elliottii* flatwoods. *Forest Ecology and Management*, 1983, **6**, 319-331.

123. Neary, D.G., Smith, J.E., Swindel, B.F., and Miller, K.V. Effects of herbicides on plant species diversity. *Proceedings Southern Weed Science Society*, 1990, **43**, 266-272.

124. Baskerville, G. Some scientific issues in cumulative impact assessment. In: *Cumulative Environmental Effects: A Bi-national Perspective*. (Eds.) G.E. Beanlands, W.J. Erckmann, G.H. Orians, J. O'Riordan, D. Policansky, M.H. Sader, and B. Sadler. Canadian Env. Assess. Res. Council/U.S. Nat. Res. Council, Ottawa, Ontario, and Washington, D.C., 1986, pp. 9-14.

125. Preston, E.M. and Bedford, B.L. Evaluating cumulative effects on wetland functions: A conceptual overview and generic framework. *Environmental Management*, 1988, **12**, 565-583.

CHAPTER 5

IDENTIFYING KEY PROCESSES AFFECTING LONG-TERM SITE PRODUCTIVITY

J.P. KIMMINS
Faculty of Forestry
University of British Columbia
Vancouver, Canada

INTRODUCTION

The world population and consumer demand for forest products continue to increase [1]. In the face of possible dramatic climate change, the carbon-sequestering and carbon storage function of forests is becoming ever more important [2, 3, 4]. The pressure to increase the area of unharvested forest reserves for wilderness, old growth and ecological reserves is growing; the 12% figure suggested by the Brundtland Commission for the area of forests to be set aside in reserves [5] is now widely accepted. In the face of these social and environmental pressures, the need to conserve, and where possible increase, the long-term productivity of those forests that are to be managed for wood fibre has never been greater. Sustainability is high on the political agenda of many nations, and nowhere is this more true than in forestry.

Sustaining forest site productivity over the long-term requires that we know how forest ecosystems function, and understand the processes that determine both the total organic production of forests and the allocation of that production to commodities valued by society. We must also know how disturbance events affect this production and allocation.

The long-term productivity of forests is affected by both human actions and natural events. These cause changes to forest ecosystems that can range from subtle, local, and often poorly documented changes in ecosystem processes, to large-scale and relatively instantaneous alterations in ecosystem structure and function. Wildfire, windstorm, insect outbreak and tree diseases can cause extensive disturbance in forests, and landslides can be a significant agent of ecosystem change in mountainous areas. In some types of forest, these natural events have played an important role in shaping the present species composition, structure and productivity. Indeed, in some forest ecosystems (e.g., northern cold forests or very humid, cool temperate rain forests), productivity is only maintained where the forests are intensively disturbed from time to time [6]. On a longer time scale, climate change has a significant effect on forest ecosystems [7]. However, unless the dire predictions of global climate change come to pass [8, 9], it is timber harvesting that constitutes the most frequent and intensive disturbance to many of the world's forests, and potentially the greatest threat to long-term site productivity on time scales relevant to forest management.

If the processes that are responsible for forest productivity are sustained, timber harvesting will not threaten net primary production (NPP) and harvest yield, and may even enhance it in comparison with an unmanaged forest. However, if the harvest is conducted in a manner that ignores the processes that determine NPP and harvestable yield, long-term site productivity may decline. Unless forest harvesting is conducted with an explicit consideration of its effect on: 1) the determinants of NPP and the allocation of NPP between commercial and non-commercial plant parts; 2) the processes of ecological succession; 3) the inherent or management-assisted recovery of the ecosystem from the harvest-caused disturbance; and 4), the frequency with which harvest disturbance is to be imposed on the ecosystem (e.g., the concept of "ecological rotation" [10]), long-term site productivity may be threatened. There are examples around the world where inappropriate harvesting and/or stand management has caused a loss of forest productivity because of the failure to consider the impact on ecosystem functioning (e.g., [11, 12, 13, 14, 15]).

The negative impacts of forest harvesting can be reduced by appropriate choice of rotation length, season of harvest, road construction methods, harvesting equipment, utilization standards (e.g., whole-tree *v.* stem-only harvesting) and silvicultural system (e.g., clearcutting *v.* a partial cutting system such as shelterwood or selection). Frequently, the impacts attributed to forest harvesting are in fact largely the consequence of post-harvest site treatment (e.g., slashburning or soil scarification; [12, 16, 17, 18a, b]), and can be avoided by using different methods of preparing the site for regeneration. Consequently, it is essential that the combined impact on site productivity of the entire stand management system be evaluated and not simply the timber harvesting process.

The ideal basis for planning the sustainable management of forest ecosystems is experience. The complexity of forest ecosystems, together with our imperfect ability to make accurate predictions about how forest stands change over time, should render experience the most reliable guide to forest policy and management. Unfortunately, we frequently lack the necessary multi-rotation experience that we need to evaluate the implications of forest management for long-term site productivity, particularly where new, previously unproven management practices and systems are imposed, or where there is a significant threat of climate change, or both. Consequently, foresters must often make management decisions based on their current knowledge of the processes affecting site productivity, or on a combination of their experience and scientific knowledge.

Forest ecosystems are complex. There is a myriad of processes that in concert determine the capture and storage of solar energy as biomass, and the transfer of this biomass from plants to herbivores and detritivores, and on to higher levels of the food web. These processes include energy and chemical exchanges between the atmosphere and the ecosystem, chemical processes in the soil and living organisms, and population and community processes that determine which species are involved in the biological component of the ecosystem, and their abundance and productivity.

It is neither possible, nor is it necessary, to consider all of the processes and components of a forest ecosystem in order to make useful predictions about the long-term consequences of forest management for site productivity of harvestable biomass. Not all ecosystem processes and components are equally important in a particular forest, and it is therefore desirable to identify those key processes, knowledge of which will permit acceptably accurate predictions concerning long-term ecosystem function for that forest. The identification of key processes is the subject of this chapter. Examples of discussions of key processes can be found in Gholz and Comerford [19], Hinckley *et al.* [20], and Perry [21].

DEFINITION OF KEY PROCESSES

"Key processes" with respect to long-term site productivity can be defined as the set of ecosystem processes that must be understood and quantified before the dynamics of net primary production and economic production in a particular ecosystem can be explained and accurately predicted, respectively, over the time period and spatial scale in question. Key processes can thus be defined in the context of understanding and explaining ecosystems, or of predicting them, or both. The spatial scale referred to in discussions of long-term site productivity is usually "the stand", an area that can vary from some fraction of a hectare to hundreds of hectares. However, it usually refers to an area of several hectares to several tens of hectares. Long-term stand productivity is thus different from long-term forest productivity, the latter involving entire management units at the landscape level: an area normally of many thousands or tens of thousands of hectares.

An alternative definition of "key processes" is the set of processes that adequately defines the differences in ecosystem function and dynamics between two different types of forest. The former is a process modelling approach to the definition of key processes; the latter is a comparative ecosystem description approach.

Irrespective of the approach, there are two major areas of ecological research that need to be considered in defining a list of key processes: the ecology of primary production (here after referred to as "production ecology") and ecological succession.

Production ecology is concerned with the capture of radiant solar energy by plants via the process of photosynthesis, the allocation of the resultant photosynthate between maintenance metabolism (respiration) and net biomass creation, and the allocation of net new biomass to various different biomass components. Production ecology examines the growth-determining site factors and the processes that affect the cascade of energy transfers from incoming solar energy to harvested products (Figure 5.1). The major growth-determining site factors are the availability of soil moisture, nutrients, and light.

Studies of ecological succession are concerned with those allogenic, autogenic, and biogenic processes that determine how different plant, animal, and microbial communities successively replace each other over time, and how resource availability, net primary production, and growth allocation (i.e., production ecology) vary as this successional replacement occurs. Succession ecology seeks to quantify those processes that result in variation in soil resources (nutrients and moisture) over

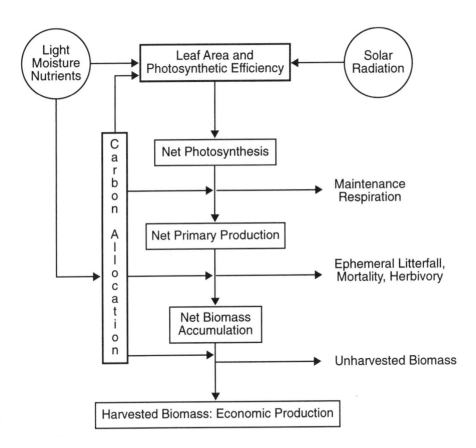

FIGURE 5.1 The major determinants of production ecology at the plant level of the ecosystem. The double boxes around Leaf Area and Carbon Allocation emphasize the importance of the processes leaf area development and efficiency, and the allocation of the resulting carbon.

the successional time sequence, how these soil resources and the plant leaf area they determine are allocated between different components of the plant community (e.g., crop *v.* non-crop plant species), and how this allocation varies with time. It is concerned with how leaf area allocation and the growth of plants determine competition for photosynthetically-active radiation and soil resources, and thus the temporal patterns of change in plant community composition.

Any list of key processes will include processes from both production ecology and succession ecology. Both these ecological subdisciplines include above-ground and below-ground processes, involving soil science, atmospheric sciences and biological sciences. Some processes will play a "key" role in any ecosystem that is studied. Other processes will be of primary importance (i.e., "key") in some ecosystems but of secondary importance in others. Similarly, the relative importance of different processes will vary at different times, both within a stand cycle (i.e., within a tree crop rotation) and over a longer successional sequence. A list that includes the key processes of all forest ecosystems is likely to be rather long. The key process list for a particular seral stage of a particular forest ecosystem may be quite short, but will change over time. Thus, a key process list that is valid for an entire tree crop rotation, or an entire successional sequence, will be much longer than that for any particular seral stage.

The quest for a single key process list that is valid for all forest ecosystems is analogous to the quest for a simple set of unifying principles that will explain the ecology of all forests. Ecological principles that are true for all ecosystems are often trivial in the context of explaining the observed differences between particular ecosystems or predicting their long-term dynamics and responses to disturbance. Similarly, the lowest common denominator list of ecosystems processes will not provide the understanding and predictive powers we seek as we strive to identify sustainable management strategies and practices for the great diversity of forest ecosystems we are managing. Conversely, a list that includes all the key processes of all seral stages of all ecosystems will be a list of most ecosystem processes.

The list of key processes that determine potential site productivity will be significantly different from the list of determinants of achieved site productivity. The former will generally focus on soil and climatic processes that determine light, temperature, soil moisture, aeration and nutrient conditions, and the length of the growing season (i.e., the basic determinants of production ecology). The latter includes biological processes involved in beneficial and antagonistic interactions between species, rates and patterns of succession, and risks from fire, insects, diseases, and wind. Anthropogenic disturbances such as air pollution and acid rain affect both potential and achieved site productivity.

These comments notwithstanding, it is still important to identify which processes are "key" to our understanding of, and our ability to predict, the form and function of particular forest ecosystems.

The Process Modelling Approach to a Definition of Key Processes
In process modelling, an ecosystem is typically represented as a series of compartments, the processes that transfer materials and energy between them, and the factors that regulate these processes (Figure 5.2). The behaviour of such models can be explored to identify the processes which have the greatest effect on the productivity of the simulated ecosystem. These are defined as "key" processes.

One problem with the model-building approach to defining key processes is that the relative importance of different processes changes over time and varies between different types of forest (e.g., papers in Kaufmann and Landsberg [23], and Dixon *et al.* [24]). This requires that ecosystem modellers represent the temporal variation in importance of different processes explicitly, and avoid the use of algorithms and coefficients (that define processes) which are unchanging over time and ecosystem condition. It also requires that models either be site-specific, or, if they are generic models, that the importance of different processes can be user-calibrated to site-specific conditions. For example, key processes in the two decades following forest harvesting (e.g., the post-harvest "assart" flush of nutrients (see review in Kimmins [25]), nutrient uptake, nitrification, denitrification, symbiotic nitrogen fixation, and competition for light and moisture between trees, herbs, and shrubs) may be different from the key processes in the final two decades of long-rotation tree crops (e.g., internal nutrient recycling, organic matter decomposition, nitrogen immobilization, non-symbiotic nitrogen fixation (e.g., [26]), and between-tree competition for nutrients). On phosphorus-fixing soils in subtropical climates, mineral soil processes that determine

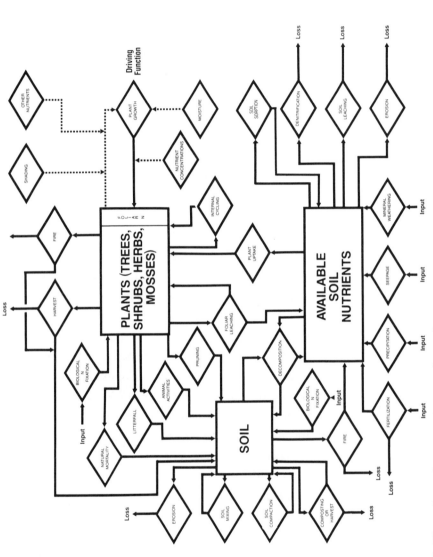

FIGURE 5.2 Flow chart of compartments and transfer processes in the ecosystem management computer simulation model FORECAST [22, 97].

phosphorus availability may be as important or more important than organic matter decomposition processes [27, 28]. In cool and humid temperate and northern forests, on the other hand, the release of nitrogen by organic matter decomposition is often critical [29, 30, 31]. Similarly, in geologically young forest soils, the key processes may be different from those in forests growing in deeply weathered ancient soil materials. There is an ever-present danger in ecosystem-modelling that a model based on key processes for a specific eco-system, or a specific seral stage or stand age for that ecosystem, will be used to predict or explain different ecosystems, or used for different seral stages or stand ages of the same ecosystem that have different key processes.

Another problem in ecosystem modelling is that so many factors which in real-ity are interrelated may be represented as being independent. For example, mois-ture affects stomatal conductance and thus photosynthesis. It also affects organic matter decomposition, nutrient mineralization, and nutrient uptake, soil animal activity, and the balance of soil flora. Both moisture and nutrient availability affect leaf area, resource allocation, internal cycling, the quality of litterfall, and the species of plant. Where moisture and nutrients are represented as independent limiting factors, or where the interactions of all these processes is not adequately represented, this may result in a mis-identification of the key processes, or inaccurate predictions of ecosystem response to disturbance, or both.

A related problem with the modelling approach to defining key processes is that models have tended to be "self-fulfilling prophesies". Naturally, models empha-size those processes that the developer of the model believes to be the most important in the system being modelled. Computer models of forest ecosystems are based on conceptual models that reflect the education, training, experience, objectives, and philosophy of the modeller [32]. Because of the nature of our educational system, these conceptual models are usually much narrower in scope and much simpler than the ecosystems they represent, and consequently sensi-tivity analysis of such models may reflect the knowledge and experience of the modeller more than the ecology of real ecosystems. For example, models that omit plant nutrition and nutrient cycling, moisture competition between species, and/or changing resource allocation and growth strategies by plants as the availability of soil resources change, are unlikely to be useful in evaluating the omitted process(es) as candidates for "key" status [33]. This limitation suggests the need to develop explanatory computer models from conceptual models that reflect the full complexity of forest ecosystem processes if these computer models are to be used to identify amongst the many ecosystem processes those that are key processes. Predictive computer models should explicitly represent all the key processes for the ecosystem in question.

Failure to develop conceptual models at appropriate levels of complexity has limited many branches of ecological science. Attempts to predict or explain the abundance and distribution of animals on the basis of simplistic mathematical models or models based solely on population processes have held back the advancement of animal population ecology for years. Similarly, the use of inappropriate ecosystem models or paradigms as the basis for resource management lies behind many failures to sustain resource values [34]. In attempting to define "key" processes through the modelling approach, great care must be taken to avoid the intellectual blind alley of simplification to satisfy a reductionist or theoretical view of nature. Although complexity may be difficult to deal with, our conceptual models of long-term ecosystem productivity must include all critical (i.e., key) processes, implicitly or explicitly.

Most models have a time step and level of complexity that reflect the intended purpose of the model: retrospective, explanatory, or predictive [35, 36]. The key processes to be included in a model will vary according to which of these purposes is to be served. An explanatory model of canopy photosynthesis may require that the effects of the evaporative demand of the atmosphere on stomatal function be represented on an hourly time step, and the effects of precipitation events on soil water (and thus on nutrient uptake and CO_2 exchange) on a daily time step (e.g., [37, 38]). The key processes in such an explanatory model will be very different from the key processes in a model designed to predict the multi-rotation consequences of different rotation lengths and utilization levels, which will probably consider processes on an annual time step. Thus, the definition of key processes in the modelling approach will be greatly influenced by the modeller's objectives and the associated time scales. There can be no "right" or "wrong" list of key processes without reference to these objectives.

Because a list of key processes that covers all types of forest ecosystem through all stages of succession might include a significant proportion of all ecosystem processes, generic ecosystem models are of necessity very complex [39]. This problem of ecosystem complexity has made it difficult to develop generalized ecosystem-level models that are universally applicable.

The Comparative Ecosystem Study Approach

This approach considers the basic functions of an ecosystem and asks the question: what processes give the maximum insight into why different ecosystems differ? Much can be learned about ecosystems through this comparative approach and it is sometimes favoured over the modelling approach. By focusing on the differences between ecosystems, fewer processes may need to be considered than in process modelling, and this is considered by some scientists to be a major advantage.

A difficulty inherent in the ecosystem comparison approach is that it may focus on identifying key processes at only one point in time: what processes are key to explaining the current differences between two or more ecosystems? Comparative studies of this type have generally had a relatively short duration (e.g., 5 to 10 years), and the key processes so identified may only reflect the stage of ecosystem development considered by the study (e.g., [16]). By focusing on the processes that define differences at only one stage of ecosystem development, key processes in a subsequent seral stage may be missed. The time dimension of ecosystem development is frequently omitted from such studies unless the chronosequence approach is employed. This method involves time-for-space substitution [40] in which a series of forests of various 'ages since disturbance' are described over a short period of time to define the long-term time trends of ecosystem development. This method assumes that all the sites in the chronosequence were very similar prior to disturbance, and that all of them have experienced a similar range of conditions during their development. Thus, time is the only variable. Unfortunately, it is often difficult to satisfy these assumptions and this has restricted the use of the chronosequence approach [41].

Comparative ecosystem description studies often employ statistical techniques (multi-variate or univariate) to identify relationships between certain measured ecosystem processes, or between these processes and ecosystem states or conditions of interest. Where such studies are based on appropriate conceptual models and are conducted over sufficiently long time periods, this approach may identify the key processes that determine differences between the form and function of different ecosystems. Where this is not the case, the correlations that are revealed may not be useful predictors of the long-term differences between the ecosystems. Correlation is not an adequate proof of causality. The fact that the rates of certain processes may be different between two ecosystems does not necessarily mean that these are key processes in understanding the ecosystem differences. The different process rates may be a consequence rather than the cause of the ecosystem difference. This problem is shared with the modelling approach: a computer model may produce the "right answer" for the wrong reason [42] under a particular range of conditions (i.e., within the domain of the calibration data set), but make erroneous predictions if used outside of this range.

Time Scale Considerations

From this discussion it should be apparent that any definition of key processes by which to understand and predict long-term site productivity in managed forests should be in the context of a conceptual model of the ecosystem that spans the time scale of forest crop production system in question. There are far too many examples of short-term evaluations of forest management treatment responses and ecological processes that have given very different answers to similar studies carried out over many years; the early results from long-term studies are often quite different from the long-term results (e.g., [16]). Short-term analytical approaches to evaluating long-term site productivity have often served our short-term understanding of these processes better than our long-term under-standing and ability to predict site productivity. Similarly, unless conceptual models at the appropriate levels of complexity are used, analysis of key processes may serve short-term scientific goals better than long-term scientific and resource management goals.

Spatial Scale Considerations

Before leaving the discussion of identification of key processes, the question of spatial scale must be considered in more detail. Forest ecologists have frequently focused on whole stands or watersheds, both in terms of response to disturbance and ecosystem development over time (e.g., the Hubbard Brook study: [43, 44]): studies at the multiple hectare and multi-decade scales. In contrast, tree physiologists, ecophysiologists, and soil scientists have tended to examine plant and soil processes at the leaf and plant or lysimeter/exchange site and soil profile levels, respectively. The different spatial and temporal scales addressed by different scientific disciplines each have their own advantages and disadvantages in the identification of key processes. However, the soil scientist and the ecophysiologist may draw different conclusions from those of the landscape level ecologist concerning which of the processes we should focus on in questions of long-term ecosystem sustainability. In reality, science requires both approaches and their marriage in the conceptual or computer models we develop to explain and predict ecosystems.

MAJOR ECOSYSTEM PROCESSES THAT PLAY A KEY ROLE IN FOREST ECOSYSTEMS

Processes Determining Leaf Area and Photosynthetic Efficiency

The key process that drives long-term site productivity is the conversion of solar energy into the energy of complex organic molecules; the interception of radiation by photosynthetically-active plant tissues and the creation of photosynthate (organic carbon). This fundamental process is the "engine" of the forest ecosystem. It is determined by a complex of processes that regulate leaf-area development over time, define the site's leaf area carrying capacity, allocate leaf area between plants of various different life form and between different species within a particular life form, and determine the photosynthetic efficiency of the foliage.

Leaf area development over time is a function of early plant growth rates, plant growth form (herbs, shrubs, trees), foliage retention (in evergreen species), and the seasonal phenology of leaf area development (in deciduous species). It also depends on the availability of soil moisture and nutrients, and of light (Figure 1; see references below and Kimmins [22]). As leaf area develops, an increasing proportion of the leaves are shaded by other leaves [45]. Depending on the relationship between light (PAR) intensity and photosynthetic rate (ie. photosynthetic light saturation curves of sun and shade foliage), this will result in a decline in the production of photosynthate per kilogram of foliage as foliage biomass increases (i.e., foliage production efficiency will decline; [46]). As light intensity within the canopy declines, specific leaf area (the surface area of leaves per kilogram of leaf biomass) may increase [47], but as a plant receives less and less light, it will produce progressively less new biomass and therefore carry less foliage (e.g., [48]). The amount of foliage developed also depends on access to the nutrients needed to build and equip the foliage. There is a close positive relationship between leaf area and nutrient availability up to the level of leaf area at which light or soil moisture become limiting [49, 50, 51].

Soil moisture poses an ultimate control on the quantity of leaves that can be carried by a plant community [52] and models that represent the processes that determine moisture availability to plants have certainly proven useful in predicting regional variations in potential site productivity (e.g., [38]). Plants will generally not hold leaves which cannot be provided with sufficient moisture to keep the stomata open long enough to maintain a positive carbon uptake balance. However, numerous fertilization experiments have shown that leaf area can be greatly increased on dry sites, at least in humid climates (e.g., [53]) demonstrating that often it is nutrient availability and not moisture availability *per se* that limits actual leaf area and determines local between-site variation in leaf area.

Carbon Allocation

There is a growing body of evidence that variation in tree stem production across the landscape is as much or more related to variation in the allocation of photosynthate between above- and below-ground biomass and between ephemeral and permanent biomass components as it is to variation in NPP.

Keyes and Grier [54] reported that two sites which varied by 100% in site index (top height of dominant trees at some index age: a measure of economic, long-term timber production potential), carried Douglas-fir stands which varied by as little as 13% in NPP. The difference was explained by differences in the allocation of NPP between stems and ephemeral fine roots caused by differences in soil resources (moisture and nutrients). Their general conclusion was confirmed for Douglas-fir by Kurz [55], and similar results have been provided for other species (e.g., [56, 57]).

The importance of carbon allocation in determining stem production requires that this be considered a key process in site productivity studies, in spite of the considerable methodological difficulties (e.g., [58, 59]).

The relative importance of moisture and nutrient availability in determining carbon allocation patterns in forest trees has not yet been fully clarified, but the evidence suggests that nutrition alone may play a more important role than moisture alone [35, 50, 60, 61, 62]. In particular, it has been suggested that root-shoot ratios and allocations are closely related to maintaining optimum foliage nitrogen concentrations, implying that nitrogen availability is a key factor in allocation to fine roots [63, 64, 65]. However, improved moisture results in improved decomposition, nutrient availability, and nutrient uptake [31], and in most attempts to separate out the individual effects of nutrients and moisture they have remained confounded [66]. The interaction between soil moisture and nutrients is so intimate that the distinction may be somewhat academic; it may be important if fertilization of dry sites achieves a significant proportional shift in carbon allocation.

From this discussion, it is apparent that nutrient availability is often the key to leaf area development and is also an important contributor to photosynthetic efficiency [67], and thus the key to site productivity. It also appears to play a key role in determining the allocation of net biomass production between permanent biomass and ephemeral biomass components, and between above-ground and below-ground biomass. Nutrient-mediated resource allocation is critical in determining the difference between net primary production and net biomass accumulation, and between net biomass accumulation and net economic production (Figure 5.1). This requires a detailed consideration of nutrient cycling and of the processes that determine nutrient availability.

Nutrient Cycling: A Key Process in Site Productivity

There are three major types of nutrient cycle: the geochemical cycle (nutrient inputs to, and outputs from, a particular ecosystem), the biogeochemical cycle (nutrient circulation between the soil and living organisms within a particular ecosystem), and the biochemical or internal cycle (nutrient redistribution between the biomass components or organs of a single organism). The relative importance of the different processes and pathways of nutrient cycling in these three major categories of cycle varies between different nutrients, plant species, seral stages, and times in a tree crop rotation cycle.

The nutrient capital in an ecosystem is ultimately dependent on the balance between geochemical inputs and outputs. Maintenance of long-term site productivity is closely related to maintaining a positive geochemical balance of growth limiting nutrients from rotation to rotation. Nutrient losses in harvested products and all the leaching, gaseous and other losses (e.g., erosion) must be compensated for by nutrient inputs to the ecosystems (or into the biogeochemical cycle from unweathered mineral reserves on site) over the rotation if site nutrient inventories are to be sustained.

Geochemical cycling is particularly important over long time scales (multiple rotations) and is often of less apparent significance over periods of a few years because the annual inputs and outputs are often small compared to total nutrient pools. However, these inputs can be very significant relative to available nutrient pools, and geochemical processes are certainly key processes for questions about long-term site productivity. They may be much less important than biogeochemical processes over shorter time scales (months to a few years) over part of the stand production cycle, but following a disturbance that has depleted the nutrients of a site (e.g., fire, or whole tree harvesting - rather large, short-term geochemical outputs), geochemical inputs such as biological nitrogen fixation, atmospheric inputs (i.e., dust, seasalts, pollution), or mineral weathering may be the major source of nutrients for plants for some years or decades. Similarly, there can be substantial short-term losses of nitrogen by leaching and denitrification following harvesting in some types of forest [43, 68]. On soils which actively immobilize phosphorus, and thereby control its availability, the short-term balance between soil sorption and desorption of phosphorus may be a key process regulating ecosystem function. Consequently, the geochemical cycle can be important over short as well as over long time scales.

The quantity of soil nutrients available annually for uptake by plants is a major determinant of leaf area development and net primary production early in the life of a stand, and it continues to be important at other times [51]. This quantity is determined first of all by the total inventory of nutrients in the biogeochemical cycle, and secondly by the rate at which nutrients move in this cycle. The former is controlled by the long-term geochemical balance. The latter is controlled by a variety of soil processes, including organic matter decomposition and mineralization (especially important for nitrogen availability and the productivity of cool temperate and boreal forests), soil weathering or sorption/desorption reactions in the mineral soil (especially important for phosphorus availability and the productivity of many tropical and warm temperate forests), dissolution/precipitation and redox reactions, and root uptake processes and their controls. In northern forests, soil temperature and energy balance processes play a key role in regulating soil processes. In warmer and drier areas, or in wet soils, soil moisture conditions may play the key role in determining nutrient availability. The important effect of soil moisture on soil organisms and on the movement of nutrients in soil renders soil moisture a key determinant of nutrient cycling, and thus of leaf area and productivity in all forest ecosystems.

In many forest ecosystems, the key to the biogeochemical cycle is the process of organic matter decomposition and the related processes of mineralization and immobilization of nutrients. Decomposition is clearly related to macro and microclimate [69] and to both the nutrient and lignin contents of the decomposing substrate and the nutritional status of the site [70 - and references therein]. The relative importance of such indicators of rates of litter mass loss as %N, % lignin, C/N ratio, and lignin/N ratio vary between litter substrate types and site types. There does not appear to be any one single regulation of decomposition processes that is valid for all situations [70].

Trees are often, but not always, the major players in the forest biogeochemical cycle. Though diminutive in stature, smaller plants such as mosses, herbs and shrubs can play a major role in nutrient circulation within the ecosystem, sometimes promoting more rapid nutrient circulation [71, 72], but sometimes competing with trees for nutrients and moisture [73, 74, 75], or otherwise slowing down organic matter decomposition and nutrient cycling (e.g., [76]). Sometimes, the negative effects of herb and shrub vegetation on tree growth may be different according to the tree species. For example, Sitka spruce trees planted on heather moorland in Scotland have frequently exhibited growth stagnation and nutritional stress, whereas pines or larch on the same site have prospered, and spruce growth has been unaffected by the heather when grown with pine or larch [77]. Successional and stand dynamics processes that determine stand structure and understorey species composition and abundance can thus have some important influences on the biogeochemical cycle, especially early in the stand cycle.

The biogeochemical cycle is relatively more important in young forests which rely on substantial nutrient uptake to develop their leaf area and canopy biomass than in older forests in which internal cycling becomes increasingly significant and may provide the majority of the nutrient needed to support new tissue production (see Cole and Rapp [78]). However, the productivity of older stands is still related to the rate of nutrient circulation within the ecosystem, and if this circulation stagnates (e.g., because of excessive accumulations or rotting wood [79] or because of reduced soil temperatures [6]), so does forest productivity. This occurs in some cool humid temperate and northern cold forest ecosystems if they are not disturbed for long periods of time.

Internal translocation of nutrients may be relatively unimportant in young evergreen plants which are not yet old enough to have senescing tissues and which have a small internal capital of nutrients relative to the nutrient demands of new growth. However, recent research which has shown active retranslocation in young as well as in senescing plant tissues requires a re-evaluation of the contributions of internal cycling [80, 81, 82]. As evergreen trees get older, internal cycling increases in importance to the point at which it may contribute more nutrients for the production of new growth than uptake from the soil [51]. In mature trees, internal cycling may compensate for year-to-year variations in availability of nutrients for uptake from the soil, maintaining a much more constant internal supply of nutrient than is supplied by the soil [83]. There is no simple relationship between the extent of retranslocation at the time of leaf senescence and the fertility of the site. In some cases, internal cycling is greater on nutrient poor sites, but in other cases the reverse may be true. For a useful review of internal cycling, see Nambiar and Fife [81].

Internal cycling is important for easily translocated nutrients like nitrogen and phosphorus, the external supply of which may not satisfy uptake demands. It is unimportant for nutrients such as calcium and boron which do not translocate efficiently. For carbon, the geochemical is the major cycle: internal cycling and biogeochemical cycling are relatively unimportant. Geochemical cycling is also important for nutrients like calcium and potassium on geologically young soils rich in unweathered minerals in areas with little input in dust and precipitation. It is much less important for these nutrients on ancient tropical soils that lack unweathered mineral reserves; on these sites, the biogeochemical cycle is of critical importance for these nutrients.

Soil Moisture As A Determinant of Site Productivity
As noted earlier, soil water plays a key role in determining both canopy function (leaf area and its photosynthetic efficiency) and the availability of nutrients. Thus, the processes that determine soil moisture conditions often play a key role in determining site productivity.

The quantity of water in a soil is determined by the inputs of moisture by precipitation, seepage and capillary rise from a water table, and the loss of water by evaporation from the soil surface, transpiration from plants, surface runoff or soil drainage. Water uptake by plants depends upon the evaporative demand of the atmosphere and the energy available to drive evapotranspiration, the occupancy of the soil by fine roots and associated mycorrhizae, and the ability of the water to move through the soil from moist soil to zones of moisture depletion (the hydraulic conductivity of the soil). As water moves through soil as a result of gravity or moisture uptake by plants, it carries nutrients with it. Water movement in the soil is thus an important process of nutrient cycling in both geochemical and biogeochemical pathways.

Soil water deficits occur when the evaporative demand of the atmosphere exceeds the ability of the soil to satisfy the plants' uptake requirement to replace lost water and maintain its water balance. When severe enough, water deficits cause the closing of stomata, terminating CO_2 exchange and photosynthesis. Lack of soil water reduces the flow of available nutrients from the soil matrix to tree roots, reduces nutrient uptake, limits decomposition and mineralization of organic matter by limiting soil animal and microbial activity, and limits the elongation of fine roots and their exploration of soil to find nutrients. Competition for water from non-crop plant species can have a similar effect on tree growth as an absolute shortage of soil water. Low soil temperatures or high concentrations of dissolved substances in soil water greatly limits water uptake by plants, creating a "physiological drought" even though water itself is present in abundance. The resistance to uptake of water by fine roots and the resistance to water movement through plants increase as temperatures drop [84, 85, 86, 87] and can lead to plant water deficits in plants exposed to moist, cold soil and air with high vapour pressure deficits.

Heat-related Processes

Heat energy is required to evaporate water and drive transpiration. Water uptake, root growth, and soil processes are all sensitive to temperature, as are the physiological processes of plants, animals and microbes; they are all affected by processes that determine heat energy budgets and heat transfer processes. The extent of heat exchange between soil and air at the soil surface affects the occurrence of growing-season radiation frost that may either kill tree seedlings or impair photosynthesis [88, 89, 90]. Such frost damage is common in clearcut areas in regions with continental climates (dry, clear air) where there is an intact organic forest floor or a mulch of dead herbaceous material. Where the mineral soil is exposed in such areas, the transfer of heat energy from the mineral soil to the air in contact with the ground can prevent or reduce radiation frost damage. Such damage in continental climates can be avoided by replacing clearcutting with partial harvesting systems.

The heat budget of a forest ecosystem is largely determined by its radiation budget and its thermal capacity. Where temperature is a critical factor controlling key ecosystem processes, heat-transfer processes will be considered as key processes.

Processes of Ecological Succession

The above-ground processes of leaf area development and light interception, the processes involved in photosynthate allocation, the soil processes of organic matter decomposition, nutrient mineralization and nutrient sorption/desorption, and the other above-ground/below-ground processes of nutrient cycling are major determinants of net primary production and potential long-term site productivity. However, achievement of economic timber production (the allocation of NPP to commercially harvestable biomass) requires that the majority of the leaf area be allocated to long-lived crop tree species. This allocation is largely determined by the processes of ecological succession, especially early in the life of a tree crop. Important successional processes include recruitment, early growth rates, and competitive displacement. These processes can have as much effect on the site productivity of crop trees that is actually achieved as the production ecology processes described above that are so important in determining potential site productivity. Pickett *et al.* [91] provide a comprehensive review of successional processes and theories.

Recruitment: This is the process by which additional plants, or plant stems, are added to the plant community. Sub-processes of recruitment are the seed rain (the rate at which spores and seeds arrive at the site), the germination of the seed bank (the population of dormant but viable seeds resident in the soil or held in reproductive structures in the canopy), and the activation of the bud bank (dormant buds on roots or stumps) [92]. If environmental conditions and the competitive environment are suitable, these new propagules will become established and begin to grow.

Different species vary considerably in their recruitment strategy [93]. Some have adapted to persist in the face of disturbance by having seed and bud banks. The recruitment of many species is entirely by seed rain; some are prolific producers of widely dispersed seed while others produce fewer, larger seed that are not dispersed very far. Very shade tolerant species may maintain a seedling bank of suppressed seedlings that are recruited (i.e., begin active growth) only when resource availability (light, nutrients, moisture) increases following disturbance to the existing community. As a result of these differences in strategy there is often a predictable sequence of plant communities that develop on a given site following a given type of disturbance. Some disturbances will favour bud bank species, some will favour seed bank or seedling bank species, and others will favour seed rain species.

In an intensively managed forest in which non-crop vegetation is controlled, recruitment processes may not have much influence on long-term site productivity. Where "weed" control is less intensive, recruitment may be a key process.

Eliminating these early seral species may not threaten long-term site productivity in some types of forests; in other forests, these species may be an important component of the ecosystem's recovery from harvest-induced disturbance (e.g., recruitment of nitrogen-fixing species).

Early Growth Rates: Some species exhibit rapid early growth rates. They can colonize a site by one or more of the recruitment processes discussed above, and achieve their maximum leaf area in a few years; e.g., 2-3 years for fireweed (*Epilobium* sp.) on a good site, or 5-6 years for pin cherry (*Prunus pennsilvanica*) in northeastern U.S. or red alder (*Alnus rubra*) in northwestern U.S. and south-western Canada. Where rapid early growth occurs, these pioneer species may monopolize the site for several years or decades. They may be able to exclude potential invading species by maintaining an unfavourable seed bed, low light conditions, or a dense root system that monopolizes the soil environment (e.g., some ericaceous shrub species; [94]). Species that spread by rhizomes may have the additional advantage that they can rapidly colonize an entire site even if only a small number of plants are able to get established initially. The rhizomatous habit can overcome problems of an inappropriate seedbed.

Because of their rapid recruitment and fast early growth, herbaceous species often form the first or "pioneer" seral stage. This may be followed by a shrub stage if there is already a bud or seed bank of shrub species, or if they are readily distributed by seed rain. Alternatively, the herbs may be displaced by invading tree species. The speed with which such displacement occurs depends on the speed with which the initial seral plant community can develop sufficient leaf area and/or root biomass, and the degree to which it can deny the seed rain of potential recruits a suitable seed bed. Where propagules of later seral species can become established before the earlier plant communities can achieve one or more of these three excluding mechanisms (or other mechanisms such as allelopathy), succession will proceed. Otherwise it may be very delayed.

Early seral herb or shrub communities may be very persistent, denying the site to invading or planted trees unless these "weed" communities are managed appropriately. In intensively managed plantation forests, weed control may render recruitment and early growth rates unimportant as factors in determining achieved site productivity, but in many forests competing vegetation can have a significant effect on rotation length yields, especially with short-rotation crops.

Competitive Displacement: Once propagules of a subsequent seral stage have become established, the recruits will generally displace the existing community by light competition. Competition for moisture and nutrients is generally more a method by which an existing community can resist displacement by invading species by restricting the leaf area development of these species.

Light competition is also the major mechanism of competition between the established individuals of an existing seral community. The ultimate cause of stand self-thinning may be lack of adequate moisture and nutrients, but the proximal cause will usually be insufficient light to sustain both aboveground growth and a root/mycorrhizal system that can compete effectively for soil resources.

Stand self-thinning, due to competition for light or soil resources, can act to limit the maximum level of biomass accumulating on a site. Where the soil is moist and fertile and the trees are long-lived and tolerant of shade, very large quantities of tree biomass can accumulate. Where the soil is nutrient-poor and/or dry, or where the trees are short-lived, maximum biomass accumulation will be much less both because of stand self-thinning and the allocation of photosynthate to ephemeral tissues (leaves, fine roots; references in Gower *et al.* [66]).

Disturbance Factors: Ecosystem disturbance and successional retrogression are as natural a part of ecological succession in most forests as are the processes of successional development described above [95]. Natural risk factors such as wind, insects, diseases, and fire play a major role in determining the rates, pathways, and duration of successional development in forest ecosystems, the rate and extent of biomass accumulation, and the achieved productivity of a site. Sometimes these agents will hasten the process of stand development by causing stand thinning; frequently they will cause stand damage or even stand destruction. Managing such risk factors may have as great an effect on achieved productivity as the successional processes that determine the interactions between crop trees and non-crop plant species, and the site processes that determine potential site productivity. Clearly, the processes that control tree mortality through these risk factors will figure prominently in the list of key processes for some forest ecosystems.

DISCUSSION AND SUMMARY

The task of defining the key processes that determine long-term forest ecosystem productivity for all the enormous variety of forests in the world is herculean at best. Key processes are those that provide sufficient understanding of the ecosystem to provide the predictive and/or explanatory powers we seek, and these processes will vary significantly between different types of forest ecosystems, and between different scientific or managerial objectives. By quantitatively defining key processes one may be able to infer the rates of a variety of other processes in the ecosystem, thereby reducing the need to quantify every individual process in order to understand and predict forest development and productivity. There is thus great merit in trying to identify such processes, in spite of the difficulties involved.

The separation of aboveground and belowground processes and of production ecology and ecological succession in this chapter is artificial because these processes are all closely linked (e.g., water flow from soil to plant to atmosphere, the interaction between soil water and soil nutrient availability, and the effect of both water and nutrient availability on the development of leaf area and the allocation of carbon between shoots and roots). The processes were discussed separately but should also be evaluated as a part of a comprehensive ecosystem model, be this a conceptual model or a computer model. Investigation of an individual process in isolation may advance our understanding of that process, but will normally give an incomplete understanding of the overall role of the process in the forest ecosystem over tree crop rotation time periods. Development of an overall ecosystem model was not the objective of this chapter, however.

The key to long-term economic ecosystem productivity is that carbon be assimilated from the atmosphere by plants and allocated to economically harvestable perennial biomass. This depends on the development of physiologically active and efficient leaf area, and maintaining stress on plants at a low enough level that they are able to allocate photosynthate to permanent, harvestable biomass components. This in turn is dependent on the ability of the soil to supply the necessary nutrients and water, and the regulation of competition for resources and other negative interactions with non-crop species. Sustaining long-term site productivity over multiple rotations depends on sustaining the processes of both production ecology and succession ecology and managing the various risk factors that can prevent the accumulation and harvest of biomass.

For many of the world's forests, we lack the empirical evidence on which to judge the degree to which the processes that determine site productivity can be altered without concomitant loss of long-term productivity. The complexity of the issues and the long time scales involved suggest that we cannot rely solely on empirical field research, especially of the reductionist, disciplinary type, to answer questions about how to sustain long-term site productivity. Such research, although vitally important, will need to be complemented by computer-based systems that are able to synthesize and integrate our knowledge of individual processes. These systems will be based on our understanding of key processes.

In order to help those responsible to make effective decisions about forest management and environmental regulations, we must harness what we know about ecosystem function in a form that is useful for managers and policy makers. Decision-support systems should be as simple as possible while still remaining effective, and therefore should be based on key processes. A systems approach to making management and policy decisions has been of benefit to agriculture, and there is reason to believe that similar approaches will be helpful to long-term forest management.

LITERATURE CITED

1. World Resources Institute. World Resources 1990-91. Oxford University Press. Oxford, 1990, 383 p.

2. Hendrickson, O.Q. How does forestry influence atmospheric carbon. *Forestry Chronicle*, 1990, **66**, 469-472.

3. Forestry Canada. The state of forestry in Canada. 1990 Report to Parliament, Minister of Supplies and Services, Ottawa, 1991, 80 p.

4. Winjum, J.K., Schroeder, P.E., and Kenady, M.J. (Eds.). *Large Scale Reforestation*. Proceedings of the International Workshop, U.S.E.P.A., Office of Research and Development, Washington D.C., EPA/600/9-91/014, 1991.

5. World Commission on Environment and Development. *Our Common Future*. Oxford University Press, 1987, 383 p.

6. Heilman, P.E. Relationship of availability of phosphorus and cations to forest succession on north slopes in interior Alaska. *Ecology*, 1968, **49**, 331-336.

7. Peters, R.L. Effects of global warming on forests. *Forest Ecology and Management*, 1990, **35**, 13-33.

8. Houghton, J.P., Jenkins, G.J., and Ephraums, J.J. (Eds.). *Climate Change. The IPCC Assessment*. Cambridge University Press, Cambridge, UK, 1990, 364 p.

9. Harrington, J.B. Climate Change. A review of the causes. *Canadian Journal of Forest Research*, 1987, **12**, 1313-1339.

10. Kimmins, J.P. Sustained yield, timber mining, and the concept of ecological rotation: a British Columbia view. *Forestry Chronicle*, 1974, **50**, 27-31.

11. Squire, R.O. Review of second rotation silviculture of *P. radiata* plantations in Southern Australia: Establishment practices and expectations. In: *IUFRO Symposium on Forest Site and Continuous Productivity*. (Eds.) R. Ballard and S.P. Gessel. USDA Forest Service, PNW Forest and Range Experiment Station, Portland, OR, General Technical Report, PNW-163, 1983, pp. 130-137.

12. Farrell, P.W. Radiata pine residue management and its implications for site productivity on sandy soils. *Australian Forestry*, 1984, **47**, 95-102.

13. Evans, J. The Usutu forest: twenty years later. *Unasylva*, 1988, **159**, 19-29.

14. Kanowski, P. *et al. Plantation Forestry*. World Bank Forestry Policy Issues Paper. Oxford For. Institute, 1990.

15. Gessel, S.P., Lacate, D.S., Weetman, G.F., and Powers, R.F. (Eds.). *Sustained Productivity of Forest Soils*. Proceedings, 7th North American Forest Soils Conference, University of British Columbia, Faculty of Forestry, Vancouver, B.C., 1990, 525 p.

16. Lundmark, J.E. [The soil as part of the forest ecosystem.] Markem som del au det skogliga ekosystemet. [The care of the soil. Properties and Utilization of Forest Soils.] Markuård, Skogsmarkens egensleaper och utuyttjande. Sveriges Skogsvårdsforbunds. *Tidskrift*, 1977, **75**, 109-122.

17. Dyck, W.J. and Skinner, M.F. Potential for productivity decline in New Zealand radiata pine forests. In: *Sustained Productivity of Forest Soils*, Proceedings, 7th North American Forest Soils Conference. (Eds.) S.P. Gessel, D.S. Lacate, G.F. Weetman and R.F. Powers. Faculty of Forestry, UBC, Vancouver, B.C., Canada, 1990, pp. 318-332.

18. Kimmins, H. *Balancing Act: Environmental Issues in Forestry*. University of British Columbia Press, Vancouver, B.C., 1990, 246 p.

19. Gholz, H.L. and Comerford, N.B. Key processes for potential monitoring in long-term field studies of forest productivity: an ecological perspective. In: *Long-term Field Trials to Assess Environmental Impacts of Harvesting*, Proceedings IEA/BE T6/A6 Workshop, Amelia Island, Florida, USA. (Eds.) W.J. Dyck and C.A. Mees. Forest Research Institute, Rotorua, New Zealand, FRI Bulletin 161, 1991, pp. 29-38.

20. Hinckley, T., Ford, D., Segura, G., and Sprugel, D. Key processes from tree to stand level. In: *Implications of Climate Change for Pacific Northwest Forest Management*. (Ed.) G. Wall. Department of Geogr. Publ. Ser. Occasional Paper No. 15. University of Waterloo, 1992, pp. 33-43.

21. Perry, D.A. Key processes at the stand to landscape scale. In. *Implications of Climate Change for Pacific Northwest Forest Management*. (Ed.) G. Wall. Department of Geogr. Publ. Ser. Occasional Paper No. 15. University of Waterloo, 1992, pp. 51-58.

22. Kimmins, J.P. (1993) Scientific foundations for the simulation of ecosystem function and management in FORCYTE-II. Forestry Canada Northwest Region, *Info. Rept. NOR-X-328*, 88p.

23. Kaufmann, M.R. and Landsberg, J.J. (Eds.). Advancing Toward Closed Models of Forest Ecosystems. *Tree Physiology*, 1991, **9(1,2)**, 1-324.

24. Dixon, R.K., Meldahl, R.S., Ruark, G.A., and Warren, W.G. (Eds.). *Process Modelling Forest Growth Responses to Environmental Stress*. Timber Press, Portland, OR, 1989, 441 p.

25. Kimmins, J.P. *The Assart Effect: Implications for Sustainable Ecosystem Management*. Manuscript in preparation, 1993.

26. Hendrickson, O.Q. Abundance and activity of N_2-fixing bacteria in decaying wood. *Canadian Journal of Forest Research*, 1991, **21**, 1299-1304.

27. Vitousek, P.M. Litterfall, nutrient cycling and nutrient limitations in tropical forests. *Ecology*, 1984, **65**, 285-298.

28. Vitousek, P.M. and Sandford, R.L. Nutrient cycling in moist tropical forests. *Annual Review of Ecology and Systematics*, 1986, **17**, 137-167.

29. Tamm, C.O. and Pettersson, A. Studies on nitrogen mobilization in forest soils. *Studia Forestalia Suecica*, 1969, **75**, 1-39.

30. Williams, B.L. Nitrogen mineralization and organic matter decomposition in Scots pine humus. *Forestry*, 1972, **45**, 177-188.

31. Swift, M.J., Heal, D.W., and Anderson, J.M. *Decomposition in Terrestrial Ecosystems. Studies in Ecology Volume 5*. University of California Press, Berkeley, California, USA, 1979.

32. Kimmins, J.P. and Scoullar, K.A. The role of modelling in tree nutrition research and site nutrient management. In: *Nutrition of Plantation Forests*.

(Eds.) G.D. Bowen and E.K.S. Nambiar. Academic Press, N.Y., 1984, pp. 463-487.

33. Kimmins, J.P., Scoullar, K.A., Apps, M.J., Kurz, W.A., and Comeau, P.G. Predicting the yield and economic returns of forest management in a changing and uncertain future: the hybrid simulation approach. In: *Forest Simulation Systems*, Proceedings, IUFRO Conference, Nov. 1988, Berkeley, California. University of California, Division of Agriculture and Natural Resources, Bulletin 1927, 1990, pp. 247-256.

34. Botkin, D.B. *Discordant Harmonies. A New Ecology for the Twenty-First Century*. Oxford University Press, 1990, 241 p.

35. Kimmins, J.P., Comeau, P.G., and Kurz, W. Modelling the interactions between moisture and nutrients in the control of forest growth. *Forest Ecology and Management*, 1990, **30**, 361-379.

36. Yarie, J. Role of computer models in predicting the consequences of management on forest productivity. In. *Impacts of Intensive Harvesting on Forest Site Productivity*. (Eds.) W.J. Dyck and C.A. Mees. Forest Research Institute, Rotorua, New Zealand, Bulletin No. 159, 1990, pp. 3-18.

37. Mohren, G.M.J. Simulation of forest growth applied to Douglas-fir stands in the Netherlands. Ph.D. Thesis. Landbouwuniversiteit. Wageningen, Netherlands, 1987, 184 p.

38. Running, S.W. Microclimatic control of forest productivity: analysis by computer simulation of annual photosynthesis and transpiration balance in different environments. *Agriculture and Forest Meteorology*, 1984, **32**, 267-288.

39. Landsberg, J.J., Kaufmann, M.R., Binkley, D., Isebrands, J., and Jarvis, P.G. Evaluating progress towards closed forest models based on fluxes of carbon, water and nutrients. *Tree Physiology*, 1991, **9**, 1-15.

40. Pickett, S.T.A. Space-for-time substitution as an alternative to long-term studies. In: *Long-term Studies in Ecology. Approaches and Alternatives*. (Ed.) G.E. Likens. Springer-Verlag, New York, 1989, pp. 110-135.

41. Cole, D.W. and Van Miegroet, H. Chronosequences: A technique to assess ecosystem dynamics. In: *Research Strategies for Long-term Site Productivity*, Proceedings, IEA/BE A3 Workshop, Seattle, WA. (Eds.) W.J. Dyck and C.A. Mees. Forest Research Institute, Rotorua, New Zealand Bulletin No. 152, 1989, pp. 5-24.

42. Bunnell, F.L. Alchemy and uncertainty: what good are models? USDA Forest Service, Pacific Northwest Research Station, Portland, OR, General Technical Report, PNW-GTR-232, 1989, 27 p.

43. Likens, G.E., Johnson, N.M., Fisher, D.W., and Pierce, R.S. Effects of forest cutting and herbicide treatment on nutrient budgets in the Hubbard Brook watershed ecosystem. *Ecological Monographs*, 1970, **40**, 23-47.

44. Bormann, F.H. and Likens, G.E. *Pattern and Process in a Forested Ecosystem.* Springer-Verlag, N.Y., 1979, 253 p.

45. Waring, R.H. and Schlesinger, W.H. *Forest Ecosystems. Concepts and Management.* Academic Press, N.Y., 1985, 340 p.

46. Ågren, G.I. Nitrogen productivity of some conifers. *Canadian Journal of Forest Research,* 1983, **13**, 494-500.

47. Klinka, K., Wang, Q., Kayahara, G.J., Carter, R.E., and Blackwell, B.A. Light growth response relationships in Pacific silver fir [*Abies amabilis* (Dougla. ex Loud.) Forbes] and subalpine fir [*Abies lasiocarpa* (Hook.) Nutt.]. *Canadian Journal of Forest Research,* in press.

48. Messier, C. Effects of neutral shade and growing media on growth, biomass allocation, and competitive ability of Gaultheria shallon. *Canadian Journal of Botany,* 1992, **70**, 2271-2276.

49. Brix, H. Effects of nitrogen fertilization on photosynthesis and respiration in Douglas-fir. *Forest Science,* 1971, **17**, 407-414.

50. Brix, H. and Mitchell, A.K. Thinning and fertilization effects on soil and tree water stress in a Douglas-fir stand. *Canadian Journal of Forest Research,* 1986, **16**, 1334-1338.

51. Miller, H.G. Dynamics of nutrient cycling in plantation ecosystems. In: *Nutrition of Plantation Forests.* (Eds.) G.D. Bowen and E.K.S. Nambiar. Academic Press, N.Y., 1984, pp. 53-78.

52. Grier, C.C. and Running, S.W. Leaf area of mature north-western coniferous forests: relation to site water balance. *Ecology,* 1977, 58: 893-899.

53. Barclay, J.H. and Brix, H. Fertilization and thinning effects on a Douglas-fir ecosystem at Shannigan Lake: 12 year growth response. Canadian Forest Service Pacific Forestry Centre Information Report BC-X-271, 1985, 36 p.

54. Keyes, M.R. and Grier, C.C. Above- and below-ground net production in 40-year-old Douglas-fir stands on low and high productivity sites. *Canadian Journal of Forest Research,* 1981, **11**, 599-605.

55. Kurz, W.A. Net primary production, production allocation, and foliage efficiency in second growth Douglas-fir stands with differing site quality. Ph.D. Thesis. Univ. of B.C., Vancouver, B.C., 1989, 224 p.

56. Vogt, K., Grier, C.C., and Vogt, D.J. Production, turnover, and nutrient dynamics of above- and below-ground detritus of world forests. *Advances in Ecological Research,* 1986, **15**, 303-377.

57. Comeau, P.D. and Kimmins, J.P. Above- and below-ground biomass and production of lodgepole pine on sites with differing soil moisture regimes. *Canadian Journal of Forest Research,* 1989, **19**, 447-454.

58. Kurz, W.A. and Kimmins, J.P. Analysis of some sources of error in

methods used to determine fine root production in forest ecosystems: a simulation approach. *Canadian Journal of Forest Research*, 1987, **17**, 909-912.

59. Persson, H. Methods of studying root dynamics in relation to nutrient cycling. In: *Nutrient Cycling in Terrestrial Ecosystems: Field Methods, Application and Interpretation.* (Eds.) A.F. Harrison, P. Inerson and D.W. Heal. Elsevier Applied Science, N.Y., 1990, pp. 198-217.

60. Axelsson, E. and Axelsson, B. Changes in carbon allocation patterns in spruce and pine trees following irrigation and fertilization. *Tree Physiology*, 1986, **2**, 205-214.

61. Grier, C.C., Hinckley, T.M., Vogt, K.A., and Gower, S.T. Net Primary production in Douglas-fir ecosystems: its relation to moisture and mineral nutrition. In: *Douglas-fir: Stand Management for the Future.* (Ed.) C.D. Oliver. 1986, pp. 155-161.

62. Nambiar, E.K.S., Squire, R., Cromer, R., Thower, J., and Boardman, R. (Eds.). Management of water and nutrient relations to increase forest growth. *Forest Ecology and Management*, 1990, **30**, 1-486.

63. Ågren, G.I. and Ingestad, T. Root:shoot ratio as a balance between nitrogen productivity and photosynthesis. *Plant, Cell and Environment*, 1987, **10**, 579-586.

64. Hilbert, D.W. Optimization of plant root:shoot ratios and internal nitrogen concentrations. *Annals of Botany*, 1990, **66**, 91-99.

65. Hilbert, D.W., Larigauderie, A., and Reynolds, J.F. The influence of carbon dioxide and daily photon-flux density on optimal leaf nitrogen concentration and root:shoot ratio. *Annals of Botany*, 1991, **68**, 365-376.

66. Gower, S.T., Vogt, K.A., and Grier, C.C. Carbon dynamics of Rocky Mountain Douglas-fir: influence of water and nutrient availability. *Ecological Monographs*, 1992, **62**, 43-65.

67. Brix, H. Mechanisms of response to fertilization. II. Utilization by trees and stands. In: *Improving Forest Fertilization Decision-making in British Columbia*, Proceedings Forest Fertilization Workshop, March 1988, Vancouver B.C. (Ed.) J.D. Lousier. B.C. Ministry of Forests, Victoria, B.C., 1991, pp. 76-93.

68. Martin, W.L. Post-clearcutting forest floor nitrogen dynamics and regeneration response in the coastal western hemlock wet subzone. Ph.D. thesis. Univ. of B.C., Vancouver, B.C., 1985, 350 p.

69. Meentemeyer, V. Macroclimate and lignin control of litter decomposition rates. *Ecology*, 1978, **59**, 465-472.

70. Taylor, B.R., Prescott, C.E., Parsons, W.J.F., and Parkinson, D. Substrate control of litter decomposition in four Rocky Mountain coniferous forests. *Canadian Journal of Botany*, 1991, **69**, 2242-2250.

71. Weetman, G.F. and Timmer, V. Feather moss growth and nutrient content under upland black spruce. Woodland Research Index No. 183. Pulp and Paper Research Institute, Canada, 1967, 38 p.

72. Yarie, J. The role of understorey vegetation in the nutrient cycle of forested ecosystems in the Mountain Hemlock Biogeoclimatic Zone. *Ecology*, 1980, **61**, 1498-1514.

73. Nambiar, E.K.S. and Zed, P.G. Influence of weeds and seedling types on the water potential and growth of young *Pinus radiata*. *Australian Forestry Research*, 1980, **10**, 279-288.

74. Sands, R. and Nambiar, E.K.S. Water relations of *Pinus radiata* in competition with weeds. *Canadian Journal of Forest Research*, 1984, **14**, 233-237.

75. Woods, P.V., Nambiar, E.K.S., and Smethurst, P.J. Effect of annual weeds on water and nitrogen availability to *Pinus radiata* trees in a young plantation. *Forest Ecology and Management*, 1992, **48**, 145-163.

76. Damman, A.W.H. Effect of vegetation change on the fertility of a Newfoundland forest site. *Ecological Monographs*, 1971, **41**, 352-70.

77. Taylor, C.M.A. and Tabbush, P.M. Nitrogen deficiency in Sitka spruce plantations. Forestry Commission Bulletin No. 89, 1990.

78. Cole, D.W. and Rapp, M. Elemental cycling in forest ecosystems. In: *Dynamic Properties of Forest Ecosystems.* (Ed.) D.E. Reichle. International Biological Programme 23. Cambridge University Press, Cambridge, 1981, pp. 341-409.

79. Prescott, C.E., McDonald, M.A., and Weetman, G.F. Differences in availability of N and P in the forest floors of adjacent stands of western redcedar - western hemlock and western hemlock - amabilis fir on northern Vancouver Island. *Canadian Journal of Forest Research*, 1992, (in press).

80. van den Driessche, R. Nutrient storage, retranslocation and relationship of stress to nutrition. In: *Nutrition of Plantation Forests.* (Eds.) G.D. Bowen and E.K.S. Nambiar. Academic Press, N.Y., 1984, pp. 181-209.

81. Nambiar, E.K.S. and Fife, D.N. Nutrient retranslocation in temperate conifers. *Tree Physiology*, 1991, **9**, 185-207.

82. Millard, P. and Proe, M.F. Storage and internal cycling of Nitrogen in relation to seasonal growth of Sitka spruce. *Tree Physiology*, 1992, **10(1)**, 33-44.

83. Turner, J. Effect of nitrogen availability on nitrogen cycling in a Douglas-fir stand. *Forest Science*, 1977, **23**, 307-316.

84. Turner, N.C. and Jarvis, P.G. Photosynthesis in Sitka spruce (*Picea sitchensis* [Bong.] Carr.) IV. Responses to soil temperature. *Journal of Applied Ecology*, 1975, **12**, 561-576.

85. Kaufmann, M.R. Soil temperature and drying cycle effects on water relations of Pinus radiata. *Canadian Journal of Botany*, 1977, **55**, 2413-2418.

86. Teskey, R.O. and Hinkley, T.M. Effects of water stress in trees. In: *Stress Physiology and Forest Productivity.* (Eds.) T. Hennessy, P. Dougherty, S. Kossuth and J. Johnson. Martinus Nijhoff Publishers, Boston, 1986.

87. Kandiko, R.A., Timmis, R., and Worrall, J. Pressure-volume curves of shoots and roots of normal and drought conditioned western hemlock seedlings. *Canadian Journal of Forest Research*, 1980, **10**, 10-16.

88. Spittlehouse, D.L. and Stathers, R.J. Seedling Microclimate. B.C. Ministry of Forests, Victoria, B.C. Land Management Report 65, 1990.

89. Adams, R.S., Black, T.A., and Fleming, R.L. Evapotranspiration and surface conductance in a high elevation grass-covered forest clearcut. *Agriculture and Forest Meteorology*, 1991, **56**, 173-193.

90. Black, T.A., Novak, M.D., Adams, R.S., Fleming, R.L., Eldridge, N.R., and Simpson, I.J. Site preparation procedures to reduce seedling water and temperature stress in backlog areas in the southern interior. Final Report on Southern Interior FRDA Project 3.02. Ministry of Forests, Victoria, B.C., 1991.

91. Pickett, S.T.A., Collins, S.L., and Armesto, J.J. Models, mechanisms and pathways of succession. *Botanical Review*, 1987, **53**, 335-371.

92. Harper, J.L. *Population Biology of Plants.* Academic Press, New York, 1977, 892 p.

93. Noble, I.R. and Slatyer, R.O.   Post-fire succession of plants in Mediterranean ecosystems.   In: *Proceedings, Symposium on the Environmental Consequences of Fire and Fuel Management in Mediterranean Ecosystems.* (Eds.) H.A. Mooney and C.E. Conrad. USDA Forest Service General Technical Report, WO-3, 1977, pp. 27-63.

94. Messier, C. and Kimmins, J.P. Above- and below-ground vegetation recovery in recently clearcut and burned sites dominated by Gaultheria shallon in coastal British Columbia. *Forest Ecology and Management*, 1991, **46**, 275-294.

95. Pickett, S.T.A. and White, P.S. *The Ecology of Natural Disturbance and Patch Dynamics.* Academic Press, San Diego, 1985, 472 p.

96. Keenan, R.J. and Kimmins, J.P. (1993) The ecological effects of clearcutting. *Environmental Reviews*, **1**, 472p.

97. Kimmins, J.P., Scoullar, K.A. and Wei, X. (1994) Incorporation of nutrient cycling in the design of sustainable, stand-level, forest management systems using the ecosystem management model FORECAST and its output format FORTOON. *Plant and Soil* (in press).

CHAPTER 6

COMPUTER SIMULATION MODELS AND EXPERT SYSTEMS FOR PREDICTING PRODUCTIVITY DECLINE

M.F. PROE
Macaulay Land Use Research Institute
Aberdeen AB9 2QJ, U.K.

H.M. RAUSCHER
USDA Forest Service, NC Forest Experiment Station
Grand Rapids, MN 55744, USA

J. YARIE
Forest Soils Laboratory, University of Alaska
Fairbanks, AK 99775, USA

INTRODUCTION

Forests are vital to the well-being of the world's population. Demand for forest products is projected to increase. To meet this demand the proportion of products utilized from each harvested area must increase and/or the area being harvested must increase. In each case there is a need to understand the consequences of harvesting forest products so that we can avoid degrading the productivity of the world's forests. Productivity can be defined as the amount of biomass produced on a site over a given period [1]. Productivity decline may reflect a reduction in the amount of biomass produced, or an increased requirement for management inputs to maintain biomass production. Attributing productivity decline to past management practices may be difficult unless satisfactory "control" areas are available for comparison. If these are not available then the cause of productivity decline may be difficult to establish. For example, long-term impacts of climate change may increase or decrease productivity on a given site. If the climate improved, then site productivity would be expected to improve - provided other factors did not become more limiting to tree growth. Hence, the maintenance of production will not, necessarily, ensure sustained productivity. Similar problems may emerge through the introduction of exotic species or genetically improved planting stock, again raising expectations of production.

Many factors are important in assessing the potential impact of harvesting upon long-term productivity. Whole-tree harvesting greatly increases the rate of nutrient offtake from a site. Whether or not this will affect productivity will depend upon the temporal balance between nutrient supply and nutrient demand by the newly developing tree crop. Nutrients may be replenished through a variety of processes including atmospheric deposition, mineralization, nitrogen fixation, and weathering. Retention of residues may simply lead to increased losses through leaching or volatilization. Removal of residues may have indirect effects upon tree growth due to changes in microclimate, exacerbated soil compaction, and altered patterns of re-vegetation leading to increased weed competition.

Computer simulation models provide a methodology to organize and evaluate our knowledge and hypotheses of both the structural and functional properties of forest ecosystems. These models can be used to predict the behaviour of key ecosystem variables and can be validated against independent data sets, where these are available. The complexity of biological processes and their interactions often make the formulation of suitable mathematical equations to describe these systems very difficult. In such circumstances it may be appropriate to capture the heuristic knowledge and understanding which humans have obtained through years of experience and scientific research in the form of an expert system.

COMPUTER SIMULATION MODELS

Introduction

The term model has two somewhat contradictory meanings. A model can be defined as a pattern of excellence worthy of being copied or, alternatively, "an imitation of something on a smaller scale" [2]. It is the second interpretation that is most relevant to computer simulation models. In mathematical terms "modelling extracts certain essential features of the system and expresses the resulting abstraction by a system of mathematical functions capable of imitating some subset of the original system's behaviour" [3].

All models are used, to a greater or lesser extent, for making predictions about the real world [4]. They can provide a powerful tool for the integration of fragmented knowledge into a coherent view of the whole system [5]. This holistic approach can be used to develop and test theories concerning the system and so gain insight into how it functions [6]. Areas in which data are lacking or theories appear to be inadequate can then be identified and research priorities determined [5]. The precise and logical nature of mathematical models also makes them an ideal medium for communicating extensive or complex pieces of information [7]. Models can, therefore, provide an integration of many concepts and hypotheses concerning the functioning of a forest ecosystem at a level which is sufficient to test aspects of each assumption without losing a perspective of the whole.

Simulation modelling is particularly useful in forestry because of both the long timescale involved and the structural and functional complexity of forest ecosystems. Models have been used for a wide range of applications including production forecasting, yield control, and the evaluation of alternative management operations. Many of these models were built upon empirical data, with little attempt to explain the underlying biological relationships. However, concern has been growing with regard to the applicability of such models to the multiple-objective, high-tech management of modern commercial forests [8]. Scientific research, economic pressures and political lobbying all influence the "environment" within which forest managers must operate. Complex interactions now operate regarding the options open to forest managers including the use of genetically improved planting stock, applications of fertilizers and herbicides, thinning and re-spacing techniques, and the implementation of whole-tree harvesting for thinnings or final fellings. It is impossible to establish field experiments to test and evaluate all these options. The development of ecologically-based computer simulation models may, however, provide the forest manager with a tool which is sufficiently flexible to be used to predict the effects of new treatments which have yet to be tested in the field [3].

The complexity of forest ecosystems makes it impossible for an individual to study the whole system in depth. Long-term projects designed to examine such systems in detail have been implemented using teams of scientists, each working in their own field of expertise. Examples include the Woodland Biome study conducted as part of the International Biological Programme (IBP) aimed at eliciting the biological basis for productivity and human welfare [9], and the Swedish Coniferous Forest Project (SWECON) set up to examine the structure and functioning of northern coniferous forests [10]. Within these large projects models were used to aid problem formulation, to synthesize and integrate data, to understand the complexity of the system, to predict its behaviour and to organize research activities [11, 12].

Models are, therefore, applicable to both forest management and research. They can be used to predict the future growth or economic performance of forests or to synthesize large quantities of fragmented data leading to an enhanced understanding of the system as a whole. The type of model selected will depend upon the context in which it is to be used.

One of the earliest discussions of models occurs in Plato's *Timaeus*, written towards the end of the fourth century B.C. In his description of the universe Plato identified three concepts: the creator, the model, and the copy [13]. The "model" was considered to be the unchangeable forms or patterns in imitation of which the creator constructed the cosmos itself. Plato realized the difficulty in understanding and describing reality and accepted that analogies and simplifications are required to facilitate both comprehension and communication of how we perceive the real world. He also recognised that our perceptions of the world change with time, but that "truth" does not. Thus, no model is ever "right". A model is developed to represent those features of a system perceived to be most relevant to the question or questions posed by the modeller and for which adequate information is currently available. Using models for purposes which differ to those for which the model was developed can be misleading. Equally, models can be updated as new information becomes available. Alternatively, models may require to be restructured if our understanding of system functioning alters substantially, or if the model is to be used to address new questions. A large number of computer simulation models of tree growth now exist but few have been developed to simulate long-term site productivity and fewer still address the question of how intensive forest harvesting may impact upon site productivity.

Single-tree *v.* Stand Models

Computer simulation models of tree growth operate across a continuum of scale from individual leaves to entire ecosystems. However, the unit of growth in forest stands is the individual tree. Models can, therefore, be categorized into those which predict growth of individual trees and those which operate at a stand level of resolution. Within this latter category, diameter distributions can be generated to provide summary statistics to describe the individual trees which comprise the stand as a whole. Equally, output from models which predict growth on an individual tree basis can be aggregated to provide summary statistics of the whole stand. The approach taken depends upon the purpose for which the model has been developed and the data available.

Single tree models often require information on the spatial distribution of each tree in the stand or simulated plot and details of the size of each tree. From such data, competition indices are calculated [14] and a "potential" growth is then modified according to the level of competition experienced by each tree. The development of powerful, modern computers allows single tree models to be implemented with little increase in cost of computation. The provision of individual tree co-ordinates can, however, be very expensive and does not form part of standard inventory measurements. If such data were required to implement the model then the cost may be prohibitive. If the data were required for development, calibration, and validation purposes only, then the cost may be moderate compared to other data requirements.

One fundamental difficulty with the use of single tree models is the adequate representation of environmental heterogeneity which operates at the single tree level, particularly belowground. To quantify such variation for the prediction of individual tree growth is difficult and very expensive. Not surprisingly, attempts at such measurements have never been made under field conditions. The scaling of a "potential" growth to individual trees is, therefore, largely theoretical. As such, there is little difference between aggregated single tree models or disaggregated stand models (using diameter distributions) when applied to the simulation of whole-tree harvesting impacts upon tree growth.

Empirical and Mechanistic Models

When assessing the type of model best suited to a particular purpose it is necessary to decide whether a better understanding of the processes operating within the system under investigation is desired, or whether the model is to be used purely to predict the output from that system. In the latter case, empirical models based upon experimental data are most appropriate, but these may yield little or no insight into how the internal mechanisms within the system operate. Empirical models rely on the process of induction in which a set of system inputs and out-puts are observed, measured, and recorded, and some or all of the mathematical model inferred. This may occur through passive experimentation where the available input and output data are accepted or active experimentation can be undertaken to apply specific inputs to the system and observe the consequent outputs [6]. The alternative to the induction technique is to use the process of deduction. This involves reasoning from known or hypothesized principles to deduce an unknown using a series of progressively more specific concepts [6]. Laws provide the basic principles governing the nature of the equations used to characterize the structure and function of a given system. For example, the principle of conservation of matter may be used to derive a series of algebraic or difference equations. Knowledge of the specific system being studied is then used to assign parameter values to the coefficients in each equation. In practice, the laws governing biological systems are rarely understood and the complexity of their structure is too great to allow purely deductive models to be developed. Simplifying assumptions and approximations are required and some degree of empiricism is necessary to estimate parameters. Mechanistic, explanatory, or process-orientated models are, therefore, ultimately based upon some degree of empiricism. It is always possible to find an empirical model that provides a better fit to a given set of data due to the constraints imposed by the assumptions incorporated within all mechanistic models [5]. Emphases placed upon processes during the construction of mechanistic models do, however, provide greater scope for the application of such models to conditions that differ from those under which the model was developed.

The extent to which an understanding of processes is incorporated into computer simulation models has been used for their classification into empirical versus mechanistic. There is, however, a continuum across the range from purely empirical to purely mechanistic (theoretical). The term hybrid has been used to cover models which employ a limited representation of processes influencing tree growth. It is useful to review a range of computer simulation models that have been developed to demonstrate the range of process representations that have been undertaken with varying degrees of success. Many of these models contain elements that could provide inputs to expert systems or computer simulation models for use in predicting long-term impacts of intensive harvesting upon site productivity.

Empirical Growth Functions

Empirical growth functions form the basis of many growth and yield tables currently used in forest management. Statistical analyses of recurrent inventory-type data are used to develop predictive models of tree growth and stand development based upon measurement of "site index" or its equivalent for a particular site. Such models are confined almost entirely to the data sets from which each model was developed, or to sets that are very similar. These limitations may be overcome, to some extent, by the use of growth functions in which the specific function represents an hypothesis related to the biology of the tree or stand. A growth function is "an analytical function which can be written down in a single equation and is used to provide a mathematical summary of time course data on the growth of a group of organisms, a single organism or part of an organism" [5]. The most usual pattern of growth in forest systems is sigmoidal, whether growth is measured in terms of height, volume, or biomass on individual stems or in the forest as a whole (Figure 6.1).

The important features of this model are that growth is initially slow, then enters a period of near linearity after which growth declines towards zero as the tree or stand matures and eventually dies. The upper limit to size is set by the asymptote and there is a point of inflection on the curve at the time of maximum current increment [15]. The simplest form of such a function is the logistic growth function [16] but more flexible forms include the Chapman-Richards [5] or the Gompertz functions [15, 17]. These growth functions have sometimes been referred to as being biologically meaningful because biological significance has been attached to their growth parameters. Others have argued that such models

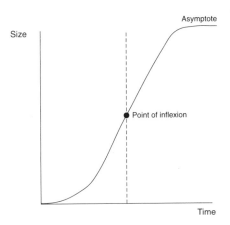

FIGURE 6.1   Sigmoidal growth curve.

should be treated as purely formalistic because the value of parameters such as maximum plant size is difficult to determine and may well depend upon environmental conditions [18]. It is clear that empirical growth functions are restricted in their capacity to describe growth and their "biological significance" is somewhat dubious.

In practice, empirical growth functions are often used to develop families of curves relating one aspect of tree growth (usually height) to age [19]. Height attained at a reference age (e.g., 50 years or 100 years) is used as an index of site quality (Figure 6.2).

Forest Site Quality Evaluation

In many cases, forest managers may wish to introduce different tree species to a site so that a direct estimate of site index for that species is not available. Forest Site Quality Evaluation uses regression techniques to relate soil, vegetation, and climatic factors to site index or some other measure of forest site productivity. Such techniques have been extensively reviewed by Carmean [19]. The potential use of site classification for predicting the effects of management on site productivity has been explored by Williams and Gresham [20] and possible future developments have been discussed by Jones [21].

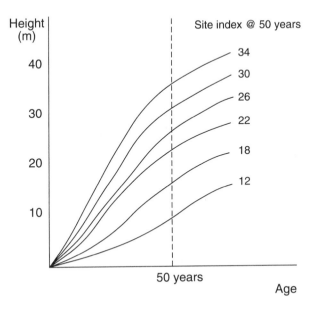

FIGURE 6.2 Site index curves.

In the context of predicting the potential impact of intensive harvesting on site quality, there are several pit-falls associated with the integration of Forest Site Quality Evaluation techniques with the use of empirically derived growth functions. One is the continued development of improved technology, particularly in the area of crop establishment. Site Index is particularly sensitive to the period over which plantations or natural stands are becoming established (Figure 6.3). Although techniques are available to attempt to "correct" for such influences [19] these are not satisfactory. Genetic research into tree improvement has also made significant advances in recent years and is likely to become increasingly influential in future rotations (Figure 6.3). In addition, management operations may degrade the site due, for example, to compaction or excessive nutrient removals (Figure 6.3). Climate change may also influence the rate of tree growth although the magnitude and direction of such changes remain unclear.

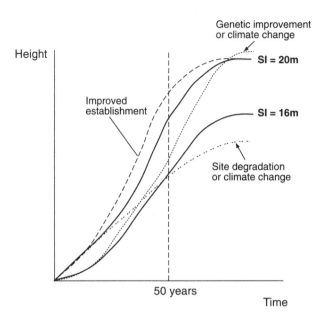

FIGURE 6.3   Potential shifts in site index curves and possible causes.

Although growth and yield tables have been widely used for many years, criticism has been growing that they provide a poor reflection of current and future patterns of forest growth. They provide a description of historical patterns of stand development, but fail to take account of the effects current management practices may have upon the empirical relationships which have been fixed within growth and yield tables. Implicit assumptions relating site factors to tree growth may no longer be valid under new management techniques and changed physical environments. In order to predict the impact of management practices such as intensive harvesting, it is necessary to develop and use "dynamic" models which represent feedback between management, site, and tree growth.

Dynamic Site-yield Models

Dynamic site-yield models couple site factors to tree growth, thereby allowing feedback in either direction. One example of such a model is FORCYTE-10 [22]. Empirical, inventory-type data are used to produce a family of Chapman-Richards growth functions used to drive the model. Feedback between the predicted demand for nitrogen and the predicted available nitrogen from soil pools (including fertilizer additions) allows tree growth to diverge from the pre-determined Chapman-Richards growth functions depending upon predicted changes in soil nitrogen supply. Feedback also exists between "litter quality" and decomposition to enable a more realistic representation of soil nitrogen availability [3].

FORCYTE-10 relies heavily upon empirical growth and yield data. The model has been calibrated for several conifer species in regions of North America [23, 24, 25, 26], and for Scots pine in Finland [27]. FORCYTE-10 was developed with nitrogen supply as a modifier to conventional site/yield data with moisture and other nutrients being implicit within the Chapman-Richards growth functions.

Continued funding of the FORCYTE series has led to the development of FORCYTE-11 [28, 29]. In this version, ecosystem process representations have been increased, although their methods of representation remain largely empirical. Growth and yield data are combined with nutritional data and light measurements to calculate foliage nitrogen efficiencies corrected for shading. These age- and site-specific values are used as the driving functions of tree growth. Other nutrients can be incorporated within the model. Water and climate remain implicit in the empirical derivation of foliage nitrogen efficiency values. A major advantage of FORCYTE is the ability to simulate growth and nutrient uptake by ground vegetation. Weed species have been shown to be important sinks contributing to nutrient retention on site after harvesting when the system may become potentially "leaky" for nutrients [30, 31]. The empirical representation of weed growth detracts from the usefulness of FORCYTE to simulate whole-tree harvesting, due to a paucity of field data.

The model JABOWA was developed to simulate competition and secondary succession in mixed species stands of the north-eastern United States [32]. In this individual tree growth model, potential diameter growth of each species is estimated as a function of tree size, with respect to a maximum size attribute, and age. Three growth multipliers based on shading, temperature, and competition are applied to determine the actual growth for a given tree. Shading by competing trees is simulated using the Beers-Lambert Law for light extinction within a canopy [33]. Actual values of the growth multipliers are modified according to species shade tolerance. The effect of temperature upon growth was included as a parabolic function of growing day degrees, modified for different species optima. Finally, a crude expression of competition for water and nutrients was derived from the ratio of current basal area to an upper limit set for each site. These deterministic growth functions for each species were coupled to stochastic routines to predict regrowth and mortality.

A large family of computer simulation models have evolved from JABOWA and can be divided into two broad groups.

Gap Models - "FORET" Group: Minor modifications were incorporated into FORET [34] which has since formed the basis of many "GAP" succession models [35]. These have concentrated primarily upon light and species competition, with very limited attempts to represent the processes of competition for other resources. Continued development of GAP models include, for example, ZELIG which has a more refined spatial representation of abiotic environmental gradients for integration with Geographic Information Systems [36]. Other workers have reduced the spatial representation, within simulated plots, from the single tree to the stand - subdivided into height classes [37].

Linked Models - "FORTNITE" Group: The growth functions developed in JABOWA have also been linked to a decomposition model [38] in the FORTNITE series of models. In FORTNITE, an additional growth multiplier was used to simulate the effect of nitrogen availability. Moisture stress was represented by a growth multiplier derived from predicted evapotranspiration [39]. Further modifications have been made in LINKAGES [40]. Moisture stress is based upon the proportion of the growing season during which soil water potential falls below the wilting point estimated for a given species. In addition, the use of growth multipliers is simplified such that an individual tree's growth is limited by nitrogen, water, light, or temperature - whichever is the most limiting [41]. The rate of litter decomposition in LINKAGES depends upon climate and litter carbon chemistry. This is a potentially useful feature of LINKAGES because the consequences of intensive harvesting may well become manifest through the impact upon forest floor microclimate, in addition to the direct removal of carbon and nutrients from a site [42, 43].

Changes in the balance between nutrient availability and demand for nutrients by the newly developing tree crop will affect the amount and quality of litterfall with profound consequences for future litter quality and, hence, site fertility.

The series of models developed from JABOWA, and to some extent FORCYTE, still rely heavily upon empirical growth functions as their drivers. Modification of these functions through growth multipliers - in whatever form - has led to some form of process representation. Although individual growth processes are not explicitly represented, those factors likely to influence these processes have been derived and applied to the empirical growth functions. One major difference between the FORCYTE and JABOWA series is important when simulating impacts of management on tree growth. The JABOWA models use empirically-derived, *fixed*, allometric relations to partition growth [44]. FORCYTE also uses allometric relations but these are taken from empirical data spanning a range of site "qualities" and the specific relationships can change during individual simulation runs. As a consequence, partitioning between tree components can differ in response to changing site characteristics. These, and similar models, have already been used to predict potential consequences of intensive harvesting upon long-term productivity [39, 45, 46, 47].

Although the division between "empirical" and "process-based" models tends to be subjective, it is possible to separate models into those which use empirically-derived growth functions - however modified - and those which attempt to drive growth directly by various levels of process representation. The complexity with which processes are represented within models forms another continuum which is no easier to resolve than that between empirical and process-based models.

"Simple" Process-based Models

McMurtrie and Wolfe [48, 49] developed a stand-level growth model with a driving function based upon a site-specific upper limit to gross photosynthesis, assuming full canopy closure and preferably derived from solar irradiance and ambient CO_2 levels. Actual gross photosynthesis was calculated according to the fraction of incident radiation intercepted by the canopy using Beers-Lambert's Law. Net photosynthesis was then derived using an explicit representation of leaf respiration. A further parameter was used to express the conversion efficiency from carbon fixation to dry matter production. Partition coefficients were then used to allocate dry matter to different tree components. By changing the assimilate partition coefficients or the photosynthetic efficiencies of the competitors the authors were able to qualitatively simulate competition effects for water or nutrients arising from the interaction between trees and grass. Such a model provides scope for a more mechanistic representation of competition effects upon growth.

A similar approach to that of McMurtrie and Wolfe, but applied at the single tree level, was taken earlier by Lindgren and Axelsson [50] in their model STAND, developed during the Swedish Coniferous Forest Project [10]. Growth was driven by foliage biomass with the upper limit to growth determined by a maximum "foliage productivity" value (biomass produced per unit foliage biomass). Potentially realizable production was then modified by three growth multipliers representing the influence of shading, nitrogen status within the tree, and tree size. Attainment of the potentially realizable production depended upon the balance between nitrogen required to fulfil this growth at the current level of nitrogen within the tree, and the predicted availability of nitrogen generated within a decomposition subroutine.

The models described in this section have illustrated a shift away from empirical growth functions, towards functions that are based on the concept of a potential canopy photosynthate production modified by factors likely to influence the realization of growth potential.

There are three main factors often identified as influencing the attainment of potential growth: self-shading, nutrient status, and moisture. Aspects of shading and nutrition have been incorporated directly into a growth function using the concept of "nitrogen productivity" initially proposed by Ingestad [51] and developed further by Ågren [52]. In order to limit growth under conditions of adequate nitrogen supply, Ågren assumed a linear decrease in nitrogen productivity with increasing foliage biomass to represent aspects of internal shading, water stress, and/or increased respiration by non-photosynthesizing tissues. It is this approach that has been incorporated into FORCYTE-11, although the nitrogen productivity values were derived from empirical Chapman-Richards growth functions coupled to data on biomass distributions and foliage nitrogen concentrations.

"Complex" Process-based Models

Trees require light, water, and nutrients to grow. Competition for these resources occurs within trees, between trees, and between trees and other biological components of the ecosystem. A large number of complex, process-based models have been developed to simulate the various processes involved in tree growth. Many of these models examine a number of key processes associated with one major aspect of growth such as photosynthesis, photosynthate allocation, or nutrient acquisition. There are, however, comparatively few models which integrate many of the key processes of tree growth into a coherent and complex representation of forest ecosystems.

McMurtrie and Wolfe's model mentioned earlier, has recently been extended into a dynamic, process-based model (BIOMASS) to predict yield as annual net canopy photosynthesis [53]. The model can also include respiratory costs and allocation of photosynthate, but these are rather simplistic representations at present. BIOMASS computes a daily water-balance and simulates a root-zone soil water storage providing a source of plant available water. Canopy photosynthesis and transpiration are, therefore, linked to soil water and an option to allocate photosynthate between tree components (including litterfall) is available. The model does not, however, incorporate nutrition explicitly, requiring foliage nitrogen concentrations as input data through time.

There have been a number of complex process-based models of competition for light and subsequent effects upon canopy processes (photosynthesis, respiration, transpiration) within single trees and at the stand levels [54, 55, 56]. These models make no attempt to predict tree growth based upon canopy processes, nor is there any link made between above- and below-ground processes. Many of the models in which such links have been made rely upon one of three representations of carbon and nutrient partitioning within trees (and other plants). The first is the functional balance approach which proposes that shoot:root ratios in plants are adjusted to maintain a balanced carbon:nitrogen ratio in the plant [57, 58]. The pipe model theory [59] proposed that each unit of foliage requires a unit of pipeline of sapwood for water transport and physical support [58, 60]. A third approach is the transport resistance model developed by Thornley [61] and recently applied specifically to trees [62]. In this approach, photosynthate is partitioned based upon two processes: substrate utilization for growth (dependent upon substrate concentration) and substrate transport between plant components (dependent upon substrate concentration gradients). Thornley's recent tree model is perhaps the most sophisticated attempt to represent the dynamics of carbon partitioning at the process level, but the model has yet to be extended to incorporate water and nutrients (with the exception of N).

One of the major problems associated with implementing a model such as that of Thornley [62] is the number of parameters which must be estimated, some of which are not directly measurable. A different approach has been taken in a model, ECOPHYS [63] developed for simulating growth of juvenile poplars. In this model the proportion of assimilate available for transport depends upon a leaf plastochron index, while photosynthate transport coefficients (the function of photosynthate transported to plant growth centres) were quantified directly using experimental radiotracer studies. The contrast between the two models illustrates how one problem of using non-measurable parameters can be substituted for another in which the level of sophistication required to determine model parameters is beyond the scope of forest managers - and many scientists!

A compromise is obviously needed where models allow a degree of process representation without the introduction of too many and frequently non-measurable parameters. One model of potential use is a recently modified version [64] of FOREST-BGC [65]. The most recent version allows carbon allocation coefficients to be altered dynamically at each annual iteration of the model according to the level of water and nitrogen limitations upon tree growth. The model uses a submodel based upon CENTURY [66] to simulate soil carbon and nitrogen pools (and hence nitrogen supply) and also incorporates soil hydrology and a water balance. Partitioning of photosynthates is distributed between different tree components and is further partitioned into maintenance respiration (as a function of temperature) and growth respiration (according to biochemical energetics).

Perhaps one of the most process-based models of tree growth currently available is CERES [67, 68]. According to Luxmoore [68] "many relationships between photosynthesis and external driving variables are confounded by internal feedback effects that reflect the influence of external factors on growth". Growth is more sensitive to water and mineral nutrient stress than is photosynthesis.

CERES incorporates source-sink feedback similar to that of Thornley [61] and operates on an hourly basis using sucrose gradients. CERES can interface with a soil-plant-atmosphere water model to examine the effect of plant water status on growth and solute transport. This ability to couple the model with a number of others to form the Unified Transport Model [67] has taken Thornley's model [62] further towards an integrated model of tree growth driven by light, water and nutrients with coupled feedbacks and dynamic partitioning of resources.

Model Validation and Error Prediction
Ideally, models should be tested, or validated, on independent data that have not been used in the construction of the model. In many cases, however, data are scarce and there are gaps even when all available data are utilized in developing the model. Under such circumstances, attempts should be made to collect new data for the testing of simulation models. In the case of models used to predict the long-term impact of whole-tree harvesting it may take many decades before sufficient information becomes available. Indeed, this is often one of the reasons for having developed the model in the first place. Even so, the model cannot be fully tested until it has been used to predict for an independent data set. "Short-cuts" may be available through the use of chronosequence [69] or retrospective research [70] (see Chapter 2), but the ultimate test will be against well-designed long-term field experiments (see Chapter 9). These experiments must have adequate control treatments to enable the cause and effect relationships between treatment and response to be determined. It is important to establish that the model provides the "right answer" for the "right reason" if it is to be used with confidence in other circumstances.

No model is ever "perfect". Approximations are made with respect to how reality is represented within the model (model structure) and in the inferences drawn from sample data and applied to other components of the population (model parameters). The latter can be particularly important where the database upon which the model was developed did not provide a true representation of the population to which the model was to be applied (model bias). There are, therefore, a number of sources contributing to errors of prediction associated with computer simulation models [71]. The whole area of error propagation in simulation models and the expression of confidence associated with predictions is an area of research that requires considerably more development.

EXPERT SYSTEMS

Expert Systems and Simulation Models

Expert systems are defined as computer programs that use knowledge and symbolic inference (reasoning) procedures to solve problems that are difficult enough to otherwise require significant human expertise for their solution [72]. Expert systems are based on symbolic logic methods which are largely qualitative rather than quantitative. The experience of developing and using expert systems is significantly different from numeric simulation modelling. We can use expert systems methods to develop decision support systems in complex domains where numerical simulation models would be inappropriate or impossible to develop.

Mathematical equations are very powerful problem-solving tools. When scientifically validated mathematical equations are available and applicable, they represent the most efficient and effective way to electronically model a biological system and hence assist in the problem solving effort. Unfortunately, mathematical equation systems describing most biological processes of any complexity are often unavailable. Human experts have, however, emerged in many fields with an effective understanding of complex biological processes through years of experience and scientific research. In such cases, despite the lack of a scientifically valid mathematical equation system, we can still extract the heuristic knowledge of human experts and develop an expert system analogue amenable to electronic manipulation. A heuristic is "a rule of thumb or a technique based on experience and for which our knowledge is incomplete. A heuristic rule works with useful regularity but not necessarily all the time" [73].

To better explain the utility of expert system methods, it is helpful to examine all the forms in which knowledge exists so that it can be manipulated by electronic computers. It is axiomatic that knowledge must first be represented as a digital symbol before it can be manipulated by computers. In the last 40 years, computer techniques have been developed to electronically manipulate many forms of non-language and language-based knowledge.

Non Language-based Knowledge: We perceive our environment through many channels: sight, sound, touch, smell, and taste. It is through these faculties that we gather almost all of our knowledge of the world. People store this knowledge in their memories and manipulate it through thoughts. Researchers in the field of artificial intelligence are working hard to create electronic analogues of ourselves that can also sense the world and generate and manipulate knowledge about it.

The electronic manipulation of human thoughts is currently impossible. However, recent medical research has shown that the brain operates using minute electrical impulses in an enormously large biological neural network. If that is the case, it may be possible at some point in the far future to directly exchange knowledge between a computer neural network and the biological neural network of our brain.

Electronic analogues to our sense of smell also do not currently exist unless devices like electronic smoke detectors are included in this category.

We have had good success in simulating the human sense of touch using electro-mechanical devices. The robotics industry has developed computer controlled arms and hands flexible enough and sensitive enough to play piano and physically manipulate delicate objects.

Computer interpreted vision is good enough to allow the development of rudimentary autonomously mobile all-terrain vehicles for military and space exploration purposes. We have had even more success in manipulating graphic, photographic, and video images using computers. With the recent advent of CD-ROM into mainstream computing, knowledge represented in these formats is almost commonplace.

The playing of music and recorded sounds under computer control is quite advanced while voice recognition is still rudimentary.

Representing knowledge in non-language based media offers several advantages for effective communication. For one thing, it is often a very compact form of knowledge, i.e., a picture is often indeed worth a thousand words. Another advantage is the ability to impart certain types of knowledge better through sound, smell, sight, or feel than through language. For example, it is difficult to describe the difference in smell of a hardwood forest versus a pine forest in autumn. Yet anyone familiar with these ecosystems can instantly differentiate between them on the basis of smell alone. The smell carries information which leads to knowledge.

To perceive and react with our environment means we must rely on our senses. To communicate with other people, we have invented language to encode our thoughts and communicate our knowledge.

Language-based Knowledge: The spectrum of language-based knowledge representation can be organized along a generality - power axis (Table 6.1). For better understanding, let us discuss the extremes first and then fill in the middle. Natural language, oral or written, is a coded representation of our thoughts. It was a tool developed so we could communicate our thoughts to others. Word processors have been developed to allow us to efficiently manipulate the written word electronically. Natural language text has great generality, i.e., it can be used to represent almost any knowledge but it is not very powerful in communicating because it is ambiguous and imprecise. To compensate for this high proportion of noise in the communication signal, natural language text is bulky - it takes many, many words to describe something a picture or an equation could represent more succinctly if only it were possible to do so.

Mathematics, at the other extreme of Table 6.1, was invented to improve the precision of expression as well as to improve problem solving power. Analytical models are composed of one or more mathematical equations with known, exact solutions. When scientifically validated analytical equations are available and applicable, they represent the most efficient and effective way to electronically model a natural system. Unfortunately, analytical equation systems describing most biological processes of any complexity are usually not available. Indeed, they may never be available for some complex systems. Therefore, in contrast to natural language text, analytical equation systems are very powerful tools but they apply only to a few, specific situations.

Until electronic computers were invented, knowledge could only be represented and manipulated as natural language text or as analytical mathematical equations. With the advent of computers and the development of the field of numerical analysis in the 1950's, it became practical, for the first time, to solve

TABLE 6.1

The language-based spectrum of knowledge representation methods.

Conceptual Word Models		Qualitative Models	Quantitative Models
Natural Language Text	Hypertext Systems	Expert Systems and Qualitative Simulation Systems	Numerical Equation Systems Analytical Equation Systems
Great Generality Ambiguous Meaning Low Problem Solving Power		<——>	Limited to Specific Problems Precise Meaning Great Problem Solving Power

systems of inter-dependent mathematical equations that could not be solved analytically. Simulation models were developed that used computer-based numerical methods to solve systems of equations approximately. Being satisfied with sufficiently precise solutions rather than demanding exact solutions to systems of mathematical equations, increased the potentially mathematically solvable set of problems by orders of magnitude. Simulation models are not as compact or powerful as analytical models but they are significantly more general in their applicability to real world problems.

Expert systems and their time variant cousins, qualitative simulation models, take another large step in the direction of generality. These systems are not based on mathematical equations at all. They grew out of the science of symbolic logic and were adapted for practical problem solving and understanding by workers in the field of artificial intelligence (AI) applications. Researchers in AI realized that much of what human experts knew could not easily be formulated into a system of mathematical equations. And yet, human experts did exist that were able to solve these problems consistently better than most other people could. By capturing this expert-based knowledge into sets of IF-THEN logical statements and then manipulating the resultant knowledge-base to try to answer some goal, expert systems were invented. Expert systems again trade off a gain in generality, i.e., an increase in the set of potentially soluble problems, with a decrease in compactness and problem solving power. Indeed, expert systems, just like their human counterparts, are not always correct; they are not even guaranteed to be able to find a solution all the time.

The movement from analytical systems, to numerical simulation systems, to expert systems all progressed from the mathematical end of the knowledge representation spectrum. In contrast, hypertext systems developed from the natural language end of the spectrum. In most natural language text, the structure of the subject is more or less hidden, camouflaged by the sequential nature of the medium and the need to bridge ideas gracefully. In natural language text, structure is subservient to the content of the material being presented. Hypertext forces the author to explicitly highlight the structure (outline or concept map) first and foremost for the user. Only secondarily is the user exposed to the content matter. It is the structure that guides the user time and again to try different paths in the hyper-base in an order that is user determined. The author can no longer rely on sequential presentation to present material, which means that each block or chunk of the hypertext must be independently understandable much like we demand that journal figures and tables be independently understandable. Hypertext gives up some generality to increase the power of natural language text to communicate understanding and problem solving. It does not go as far as expert systems do, however.

Refining the knowledge encoded in natural language text first into hypertext, then into expert systems, and finally into mathematical models produces a significant compaction of knowledge bulk, increases understanding through progressive organization, evaluation, and synthesis, and promotes more effective problem solving and communication. The cost is a step-by-step reduction in the set of problems potentially solvable and the effort of refining knowledge at each step.

Structure and Function of Expert Systems

The structure of an expert system, also known as a knowledge-based system, may be seen to consist of many component parts (Figure 6.4) [74]. Not all of these elements are necessarily included in any one system. Rather, we have illustrated a relatively comprehensive collection of components.

The process of running an expert system is commonly referred to as a "consultation", alluding to the analogous scenario of consulting a human expert. The reasoning engine is a computer program (consisting of algorithms) that navigates through the knowledge base using inference methods and control

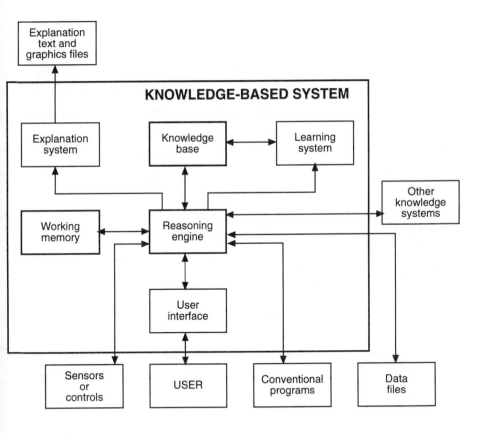

FIGURE 6.4 The components of an expert (knowledge-based) system.

strategies to solve problems. To make the best decisions possible in an imperfect world, we must often use the best available resources at our disposal. Sometimes certain desirable resources are unavailable, sometimes the ones we do have are unreliable, and sometimes our understanding is incomplete and can only provide us with vague answers. For expert systems to be useful, they, like us, also must possess an ability to reason in this type of "dirty" environment. Therefore, most expert systems allow for inexact reasoning or approximate reasoning in the course of the problem solving consultation.

The knowledge base contains the domain-pertinent knowledge. The expert knowledge encoded in the knowledge-base of expert systems comes from human experts either directly through interviews or indirectly through condensation of the literature. Experts perform better than non-experts because they have a large amount of compiled, domain-specific knowledge stored in long-term memory and they have become skilled at using that knowledge to solve specific problems [72]. A world-class expert is said to have between 50,000 and 100,000 chunks of knowledge stored in long-term memory. When young, we begin to organize our experiences into chunks centred around objects, either concrete or abstract. As we learn more, we cluster more and more knowledge around successively more abstract chunks. But no matter how complex our chunks become, our ability to manipulate any one chunk seems to stay about the same. Experts ordinarily emerge after about 10 years or more of study and work in their fields because it takes that long to acquire, chunk, and organize the necessary knowledge and experience.

One of the distinguishing characteristics of knowledgeable people (experts) is the inaccessible nature of their knowledge. Their knowledge is inaccessible in the sense that an expert encounters difficulty when relating the how's and why's of his or her knowledge to others. Much of an individual's knowledge is implicitly [75] attached to and integrated with prior experiences and ways of doing things - subsequent extraction is difficult. This phenomenon has also been referred to as the compiled nature of expertise. This characterization of human expertise arises from an analogy between knowledge and computer programming languages. It's suggested that expertise is similar to a compiled computer program. A compiled program is useful because it can be executed on some machine to perform some computation. If one looks at the compiled code, however, it's just machine code

(0's and 1's), and it's impossible to understand how the program's logic works. If we can "reverse compile" it back into the source code of some programming language which we understand, then the rationale becomes much more apparent. It is similar with human expertise. The most difficult and time-consuming part of constructing expert systems deals with "reverse compiling" human expertise.

A working memory serves as the work area (scratch pad) of the inference engine where all currently known, problem-specific information is stored.

User interface software enables the system to elicit answers to questions and to display screen reports. Consultation with the expert system usually occurs through some user-interface program, which may be menu driven or may incorporate some sophisticated natural language capability or, possibly, visual-aid graphics. Facts about a current consultation problem are provided by the user and entered into the working memory. A reasoning engine then attempts to validate other facts (hypotheses) from working memory facts and the knowledge base "understanding" of the application area.

The explanation system along with the user interface constitutes the human-machine interface. Explanations may answer "why" a question is asked of the user, "how" an answer has been derived, the meaning of terms, and the intent and source of reference for the knowledge in the knowledge base. An expert system may input data from simulation models (in external programs) or database/spreadsheet files as well as output data to simulation models and change database/spreadsheet files. Non-language based information is acquired by the expert system through connections to digital or analog sensors. Output goes to electro-mechanical controls.

Finally, when an expert system finds a dead-end in its chain of logic (i.e., a point in the line of reasoning from which no solution can be reached), a learning system might report this finding to the user and help the user remedy the situation by making changes to the knowledge base during the running of the consultation.

Benefits and Limitations of Expert Systems

Many potential benefits can be derived from the application of expert systems in science and society. Expert systems, in conjunction with hypertext systems, can be used to bring together all the knowledge in a particular domain in a formulated, testable, and maintainable way. This will provide an explicit record of expertise to ensure some protection from catastrophic loss, such as the retirement of a key person, and to provide a concrete record of current decision strategies. An explicit record affords the decision-maker a measure of accountability - a means for, at least partially, justifying decisions. This can be especially desirable when decisions may be interpreted as controversial in nature. Also, strategies used in solving problems and making decisions become less mysterious. Artificial expertise provides a focus for upgrading and improving management strategies over time [76]. As new information is produced, it will be incorporated into our understanding (knowledge) of that field, and lead to new strategies for the application of that knowledge in the real world.

One of the primary advantages of the expert system approach is the ability to create systems which fuse the knowledge of several separate disciplines (e.g., [77]). This breadth of expertise endows them with a greater potential to solve complex problems than human experts who are often knowledgeable only in a single domain. In a sense we can create "super" experts that have a greater breadth of knowledge and can address a wider range of problems.

Most knowledge found in textbooks, handbooks, research papers, and other "public" sources address general principles and make recommendations for the average case. Expert systems are designed to translate this general understanding into a solution process for specific problems. Differences between routine situations and those requiring more analysis or data collection or attention by a human expert can be identified and treated appropriately [78].

Expert systems, being part of and reliant on artificial intelligence, have all the same limitations associated with an inability to field truly intelligent systems. As Artificial Intelligence (AI) develops and continues to create more sophisticated and intelligent systems, expert systems will improve to the point where many limitations may eventually disappear. Bowerman and Glover [79] note several categories of expert system weaknesses:

1. Systems function in a subdomain of the full human potential: i.e., cognitive and logical. They are not adept in managing sensory input or output, or dealing with less analytical aspects of human reasoning abilities, e.g., ellipsis, intuition.

2. The total knowledge available to a human expert appears in many forms, of which expert systems have no understanding, e.g., social knowledge, analogical reasoning, random memories, feelings, emotions, common sense, and irrational information. Their problem-solving skills have a very narrow focus.
3. Expert systems degrade almost immediately when confronted with situations outside their domain. The ability to recognise and recover from errors is non-existent.
4. Human intellectual development relies extensively on learning abilities; few current systems have any capacity for learning.
5. They exhibit no sense of self; therefore, they possess no understanding of what they know - or what they don't know.
6. The current state of scientific understanding limits their abilities; they can acquire no new knowledge. Some are able to solve problems more effectively than humans because of their exhaustive search methods, but they have not increased knowledge in the domain.

Limitations do not imply any theoretical defect, currently apparent, that would prevent expert system developers from eventually overcoming these problems. This technology remains in its infancy. Vigorous research in AI has as its goal the solution of several of these problems; in particular, learning methods, alternative logics, and common sense reasoning are receiving substantial effort. The critical breakthrough for solving many of these problems may likely involve some capacity for learning [80]. Learning will enable systems to recognise familiar situations and react quickly, to adapt to new and unfamiliar circumstances, and to expand their narrow focus to larger, more general problems.

Current Development in Expert Systems

Articles on AI and expert systems in natural resource management began to appear in significant numbers in 1983 [81]. The scientific journal *AI Applications* covering natural resources, agriculture, and environmental science began publishing in 1987. A review article reporting on 74 AI projects in natural resources world-wide was authored by Rauscher and Hacker [82]. Another world-wide review of the literature on AI and environmental protection was prepared by Simon *et al.* [83] in German.

Rauscher and Hacker [82] identified a total of 74 projects worldwide. About half the developers had invested substantial resources in learning a complex new programming methodology - the most frequently used language being ProLog. Most other applications were developed using Artificial Intelligence "shells" or "toolkits". Only 16% of projects used standard computer languages as a main or supporting language for their AI applications. The majority of these AI applications have been developed for IBM PC-compatible microcomputers.

The dominant domain of application for AI systems in forestry appears to be silviculture, growth, and yield:

Silviculture/Growth and Yield	27%
Fire Management	18%
Pest Management	16%
Soil/Site/Environment	14%
Land Management Planning	12%
Harvesting/Products	7%
Others	6%

The range of applications is very broad. Almost all specialities within natural resource management have projects within them that may benefit from the adoption of AI methods for problem solution.

Since 1989, the C/C++ programming language and the object oriented programming philosophy have gained in popularity. Lacking a comprehensive survey more recent than that of Rauscher and Hacker [82], it is difficult to assess the current state of affairs objectively.

To summarise, Expert Decision Support Systems (EDSS) methodology is well developed and, in many fields, well tested [84]. Outside natural resource science and management, the power and usefulness of expert systems is well recognised [85]. The situation is somewhat different within natural resource management. The adoption of expert systems methods by the forest science research community has steadily expanded over the last 5 years. Prototypical systems (i.e., developed systems that exist, have been reported in the scientific literature, but have not yet seen extensive field testing) are now common [77, 86, 87, 88, 89].

On the other hand, in the natural resource field, industrial strength, fully tested and deployed management EDSS are still rare. The financial support for the development of expert decision support systems and their subsequent use by the forest management community has been slow in developing. Some constraints have been identified that account for this condition. First, the methodology is new and untested. This is usually an advantage for researchers but a constraint for managers. Second, the terminology, such as "artificial intelligence" and "expert system", is unusually threatening, especially to forest managers with little computer experience, and this prevents an objective evaluation of the utility of this technology. Third, several cycles of media-driven "hype" have raised fears and expectations among the management community to unrealistic levels. Fourth, development of expert decision support systems requires fairly advanced micro-computer systems with powerful graphic programming environments. Many forest management institutions have not yet invested in these systems and there-fore cannot execute and evaluate research systems which have already been developed. Finally, developing expert decision support systems for research demonstration is much less costly than developing them for management use. The user involvement, deployment of the system for initial testing, monitoring the tests, making numerous changes in user interface as well as content, and finally, developing effective maintenance strategies raises the cost by one to two orders of magnitude. To progress from prototype research systems to fully developed management systems will require substantial further investment.

LITERATURE CITED

1. Dyck, W.J. and Skinner, M.F. Potential for productivity decline in New Zealand radiata pine forests. In: *Sustained Productivity of Forest Soils*. Proceedings, 7th North American Forest Soils Conference, Vancouver, Canada. (Eds.) S.P. Gessel, D.S. Lacate, G.F. Weetman, and R.F. Powers. Faculty of Forestry, UBC, Production, Vancouver B.C., 1990, pp. 318-332.

2. MacDonald, A.M. *Chambers Twentieth Century Dictionary*. W & R Chambers Limited, Edinburgh, 1982.

3. Kimmins, J.P. and Scoullar, K.A. The role of modelling in tree nutrition

research and site nutrient management. In: *Nutrition of Plantation Forests*. (Eds.) G.D. Bowen and E.K.S. Nambiar. Academic Press, London, 1984, pp. 463-487.

4. Kowal, N.E. A rationale for modelling dynamic ecological systems. In: *Systems Analysis and Simulation in Ecology*. (Ed.) B.C. Patten. Academic Press, London, 1971, pp. 123-194.

5. France, J. and Thornley, J.H.M. *Mathematical Models in Agriculture*. Butterworths, London, 1984.

6. Karplus, W.J. The spectrum of mathematical modelling and systems simulation. *Mathematics and Computers in Simulation*, 1977, **XIX**, 3-10.

7. Jeffers, J.N.R. *An Introduction to Systems Analysis: with Ecological Applications*. Edward Arnold, London, 1978.

8. Kimmins, J.P. Future shock in forest yield forecasting: the need for a new approach. *Forestry Chronicle*, 1985, 61: 503-512.

9. Rosswall, T. *Systems Analysis in Northern Coniferous Forests*. Swedish Natural Science Research Council - Bulletins from the Ecological Research Committee, Number 14, 1971.

10. Persson, T. Structure and Function of Northern Coniferous Forests - An Ecosystem Study. *Ecological Bulletin (Stockholm)*, 1980, Number 32.

11. Reichle, D.E. Systems analysis as applied to ecological processes: a mechanism for synthesis, integration and interpretation of I.B.P. woodlands ecosystem research. In: *Systems Analysis in Northern Coniferous Forests*. (Ed.) T. Rosswall. Swedish Natural Science Research Council - Bulletins from the Ecological Research Committee, 1971, **14**, 12-28.

12. Andersson, F. Ecosystem research within the Swedish coniferous forest project. In: Structure and Function of Northern Coniferous Forests - An Ecosystem Study. (Ed.) T. Persson. *Ecological Bulletin (Stockholm)*, 1980, **32**, 11-33.

13. Plochman, G.K. *Plato*. Dell Publishing Company, New York, 1973.

14. Munro, D.D. Forest growth models - a prognosis. In: *Growth Models for Tree and Stand Simulation*. IUFRO Working Party S4.01-4, Proceedings of Meetings in 1973. (Ed.) J. Fries. Royal College of Forestry, Department of Forest Yield Research (Sweden) Note No. 30, 1974, pp. 7-21.

15. Rawat, A.S. and Franz, F. Detailed non-linear asymptotic regression studies on tree and stand growth with particular reference to forest yield research in Bavaria (Federal Republic of Germany) and India. In: *Growth Models for Tree and Stand Simulation*. IUFRO Working Party S4.01-4, Proceedings of Meetings in 1973. Royal College of Forestry, Dept of Forest Yield Research (Sweden), Note Number 30, 1974, pp. 180-221.

16. Cooper, C.F. Carbon storage in managed forests. *Canadian Journal of Forest Research*, 1983, **13**, 155-166.

17. Nokoe, S. Demonstrating the flexibility of the Gompertz function as a yield model using mature species data. *Commonwealth Forestry Review*, 1978, **57**, 35-42.

18. Zelawski, W. Modelling of the dry matter accumulation in plants by means of asymptotic (logistic) and exponential functions. *Studia Forestalia Suecica*, 1981, **160**, 31-38.

19. Carmean, W.H. Forest site quality evaluation in the United States. *Advances in Agronomy*, 1975, **27**, 209-269.

20. Williams, T.M. and Gresham, C.A. Predicting Consequences of Intensive Forest Harvesting on Long-term Productivity by Site Classification. IEA/BE Project A3, Report No. 6, Baruch Forest Science Institute of Clemson University, Georgetown, S.C., 1988, 180 p.

21. Jones, R.K. Role of site classification in predicting the consequences of management on forest response. In: *Impact of Intensive Harvesting on Forest Site Productivity*, Proceedings, IEA/BE A3 Workshop, South Island, New Zealand, March 1989. (Eds.) W.J. Dyck and C.A. Mees. IEA/BE T6/A6 Report No. 2, Forest Research Institute, Rotorua, New Zealand, FRI Bulletin No 159, 1990, pp. 19-38.

22. Kimmins, J.P. and Scoullar, K.A. *FORCYTE-10: A User's Manual (Second Approximation)*. Faculty of Forest Ecology, University of British Columbia, Canada, 1983.

23. Feller, M.C., Kimmins, J.P., and Scoullar, K.A. FORCYTE-10: calibration data and simulation of potential long-term effects of intensive forest management on site productivity, economic performance and energy benefit/cost ratio. In: *IUFRO Symposium on Forest Site and Continuous Productivity*. (Eds.) R. Ballard and S.P. Gessel. USDA Forest Service General Technical Report PNW-163, Portland, Oregon, 1983, pp. 179-200.

24. Sachs, D. and Sollins, P. Potential effects of management practices on nitrogen nutrition and long-term productivity of western hemlock stands. *Forest Ecology and Management*, 1986, **17**, 25-36.

25. Yarie, J. A preliminary comparison of two ecosystem models, FORCYTE-10 and linkages for interior alaska white spruce. In: Predicting consequences of intensive forest harvesting on long-term productivity. (Ed.) G.I. Ågren, Swedish University of Agricultural Science, Dept of Ecology and Environmental Research Report No. 26, 1986, pp. 95-103.

26. Meades, W.J. The integration of forest site classification and FORCYTE-10 calibration for balsam fir in western Newfoundland. In: *Predicting Consequences of Intensive Forest Harvesting on Long-term Productivity by*

Site Classification. (Ed.) T.M. Williams and C.A. Gresham. IEA/BE Project A3, Report No. 6. Baruch Forest Science Institute of Clemson University, Georgetown, S.C., 1988, pp. 29-40.

27. Kellomäki, S. and Seppala, M.   Effects of timber harvesting and forest management on the nutrient cycle and productivity of forest ecosystems with reference to the selection of a proper management regime for timber production.   Final report of the preliminary study supported by the grant (No 1133) from the Foundation from the Research and Natural Resources in Finland. University of Joensuu, Joensuu, Finland, 1986, 67 p.

28. Kimmins, J.P. and Scoullar, K.A.   *FORCYTE-11.   User's Manual (3rd edition).* Report to Forestry Canada, Northern Forest Research Centre, Edmonton, Alberta, 1989, 430 p.

29. Kimmins, J.P., Comeau, P.G., and Kurz, W.   Modelling the interactions between moisture and nutrients in the control of forest growth.   *Forest Ecology and Management*, 1990, **30**, 361-380.

30. Likens, G.E., Bormann, F.H., Johnson, N.M., Fisher, D.W., and Pierce, R.S. Effects of forest cutting and herbicide treatment on nutrient budgets in the Hubbard Brook Watershed ecosystem. *Ecological Monograph*, 1970, **40**, 23-47.

31. Emmett, B.A., Anderson, J.M., and Hornung, M.   Nitrogen sinks following two intensities of harvesting in a Sitka spruce forest (N. Wales) and the effect on the establishment of the next crop.   *Forest Ecology and Management*, 1991, **41**, 81-93.

32. Botkin, D.B., Janak, J.F., and Wallis, J.R.   Some ecological consequences of a computer model of forest growth. *Journal of Ecology*, 1972, **60**, 849-873.

33. Monsi, M., Uchijima, Z., and Oikawa, T.   Structure of foliage canopies and photosynthesis. *Annual Review of Ecological Systems*, 1973, **4**, 301-327.

34. Shugart, H.H. and West, D.C.   Development of an Appalachian deciduous forest succession model and its application to assessment of the impact of the chestnut blight. *Journal of Environmental Management*, 1977, **5**, 161-179.

35. Shugart, H.H.   *A Theory of Forest Dynamics: The Ecological Implications of Forest Succession Models.*   Springer, New York, 1984, 278 p.

36. Urban, D.L., Bonan, G.B., Smith, T.M., and Shugart, H.H.   Spatial applications of gap models. *Forest Ecology and Management*, 1991, **42**, 95-110.

37. Fulton, M.R.   A computationally efficient forest succession model: design and initial tests. *Forest Ecology and Management*, 1991, **42**, 23-34.

38. Aber, J.D., Botkin, D.M., and Melillo, J.M.   Predicting the effects of different harvesting regimes on forest floor dynamics in northern hardwoods. *Canadian Journal of Forest Research*, 1978, **8**, 306-315.

39. Aber, J.D., Melillo, J.M., and Federer, C.A. Predicting the effects of rotation length, harvest intensity and fertilization on fibre yield from northern hardwood forests in New England. *Forest Science*, 1982, **28**, 31-45.

40. Pastor, J. and Post, W.M. Influence of climate, soil moisture and succession on forest carbon and nitrogen cycles. *Biogeochemistry*, 1986, **2**, 3-27.

41. Pastor, J. Reciprocally linked carbon-nitrogen cycles in forests: biological feedbacks within geological constraints. *Predicting Consequences of Intensive Forest Harvesting on Long-term Productivity*. (Ed.) G.I. Ågren. Swedish University of Agricultural Sciences, Dept. Ecology and Environmental Research Report No. 26, 1986, pp. 131-140.

42. Hendrickson, O.Q., Chatarpaul, L., and Robinson, J.B. Effects of two methods of timber harvesting on microbial processes in forest soil. *Soil Science Society of America Journal*, 1985, **49**, 739-746.

43. Jansson, P. Simulated soil temperature and moisture at a clearcutting in central Sweden. *Scandinavian Journal of Forest Research*, 1987, **2**, 127-140.

44. Riggan, P.J. and Cole, D.W. Simulation of forest production and nitrogen uptake in a young Douglas fir ecosystem. In: *Modelling Wastewater Renovation - Land Treatment*. (Ed.) I.K. Iskander, 1981, pp. 410-433.

45. Swank, W.T. and Waide, J.B. Interpretation of nutrient cycling research in a management context: Evaluating potential effects of alternative management strategies on site productivity. In: *Forests: Fresh Perspectives from Ecosystem Analysis*. (Ed.) R.H. Waring. Oregon State Univ. Press. 1980, pp. 137-158.

46. Proe, M.F. Predicting the effects of whole-tree harvesting on long-term site productivity for stands of Corsican pine. In: *Predicting Consequences of Intensive Forest Harvesting on Long-term Productivity*. (Ed.) G.I. Ågren. Swedish University of Agricultural Science Report Number 26, 1986, pp. 117-129.

47. Yarie, J. A comparison of the nutritional consequences of intensive forest harvesting in Alaskan taiga forests as predicted by FORCYTE-10 AND Linkages 2. IEA/BE Project A6 Report No 1. Forest Research Institute, Rotorua, New Zealand, 1989, 59 p.

48. McMurtrie, R. and Wolfe, L. Above and below ground growth of forest stands. A carbon budget model. *Annals of Botany*, 1983, **52**, 437-448.

49. McMurtrie, R. and Wolfe, L. A model of competition between trees and grass for radiation, water and nutrients. *Annals of Botany*, 1983, **52**, 449-458.

50. Lindgren, A. and Axelsson, B. STAND - A simulation model of the long-term development of a pine stand. Swedish Coniferous Forest Project Technical Report Number 28. Swedish University of Agricultural Sciences, Uppsala, Sweden, 1980.

51. Ingestad, T. Nitrogen and plant growth; maximum efficiency of nitrogen fertilizers. *Ambio*, 1977, **6**, 146-151.

52. Ågren, G.I. Theory for growth of plants derived from the nitrogen productivity concept. *Physiologica Plantarum*, 1985, **64**, 17-28.

53. McMurtrie, R.E., Rook D.A., and Kelliher, F.M. Modelling the yield of *Pinus radiata* on a site limited by water and nitrogen. *Forest Ecology and Management*, 1990, **30**, 381-413.

54. Oker-Blom, P. and Kellomäki, S. Effect of grouping of foliage on the within-stand and within-crown light regime: comparison of random and grouping canopy models. *Agricultural Meteorology*, 1983, **28**, 143-155.

55. Oker-Blom, P., Kaufmann, M.R., and Ryan, M.G. Performance of a canopy light interception model for conifer shoots, trees and stands. *Tree Physiology*, 1991, **9**, 227-243.

56. Wang, Y.P. and Jarvis, P.G. Description and validation of an array model - MAESTRO. *Agricultural and Forest Meteorology*, 1990, **51**, 257-280.

57. Davidson, R.L. Effect of root/leaf temperature differentials on root/shoot ratios in some pasture grasses and clover. *Annals of Botany*, 1969, **33**, 561-569.

58. Mäkelä, A. Modelling structural-functional relationships in whole-tree growth: resource allocation. *Process Modelling of Forest Growth Responses to Environmental Stress*. Section II, Tree structure and function, Chapter 7. (Eds.) R.K. Dixon, R.S. Meldahl, G.A.Ruark, W.G. Warren. 1990, pp. 81-95.

59. Shinozaki, K., Yoda, K., Hozumi, K., and Kira, T. A quantitative analysis of plant form - the pipe model theory. I. Basic analysis. *Japanese Journal of Ecology*, 1964, **14**, 97-105.

60. Valentine, H.T. Tree growth models: derivations employing the pipe-model theory. *Journal of Theoretical Biology*, 1985, **117**, 579-585.

61. Thornley, J.H.M. A balanced quantitative model for root:shoot ratios in vegetative plants. *Annals of Botany*, 1972, **36**, 431-441.

62. Thornley, J.H.M. A transport-resistance model of forest growth and partitioning. *Annals of Botany*, 1991, **68**, 211-226.

63. Rauscher, H.M., Isebrands, J.G., Host, G.E., Dickson, R.E., Dickmann, D.I., Crow T.A., and Michael, D.A. ECOPHYS: An ecophysiological growth process model for juvenile poplar. *Tree Physiology*, 1990, **7**, 255-281.

64. Running, S.W. and Gower, S.T. FOREST-BFG, A general model of forest ecosystem processes for regional applications. II. Dynamic carbon allocation and nitrogen budgets. *Tree Physiology*, 1991, **9**, 147-160.

65. Running, S.W. and Coughlan, J.C. A general model of forest ecosystem

processes for regional applications. I. Hydrologic balance, canopy gas exchange and primary production processes. *Ecological Modelling*, 1988, **42**, 125-154.

66. Parton, W.J., Stewart, J.W.B., and Cole, C.V. Dynamics of C, N, P and S in grassland soils: a model. *Biogeochemistry*, 1987, **5**, 109-131.

67. Dixon, K.R., Luxmoore, R.J., and Begovich, C.L. CERES - A model of forest stand biomass dynamics for predicting trace contaminant, nutrient and water effects. I. Model description. *Ecological Modelling*, 1978, **5**, 17-38.

68. Luxmoore, R.J. A source-sink framework for coupling water, carbon, and nutrient dynamics of vegetation. *Tree Physiology*, 1991, **9**, 267-280.

69. Cole, D.W. and Van Miegroet, H. Chronosequences: a technique to assess ecosystem dynamics. In: *Research Strategies for Long-term Site Productivity*. Proceedings, IEA/BE A3 Workshop, Seattle, WA, August 1988. (Eds.) W.J. Dyck and C.A. Mees. IEA/BE A3 Report No. 8. Forest Research Institute, New Zealand, Bulletin 152., 1989, pp. 5-23.

70. Powers, R.F. Retrospective studies in perspective: strengths and weaknesses. In: *Research Strategies for Long-term Site Productivity*. Proceedings, IEA/BE A3 Workshop, Seattle, WA, August 1988. (Eds.) W.J. Dyck and C.A. Mees. IEA/BE A3 Report No. 8, Forest Research Institute, New Zealand, Bulletin 152, 1989, pp. 47-62.

71. Gardner, R.H., Dale, V.H., and O'Neill, R.V. Error propagation and uncertainty in process modeling. In: *Process Modeling of Forest Growth Responses to Environmental Stress*. (Eds.) R.K. Dixon, R.S. Meldahl, G.A.Ruark and W.G. Warren. Timber Press, Portland, Oregon, 1990, pp. 208-219.

72. Harmon, P. and King, D. *Expert Systems: Artificial Intelligence in Business*. Wiley and Sons, New York, 1985.

73. Kurzweil, R. *The Age of Intelligent Machines*. MIT Press, Cambridge, MA, 1990, 565 p.

74. Schmoldt, D.L. and Rauscher, H.M. *Building knowledge-based systems for natural resources management*. Chapman and Hall, New York, NY, In Press, (accepted for publication Oct. 1992).

75. Barr, A. and Feigenbaum, E.A. (eds.). *The Handbook of Artificial Intelligence*. 2 volumes, William Kaufmann, Inc., Los Altos, CA, 1981-1982.

76. Starfield, A.M. and Bleloch, A.L. Expert systems: an approach to problems in ecological management that are difficult to quantify. *Journal of Environmental Management*, 1983, **16**, 261-268.

77. Schmoldt, D.L. and Martin, G.L. Construction and evaluation of an expert system for pest diagnosis of red pine in Wisconsin. *Forest Science*, 1989, **35(2)**, 364-387.

78. Starfield, A.M. and Bleloch, A.L. *Building Models for Conservation and Wildlife Management.* MacMillan Publishing Company, 1986.

79. Bowerman, R.G. and Glover, D.E. *Putting Expert Systems into Practice.* Van Nostrand Reinhold, New York, 1988.

80. Schank, R.C. *The Cognitive Computer: On Language, Learning and Artificial Intelligence.* Addison-Wesley Publishing Company, Inc., Reading MA, 1988, 268 p.

81. Davis, J.R. and Clark, J.L. A selective bibliography of expert systems in natural resource management. *AI Applications,* 1989, **3**, 1-18.

82. Rauscher, H.M. and Hacker, R. Overview of artificial intelligence applications in natural resource management. *Heuristics: The Journal of Knowledge Engineering,* 1989, **2(3)**, 30-42.

83. Simon, K.H., Manche, A., and Uhrmacher, A. "Expertensysteme aus dem Umweltsektor," Umweltbundesamt, Bismarckplatz 1, 1000 Berlin 33, Germany, TEXTE 47/92, 1992.

84. Klein, M. and Methlie, L.B. *Expert Systems - A Decision Support Approach with Applications in Management and Finance.* Addison-Wesley Publishing Company, 1990, 539 p.

85. Suzuki, N. Application-specific shell. *New Generation Computing,* 1988, **5**, 317-318.

86. Rauscher, H.M., Benzie, J.W., and Alm, A.M. A red pine forest management advisory system: knowledge model and implementation. *AI Applications in Natural Resource Management,* 1990, **4(3)**, 27-43.

87. Berry, J.S., Kemp, W.P., and Onsager, J.A. Integration of simulation models and an expert system for management of rangeland grasshoppers. *AI Applications,* 1991, **5(1)**, 1-14.

88. Bolte, J.P., Hannaway, D.B., Shuler, P.E., and Ballerstedt, P.J. An intelligent frame system for cultivar selection. *AI Applications,* 1991, **5(3)**, 21-31.

89. Goforth, G.F. and Floris, V. OASIS: an intelligent water management system for South Florida. *AI Applications,* 1991, **5(1)**, 47-55.

CHAPTER 7

SITE CLASSIFICATION: ITS ROLE IN PREDICTING FORESTLAND RESPONSES TO MANAGEMENT PRACTICES

R.K. JONES
ESRI Canada Ltd
Victoria, B.C., Canada

SITE CLASSIFICATION AND FOREST MANAGEMENT IN THE 1990s

Organizations involved in natural resource management are increasingly being affected by the state of their knowledge about ecological conditions characterizing their land base and how these conditions influence land use, resource planning, and business decisions. This need for a greater understanding of the ecological nature of the forest is being driven by a number of environmental, economic, and socio-political factors. Forest management is now really forestland management, in the broadest sense of the word "land". With this changing outlook, we are seeing an emergence of more ecological management approaches which are better able to accommodate non-timber values, anticipate and mitigate environmental impacts of forest operations, and which are beginning to address the principles of sustainable development and integrated resource management.

These new perspectives on natural resource management are having a significant effect on how foresters and land resource professionals conduct land inventories, organize land information systems, incorporate public viewpoints, develop planning alternatives, make management decisions, and monitor the use and health of land resources. In terms of land inventory programs, these changes have brought about a need for more comprehensive and higher resolution land information. Timber inventories are evolving into more complete vegetation resource inventories and greater attention is being paid now to other land-related information on climate, soils, geology, wildlife, and water resources. Finally, consistent with these trends, there is once again renewed interest in more integrated, ecological approaches to classifying and mapping sites — i.e., site classification. In this chapter we will focus our attention on site classification systems and how they are used as tools to help foresters and natural resource professionals make more informed land management decisions.

INTRODUCTION TO SITE CLASSIFICATION

As a term, site classification connotes different things to different people working in the natural resources area. Variation in the interpretation of its meaning results from differences in its particular resource application and from distinctions in the underlying philosophy. Site classifiers have debated over their methods for decades. While much of this dialogue has helped to move the science forward, in some instances it has not encouraged the acceptance and use of classifications in forest management planning. As well, the existence of many different approaches has not always promoted the development of closer bonds with empirical and process research programs. Regardless of these contrasts, all site classification systems attempt to serve the practical need to stratify an often complex forestland mosaic into more or less homogeneous units. A classification system can be used to characterize conditions about any one point on the ground or to serve as a legend, or key for describing land features delineated on a map.

Site classification systems have made important contributions to our understanding of the nature and distribution of ecological-land conditions and the influence these properties have on forestland use. Site classification is a descriptive and genetic science for land in the same way as taxonomy and systematics is applied to biological organisms. The primary users of site classification systems include land resource planners, managers, and researchers. Growing demands on the use of the forest resource is bringing about an increasing need to have site classification systems in place as frameworks for planning and decision-making.

This trend also suggests that traditional methods for predicting or interpreting forest response based on different classes or site types needs to become more directly linked with empirical and process research knowledge.

The term site classification will be used to refer to any form of classification system which stratifies biotic and/or abiotic land features and which may include a mapping function. Divisions of different ecological-land features or groupings of similar ecological-land features within a classification structure will be referred to simply as "classes".

In this chapter, site classification is reviewed from a generic viewpoint in order to emphasize similarities, rather than differences, in the many systems that have been developed throughout the world. In taking this approach, no one system is necessarily better than another, although any one approach may be better suited to a particular application than another. Readers interested in reviewing the characteristics of specific site classification systems are directed to publications such as Burger [1], Jones [2], Williams and Gresham [3], Bockheim [4], Wickware and Stevens [5], Barnes *et al.* [6], and Sims [7]. Following a review of some basic site classification principles, this chapter will then look at the interpretive or predictive nature of these systems and their relation to the empirical and process research domains. The penultimate section discusses how site classification systems might be better linked to empirical and process models used to forecast and evaluate the effect of management interventions on the forest resource. The final section focuses on information and data acquisition technologies and highlights some of the new directions being explored in site classification through the application of computer-based information and knowledge-based technologies.

GENERIC COMPONENTS OF SITE CLASSIFICATION

Definition

Site classification systems provide a framework to organize and communicate our knowledge about the nature of physical and biotic features of land. By way of this structure, the fundamental premise is that future responses of forestland to management activities and natural events can be predicted - i.e., that similar forestland conditions respond in a similar manner to similar perturbations. This predictive function presumes that experience with past events and subsequent forest behaviour, referenced to these classes, will be mirrored again in the future.

Most recent site classification systems are ecological in nature in that they address both physical and biotic features of land, or at least recognize either explicitly or implicitly, relationships between physical and biotic ecosystem components. Some classifications emphasize physical features and then infer related forest-plant associations, while others stress the vegetation and infer related soil-landform conditions. Some classifications pay equal attention to both vegetation and physical land features. In this latter case, the landform, soil, and vegetation features may be dealt with as discrete components and classifications or they may be integrated to varying degrees into "ecosystem types" (a taxonomic and non-spatial classification) or "landscape types" (a mapping or cartographic and spatial classification). Whatever the situation, nearly all site classification work hinges on the concepts of soil formation by Jenny [8] and vegetation formation by Major [9]. As integrators, soil, vegetation, soil and vegetation, or ecosystem entities represent the combined effects of climate, relief, parent material, organisms, and time and are felt to function as individuals or entities in themselves, suitable for classification, interpretation, and as a basis for research. As Williams [10] states: "classification could be considered the science of defining subsets of the universe in which a given process or processes act in a common manner".

In terms of application, most site classification schemes attempt to provide forest managers with a means to stratify a complex forest mosaic into more or less homogeneous land units. The classification procedure may employ a number of methods whereby the components of climate, landform, soil, and vegetation are aggregated, divided, ordinated, synthesized, and/or integrated into classes. As Pfister [11] has pointed out: "[site] classifications can be based on site productivity (site index), landform or physiographic features, soil taxonomy, cover types, potential vegetation, or integrated systems". The classes are given many different terms including types, units, groups, coenoses, series, or associations, just to name a few.

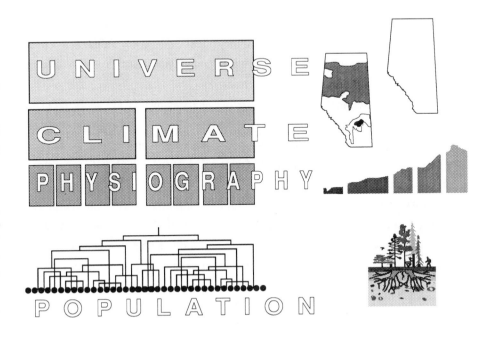

FIGURE 7.1 Building a site classification system involves both the aggregation of detailed on-the-ground individual site data and the partitioning of large geographic regions into sub-regions based on climate and landform-physiographic features.

Methods

Typically, the classification process begins by partitioning some geographic region of interest into climatic zones and/or large landform-physiographic units at relatively small scales (Figure 7.1). At this stage, these zones may or may not be delineated formally on maps. As well, existing forestland resource maps and earlier classification schemes for the same or adjacent areas are reviewed and are often used to aid in the development of these broad strata.

The next step in the process concentrates on the description and sampling of soil and vegetation conditions on the ground using conventional plots or releves.

Site selection methods are often quite subjective, although sampling along transects at set intervals has been used to help reduce bias, determine within-class variability, and to evaluate the "mappability" of the classes. In almost all cases, some form of pre-stratification of the region is used as a basis to distribute sample plots and ascertain replication requirements.

Plot data, often including soil laboratory analysis results, are analyzed using a number of techniques ranging from the simple, subjective sorting of rough field notes to more objective, computer-assisted classification and ordination analysis of coded data. Whatever the method of analysis, inevitably a classification structure emerges with the classes being defined on the basis of differentiating and accessory characteristics and so-called modal conditions. In recent years there has been an increasing use of computers and multivariate analytical techniques to assist in the classification process. Using these tools, a classifier is able to organize, manage, analyze, and display larger amounts of data than was ever possible before. Numerous alternative classification structures and identification keys can be created in a matter of minutes using sophisticated software packages and high-speed mini- or microcomputers. While these advances in data analysis capability have aided greatly in the classification analysis phase of the work, it is important to recognize that these analyses are still only tools to the ecologist. Ultimately, the classifier must choose which analytical output is most suited to the purpose at hand and makes "ecological sense".

Structure: Taxonomic and Cartographic

Hierarchical structures are common to most classification systems. They are useful in showing how a region has been divided into successively smaller units and how individual classes aggregate to form more general levels in the classification system. There are two principal forms of hierarchies: taxonomic and cartographic, and there are some structures which are hybrids of each. Taxonomic hierarchies tend to focus on showing so-called "genetic" relationships, while cartographic hierarchies attempt to show geographic relationships between the classes, from a mapping perspective. Hybrid structures include both the taxonomic affinities and their mappable counterparts for some or all levels of the hierarchy (e.g., [12, 13]).

While classification is a prerequisite and integral component of all land resource mapping, problems with the mappability of classes at different levels of the hierarchy have arisen and have caused some confusion between classifiers, mappers, and users. Users of site classification systems should acquaint themselves with the relationship between classification and mapping, and with the effect of hierarchical structures and scale on land resource inventories. Moon [14, 15] and Arnold [16] provide a good discussion of these important issues.

Identification and Mapping

Site classification systems are applied in forest management planning in two ways. Firstly, they are used on-site to allocate or classify an immediate forest area of interest. The amount of area involved with this on-the-ground assessment might range from as small as one hectare to as large as several hundred hectares. Secondly, site classifications are used as a framework for mapping and describing larger forest management areas, comprising perhaps several thousand hectares. In this case, the mapping programs usually include the use of some form of aerial imagery for polygon delineation followed by on-the-ground sampling to verify and improve image interpretation methods. In the foreseeable future, it is likely that identification and mapping procedures will continue to rely on long-standing, recognizable features of the forest vegetation and landscape (e.g., on the ground: species assemblages, soil texture, soil drainage, etc.; and on air photos: stand pattern, tone and texture, slope, slope position, slope configuration, etc.). This limited number of observable site features will continue to restrict the site classifier in terms of the properties that can be used for ground and image identification and the nature of the classes defined.

On-site allocation procedures are frequently in the form of identification keys which employ field recognizable, diagnostic, or dominant class features (e.g., [17, 18]). An allocation decision using a key is usually verified by comparing the immediate site conditions observed in the field with a list of typical features described for each class. More recent classification programs have attempted to develop identification procedures which use only a restricted set of basic site features. This approach frees the user from having to become an expert in site classification and allows a site to be allocated quickly, often within five minutes. The use of practical identification methods together with the use of handy, field guide publication formats has aided greatly in a wider acceptance and use of classification systems in natural resource management programs.

Mapping applications of site classification have been infrequent at the lowest, most detailed level of the system. This is because the most suitable mapping scale for this level (e.g., an ecosystem type) is often large (e.g., 1:5000 - 1:20000) making the mapping of an entire forest management area at any one time impractical. In some classification systems, the mapping function has been addressed with the inclusion of photo-interpretation materials and related mapping aids (e.g., [19, 20]). Mapping can occur with varying degrees of formality, ranging from a simple annotation of airphotos or existing forest cover maps with the classification on the one hand, to the colour publication of high-quality ecosystem maps on the other hand. In most instances, mapping is done for priority areas that are to be operated on over the next, say, five years (e.g., 10000 ha). In this manner, maps for an entire forest management area will be obtained gradually and more affordably. Later on in this chapter, the future role of information and data acquisition technologies will be discussed as one way of addressing these mapping difficulties.

SITE CLASSIFICATION AS AN INTERPRETIVE OR PREDICTIVE TOOL

A site classification system is a model of the ecological features and relationships which characterize the ecosystems and their distribution within a region, usually from some management perspective such as harvesting and regeneration silviculture or forest productivity. A site classification system is seldom used as predictive tool in itself. Rather, a site classification is a vehicle for organizing and allocating current and future knowledge and experience on forestland response to certain management practices (Figure 7.2). The interpretive or predictive power of a classification depends on many factors, including the degree to which:
1. specific interpretive requirements were identified prior to the classification development and were able to influence the classification structure (e.g.,

organic matter depth not to exceed 20 cm as differentiating criteria);

2. existing knowledge and experience (e.g., growth and yield relationships, rules of thumb developed from managing the forest) are incorporated within the classification system;

3. existing information can be allocated and evaluated within the context of the classification framework (e.g., regeneration performance surveys and other field or laboratory trials);

4. future knowledge, experience, and response data can be structured within the classification system;

5. the classification reflects an organization of fundamental ecosystem processes and hence, their influence on forestland response to management practices; and

6. there is flexibility within a classification system to accommodate new knowledge (e.g., the ease with which the classification structure can be extended or modified).

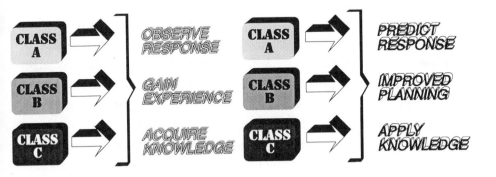

FIGURE 7.2   A site classification system comprising classes A, B, C, etc., provides a practical way to associate and transfer knowledge and experience about forestland responses to management activities.

The interpretive systems developed for most site classification systems have employed mixtures of some or all of the above factors. As is the case with all forms of predictive models — classification, empirical, process, or expert-based — the true test of predictive capability is with field validation, usually of a long-term nature. Some classification systems include interpretive evaluations for the users of the system. As with the classification systems themselves, a wide range of approaches have been used to express the interpretations. Typically, each class or grouping of classes is given a rating or interpretive indice for the evaluation of such things as productivity (e.g., site index, or range of mean annual increment), operability, potential for competitive vegetation following disturbance, soil compaction and erosion hazard, windthrow hazard, species suitability, frost heave hazard, preferred season for harvest, etc. Examples of some site classifications which have incorporated these types of evaluations include Corns and Annas [21], Zelazny *et al.* [22], and Racey *et al.* [23].

RELATIONSHIP OF SITE CLASSIFICATION TO EMPIRICAL AND PROCESS RESEARCH AND MODELS

Site Classification, Empirical, and Process Research Paradigms

Site classification, empirical modelling, and process modelling represent three areas of research, or "study paradigms", which collectively contribute to the prediction of the effects of management on forestland response. Each paradigm brings with it a certain perspective on the nature of the problem, the methods most suited for its investigation, and the character of the predictive products developed. Some site classification studies have attempted to incorporate empirically-based predictive equations (e.g., site quality evaluation studies) in the development of productivity interpretations. In turn, some predictive modelling programs have employed both empirical and process knowledge. Typically, however, research programs tend to focus on methods of investigation most suited to one research paradigm.

Site classifiers, empirical modellers, or process modellers seldom associate with one another. Each scientist views his or her particular domain differently in terms of spatial scale, temporal scale, parameters, and processes. A few scientists began their research in site classification and moved later into empirical and process research, while others have shifted their research emphasis from empirical to more process-based approaches. Leary [24] has explained this migration as a natural progression from "description" to "prediction" activities and from "statistical" to "deterministic" science; he states that the change from one paradigm to another "stretches scientists' disciplinary training, so they must learn a new discipline or join with colleagues expert in the needed fields to form interdisciplinary research groups". It is the view of this author that there is much opportunity for improving the interaction between the three research paradigms. The more scientists are able to "stretch" their individual disciplines, the greater chance there will be of developing practical predictive tools for forestland managers.

Relationships Between the Paradigms
In order to support the notion that all three research paradigms contribute collectively to predicting the consequences of management on forest response, it is important to understand the nature of relationships between the research areas. In the past, each paradigm has progressed and accumulated knowledge relatively independent of the other two. Interaction between paradigms has been infrequent, although there has been some indirect exchange of knowledge, primarily through the literature.

Although research findings from each paradigm have contributed individually to the base knowledge for the development of predictive tools, new opportunities exist for integrating the knowledge from all three paradigms. Figure 7.3 shows schematically the relationships — vertical arrows — between the three paradigms along their individual development paths — horizontal bars. In the future (right-hand side of the figure), land information systems and geographic information system technology are important spatial integrative tools for linking, analyzing, and displaying current and future knowledge (i.e., predictive models) with site classification systems and related forestland resource inventories.

Among the earliest studies in site classification were those which concentrated on the basic description of land resources. Early land surveys, particularly in the "new world" provided extensive notes on land features of importance to

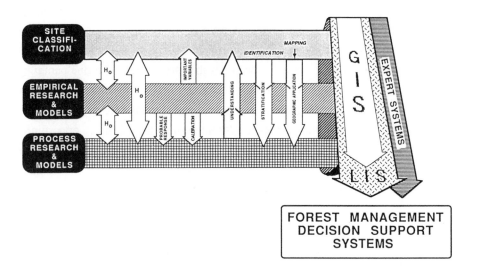

FIGURE 7.3   Relationship between three principal land resource-ecosystem study paradigms. Information technologies such as GIS and expert systems are making for more effective land information systems (LIS) and more meaningful decision support tools to support forest and land management business functions.

human settlement. These surveys formed the basis for much of the early site classification work. Classification studies have become more systematic over the years, but still concentrate primarily on describing how forest conditions vary in space and with time. Some studies have generated hypotheses, formal or otherwise, about ecological relationships and ecosystem function (i.e., why and how forestland conditions vary in space and with time). The extent to which these hypotheses have provided a basis for further study in the other two research paradigms is uncertain, but undoubtedly some of these propositions have inspired experimental testing by the empiricists and stimulated fundamental thinking by the theoreticians. The migration of hypotheses to other paradigms is shown by vertical "H_o" arrows in Figure 7.3.

Empirical studies of the forest are at least as old as those in site classification. In fact, activities in the two domains occurred more or less concurrently, but were often isolated. In Canada, for example, two factions were formed early on: the mensurational empiricists on the one hand and the site classifiers on the other. Large, forest cover type mapping programs were usually associated with empirical growth and yield programs. Site classification studies, despite the inclusion of some growth and yield estimates, were viewed largely by most foresters as being unnecessary.

Empirical research progressed in scope to include the study of managed stands and plantations. Further, silvicultural trials were, and continue to be important ways for investigating the effects of spacing, seedling quality, site preparation techniques, fertilizers, thinning regimes, etc. Many of these data provide useful calibration information for process model research. The probable response findings from empirical studies have, in turn, generated hypotheses suitable for investigation in the two neighbouring study paradigms. For example, site quality evaluation studies (e.g., [25]) identify potentially significant site variables for predicting site productivity (e.g., site index), and some of these factors have been included as diagnostic criteria in site classification programs. As well, empirically derived functions and significant variables can often inspire theoretical investigators to experiment further to rationalize these findings.

The basis for much process research in forestry comes from basic understandings gained in discipline areas like biology, physics, and chemistry. The process research domain focuses on explaining mechanisms related to such things as the capture and storage of sunlight energy, the storage, movement, and availability of water and nutrients, and forest floor decomposition. Understandings developed here often provide causal explanations to "black box" functions from the empirical realm and to observed forest landscape associations noted during the development of a site classification.

Site Classification as a Tool for Stratification and Geographic Application
Much research in the past has paid only minor attention to inherent variation in forestland conditions and the influence these conditions have on experimental results. Site classification systems provide empirical and process scientists with an ability to address the natural variation in forestland conditions by using the classification to stratify the sampling program and to identify and establish priority ecosystems for research (see "identification/stratification" in Figure 7.3). The classification can be used initially to screen candidate study sites and then used later to identify other similar sites (classes) for calibrating and testing research findings elsewhere.

The second important role of site classification is its use, by way of mapping, to apply and extend empirical and process research knowledge and predictions to actual forestland areas (see "mapping/geographic application" in Figure 7.3). When a classification system has been used to stratify the sampling program, then model predictions can be applied, *by class*, to forestland areas defined in terms of the classification. When a site classification has not been used to stratify the process research however, then a linkage between the model and the site classification will be needed to provide a geographic expression of the prediction for the forest area at large. The development of linkages between models and classification systems is discussed below. Ultimately, predictive models must have the facility for geographic expression to definable forestland areas if they are to be of any practical value to forest management.

LINKAGE OF SITE CLASSIFICATIONS TO EMPIRICAL AND PROCESS MODELS

The need to link forest site classification to empirical and process research models has been expressed often in recent years (e.g., [26,27,28]). It is only more recently however, that studies have begun to concentrate on the application of existing classifications and land surveys as input data sources for predictive models. Most of this research has been in the area of "land evaluation" using conventional (agricultural) soil surveys. Land evaluation programs are beginning to demonstrate the value in having an exchange of ideas between classifiers (mappers) and modellers (evaluators). For example, the opening address at a symposium on "Land Qualities in Space and Time" expressed the desire that: "A dialogue will be stimulated among field workers, involved with soil survey and land evaluation, and researchers involved with various modern techniques such as simulation modelling, including agronomic, soil physical and soil chemical aspects, geostatistics, remote sensing, geographical information- and data-management systems." [29].

When broken down into its simplest elements, the mechanisms for linking a site classification with a model are quite straightforward. First, there is a model which has certain input requirements and which generates particular predictive outputs. Second, there is a classification system comprising classes; each class with a definable range of property values. Third, there is a forestland map which delineates and describes different forestland areas on the basis of the classification (or at least there is the capability to map). Since it is necessary for a forest manager to make predictions ultimately for the whole forestland area, model input requirements need to be somehow linked to the properties defining the mapped classes. Having made the linkages, model predictions are made for each class and are portrayed geographically for the whole forest using the site classification map.

The linkage of a site classification to a model will be addressed from three perspectives: 1) linking an existing model to an existing site classification (Figure 7.4); 2) developing a new site classification oriented in its purpose towards the input requirements of an existing model (Figure 7.6); and 3) developing a new model oriented towards the use of the differentiating and accessory class attributes of an existing classification as input variables (Figure 7.7). The discussion following will focus on the model – site classification interface for each perspective.

The important component in each scenario is the linking of the class properties to the model's input requirements. As discussed above, the classifier is limited often by the type and number of properties that can be used on the ground and on imagery (e.g., air photos) for class identification. This restriction frequently influences the nature of a site classification if it is to be of any practical use. If differentiating class properties are not field recognizable or image interpretable then practical, consistent identification surrogates must be found. These realistic considerations mean that class attributes and model input requirements are seldom the same.

Situation 1: Existing Site Classification ↔ Existing Model

In linking existing site classification to an existing model the following procedure is proposed (see also Figure 7.4).

1. An initial screening of differentiating and accessory class attributes in terms of their direct suitability for model input requirements (e.g., Figure 7.4: Attribute 1 -unsuitable, Attribute 2 - direct attribute suitable for model input).

2. An evaluation of differentiating and accessory class attributes with respect to their potential for *estimating* individually, or in combination, the required model inputs -i.e., a transformation to form derivative class attributes (e.g., Figure 7.4: Attributes 3 and 4).

3. If required, a determination of new accessory class attributes, through additional field description, sampling, or laboratory analyses (e.g., Figure 7.4: data collection by class).

The occurrence of attributes that are directly suitable from a classification (Step 1) is rare for most models unless the model was designed to use basic, descriptive ecological data (e.g., average stand age and height, understorey species composition and abundance, organic matter thickness and type, etc.). Step 2, labelled "transform" in Figure 4, is the most likely method for creating a linkage. These transformations have been given particular attention in the land evaluation studies described earlier. "Pedo-transfer-function" is the term used in these studies to describe algorithms which process classification attributes to generate derivative properties that can be used by a predictive model [29]. Continuous classification attributes and nominal or ranked classification attributes are treated by continuous and class pedo-transfer-functions respectively. In Figure 7.5, Bouma [29] illustrates an example of a pedo-transfer procedure for estimating soil water deficits from typical soil survey data.

If neither Step 1 or 2 yield useful model input requirements, then the only remaining option to develop a linkage is to proceed with the determination of new

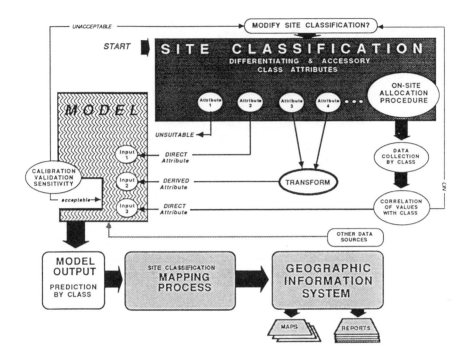

FIGURE 7.4   Procedure required to link an existing model with an existing site classification system.

accessory class properties needed by the model (Step 3). This may require additional field sampling of the original plots used in the development of the classification or of other sites characteristic of the class(es) (Figure 7.4: data collection by class). It should be confirmed that the new class data values do, in fact, covary with the classes (Figure 7.4: correlation of values with class).

Situation 2: Existing Model ↔ New Site Classification
In developing a new site classification system oriented in its purpose towards pro-

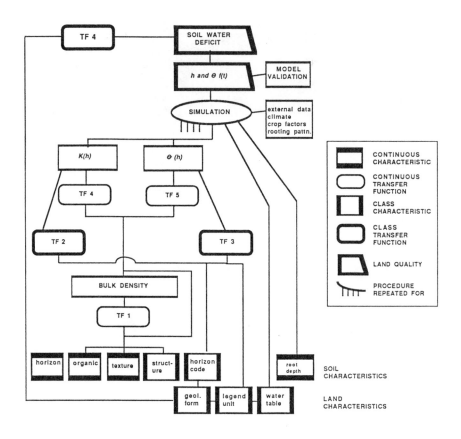

FIGURE 7.5   Example of a pedo-transfer procedure for estimating soil water deficits from typical soil survey data [29].

viding the input requirements of an existing model the procedure starts with the model input requirements and identifies a list of candidate field and/or image interpretable attributes that might be used to form a site classification system (see also Figure 7.6). The procedure proposed is as follows.

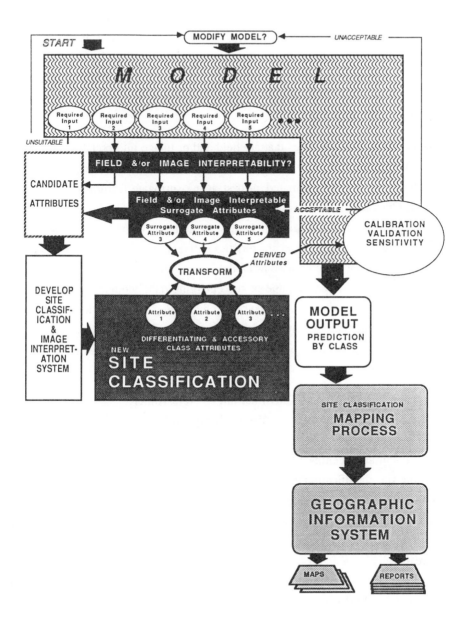

FIGURE 7.6   Procedure required to develop a site classification focused on the
input requirements of an existing model.

1.   An initial screening of the model input attributes in terms of their field and/or image interpretability (e.g., Figure 7.6: Required input 1 - unsuitable, required input 2 -directly suitable).

2.   An evaluation of those model input requirements which can be *estimated* reliably from surrogate field and/or image interpretable features (e.g., Figure 7.6: surrogate attribute 3 - 5).

The occurrence of attributes that are field and/or image interpretable, and therefore directly suitable, will be rare for most models. In most cases, model input attributes can only be estimated, at best, using surrogate features (e.g., by estimating C:N ratios from forest humus types, thicknesses, and horizon sequences or by estimating leaf area index from large scale aerial photography). Similar to the first situation, any surrogate characteristics will require various transfer functions, with testing, prior to being used as model inputs.

As discussed already, some of the emerging data acquisition technologies may have some potential for measuring and "reading" some of the less conventional forestland features. Some of these features could be more directly applicable in modelling or could provide more accurate and consistent alternatives for surrogate attributes.

Situation 3: Existing Site Classification ↔ New Model

In developing a new model oriented towards the use of a existing classification attributes as input variables, the following procedure is proposed (see also Figure 7.7).

1.   An initial screening of differentiating and accessory characteristics of an existing site classification in terms of their significance to forestland quality evaluation and to underlying empirical or process model concepts — i.e., the selection of attributes which individually, or in combination, might describe available moisture and nutrient supply (Figure 7.7: attribute 1 - unsuitable, attributes 2 and 3 are directly suitable for being added to the pool of candidate model inputs).

2.   An evaluation of differentiating and accessory attributes in terms of their potential for *estimating* (inferring), individually or in combination, properties of significance to forestland quality and to underlying empirical or process model concepts — i.e., a transformation to form derivative class attributes (e.g., Figure 7.7: attribute 4).

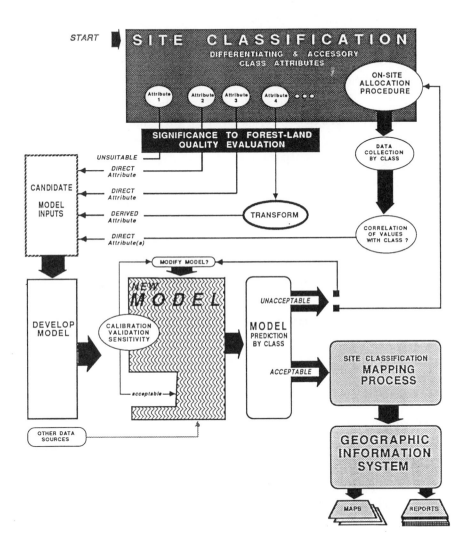

FIGURE 7.7   Procedure required to develop a new model which uses an existing site classification to identify candidate model input variables.

From the candidate list of site attributes a model is created and tested. The "by class" predictions are evaluated and may be unacceptable. In this instance, either the model could be modified or new accessory class attributes could be ascertained through additional field sampling of the original plots used in the development of the classification or of other sites characteristic of the class(es) (Figure 7.7: data collection by class).

Models developed under this last scenario are likely be less complex than some of those described elsewhere, because they will be limited, at least initially, by having to use typical site classification features. However, they will have the advantage of being easily extended to a larger forestland base via the site classification mapping process. The form of these models might be similar conceptually to the soil productivity index model developed by Gale and Grigal [30].

INFORMATION SYSTEM AND DATA ACQUISITION TECHNOLOGIES AND OPPORTUNITIES

With the demand for site classification information growing, more efficient, reliable, and economical inventory methods need to be developed soon. As Burrough [31] states: "We have arrived at a state of affairs that has been termed 'the parameter crisis' — as models get increasingly realistic they need more and better data to control them and to drive them."

Geographic information systems (GIS), artificial intelligence (e.g., expert systems, neural networks), and data acquisition technologies, individually and through their integration, are now playing an important role in the integration of data, information, and knowledge. Many of these readily evolving computer-based tools allow scientists to capitalize more fully on existing land information (e.g., forest cover maps, soil surveys, digital terrain data, spectral imagery, hydrography, etc.), known ecosystem processes and response knowledge, and site classification frameworks. Through the integration of these technologies, site classification systems can now be linked to forest resource information systems and to forecasting models and diagnostic tools. Collectively, these integrated components in turn form the essential elements to a forest management decision support system (Figure 7.3). An attractive aspect of this line of research is that it integrates:

- empirical, process, and heuristic-based knowledge of ecosystem processes;
- an understanding of forest landscape features important in controlling these ecosystem processes;
- a classification knowledge and distribution pattern knowledge of forestland conditions (i.e., site classification and site classification maps); and
- powerful spatial, temporal, and statistical analytical tools.

Paralleling these developments, continued advancements are occurring in the area of data acquisition technologies. A number of remote data collection technologies are being developed and integrated with the intent to either supplement or replace conventional inventory methods. For example, devices like multispectral image scanners, videos, radars, lasers, and accurate navigational instruments are now being mounted onto various aerial and terrestrial vehicles for local, quick turn-around survey missions. These data acquisition methods are able to provide data of high information value in an efficient and relatively inexpensive manner. To complement these developments, improved interpolation procedures have been developed which can extend point data values to those land areas not sampled more reliably. These procedures are particularly useful when typical forestland feature boundaries are less evident. With additional tools like Global Positioning Systems (GPS), advanced information systems will be able to provide the necessary spatial, temporal, and virtually real-time infrastructure for receiving, geo-referencing, structuring, and processing a wide range of data sources. All of these technology developments suggest that conventional approaches to site classification and mapping are soon to be complemented, possibly even challenged, by a number of readily emerging spatially-oriented, digital land characterization tools.

Digital Site Classification Mapping Technologies
As mentioned earlier, traditional land resource inventories, whether as individual resource themes or as integrated ecological surveys, are expensive and are formidable. Typically, the final inventory maps and reports have resulted in a single

point-in-time, static product. Further, the reliability of land resource inventories have seldom been addressed either as a part of the survey or in the form of a post-survey audit. Partly in response to these problems and the need for current, high-quality land information, there has been much activity during the 1980s in the application of computer-aided mapping and tabular analysis tools to support many land inventory business functions.

Figure 7.8 portrays the natural resource inventory business area during the 1980s as three relatively distinct program areas: 1) the continued conventional inventory of land to generate land information (e.g., forest cover, soil and base mapping programs); 2) continued research and the acquisition of knowledge and experience about ecosystems and landscapes and how we classify, map, and manage them; and, 3) the development and application of information technologies to both land information and land-related knowledge. Much of the effort to date has been focused on the acquisition of hardware and software followed by the laborious loading of large volumes of map and tabular land information (large arrow and "GIS" in Figure 7.8). For example, during the 1980s most provinces in Canada initiated a digital base mapping program, many provincial forest inventory programs adopted a digital format, and the national soil survey program (i.e., CanSIS - CANadian Soil Information System) continued to develop computer-based data management and GIS-based cartographic systems.

The linking of land resource information systems, both spatial and tabular, with knowledge bases to produce dynamic, integrated mapping systems represents a major opportunity for research and development in the 1990s (Figure 7.9). Such "ecologically-oriented, predictive mapping systems" will have the capability to:

• incorporate and link the computer-based functionalities of both information technologies (e.g., GIS, data bases) and knowledge-based technologies (e.g., conventional models or artificial intelligence approaches like expert systems);
• function in an automated or semi-automated mode, but allow for intervention by the land resource expert;
• generate maps and associated spatial attributes for large areas which have land information and knowledge bases in place;
• be dynamic, such that maps and associated spatial attributes can be regenerated when changes occur with either the information or knowledge base; and

- analyze the quality or reliability of the map and associated spatial attribute information products by incorporating uncertainty modelling procedures.

In terms of the land information activities, these predictive mapping systems will be able to capitalize fully on information contained in existing resource inventories of the forest cover, soils, hydrography, contours, and digital terrain model derivatives (e.g., slope, aspect, slope configuration). In terms of the knowledge base, these systems will be able to take advantage of existing ecological classification systems (i.e., relationship knowledge), the mapping experience of the original surveyors (e.g., foresters and pedologists), and knowledge about land responses to management practices.

Already there is considerable evidence of research programs based in GIS, expert systems, or the natural resource sciences exploring aspects of the problem. For example, Band *et al.* [32] working with mountainous landscapes in Montana

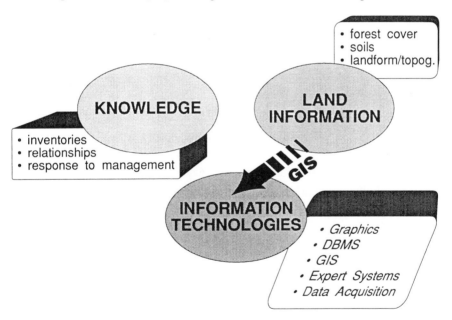

FIGURE 8. The natural resource inventory business area during the 1980s comprised three relatively distinct program areas. The main interaction between program areas was the loading of large volumes of map and tabular data into GIS.

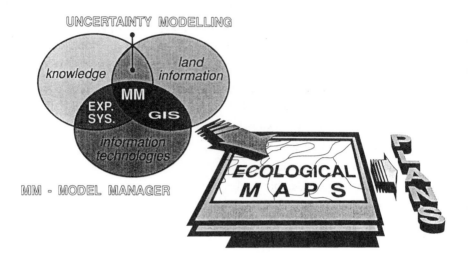

FIGURE 7.9 In the 1990's, knowledge, land information, and information
system program areas will become more closely integrated allowing for the
development of new approaches to mapping ecosystem conditions to support
contemporary planning needs.

have developed a process-oriented, forest ecosystem approach to watershed mapping using GIS and remote sensing. Moore *et al.* [33] have explored the use of
decision tree analysis, linked to a GIS as a method of predicting vegetation distribution in the south coast of New South Wales, Australia. Moore and his collaborators concluded that their predictive modelling procedure resulted in the
generation of maps of sufficiently fine resolution to be useful for monitoring management policies in natural or semi-natural ecosystems. They felt that the maps
"contain significantly more ecological information than would be obtained in a
comparable period using more conventional survey techniques". Also in
Montana, Coughlan and Running [34] have developed an ecological process-oriented expert system that creates biophysical polygons by aggregating areas having similar ecological properties (leaf area index, soil water holding capacity, or

climatic efficiency). Twery *et al.* [35] describe research which focuses on incorporating intelligent functions in a geographic information system as a means to generate (predict) new information, specifically, predicting species composition from topography. Using their system for an area in West Virginia, they determined, for example, what topographic characteristics could be used in an expert system to predict forest cover, thus negating the need in those areas for geological, soil, or hydrological data.

In Canada, Mackey and McKenney [36] have initiated a large area "bio-environmental indices" and economic trade-off research program in Ontario which uses as its basis a series of spatial modelling routines to obtain reliable estimates of the distribution patterns and environmental relations of plants and animals. These estimates can then be used to assess a number of landscape and environmental qualities or values such as biodiversity, habitat suitability, and representativeness. Many of the principles being applied in this investigation have been transferred from similar studies conducted out of the Centre for Resource and Environmental Studies, Australian National University (e.g, [37]). In Alberta, related research applies knowledge-based tools together with spatial and attribute uncertainty functions as a means to spatially extend the use of conventional site classification systems and land resource inventory maps (Figure 7.10) [38].

Although these approaches do not negate the need totally for the collection of ground truth information, they do have the potential for generating reasonable quality, land resource inventory information for large forestland areas relatively quickly and inexpensively. At a minimum, the map products developed can serve as a first approximation for subsequent field survey and enhancement and can provide useful information for whole forest level planning.

SUMMARY

A review of the generic characteristics and objectives of site classification systems suggests that there are many fundamental commonalities between the various approaches. Mapping, mappability, and ground identification are important functions that must be integral to any classification system if it is to be of any practical value to forest management. Site classification, empirical research, and process research domains historically have pursued their own interests, but are beginning to integrate more as classification programs become more closely scrutinized for their interpretive or predictive value, and as model research aspires to predict responses beyond the laboratory or single plot to the forest estate at large.

Each research paradigm continues to make important contributions to our knowledge of the forest and its management. Some future research should focus on studies which build better systems that integrate the paradigms. Geographic information systems, land information systems, expert systems and emerging data acquisition technologies are all tools which can assist in this integration of knowledge from each research paradigm and in the development of better resource inventories.

Site classification systems, through their mapping function, provide the means ultimately by which predictive models are given geographic expression. A necessary prerequisite to this geographic extension however, is the linking of the model input requirements to a classification system. This is not a trivial task, since seldom, if ever, do the required properties of both the model and the classification align. Class properties must be linked in some way to model input requirements and/or model input requirements must be linked to field or image interpretable forestland features. Recent studies in the area of land evaluation may provide some insights into how forestland (site) classification systems can be used to apply predictive models. In all of this work, classifiers and modellers must ensure that class, map, and model reliabilities and assumptions are stated clearly to the user.

Site classification as we have known it over the last 80 years is changing in its form. High speed computers coupled with sophisticated software tools like GIS and increasingly larger amounts of digital land information allows ecologists to access, process, and analyze resource data in ways that were previously not possible. Now it is feasible to generate large area ecological inventories with far less effort than conventional methods employed in the past. With these new capabilities in place, many of the principles underlying site classification research will continue to play an important role in guiding land resource analytical strategies and in how we ultimately interpret the results from a forestland management perspective. No technology can replace an ecologist's understanding of "the woods".

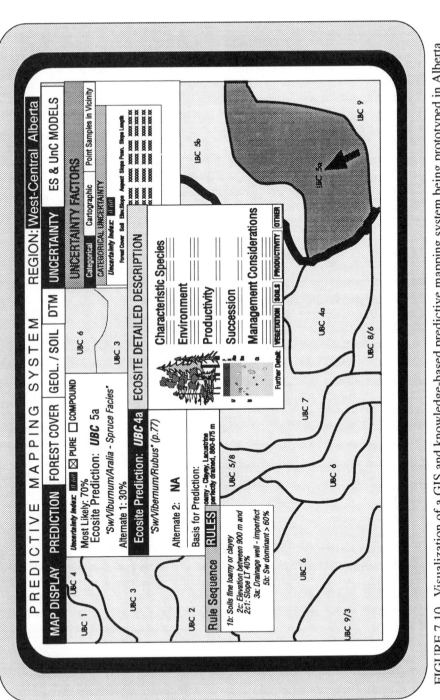

FIGURE 7.10 Visualization of a GIS and knowledge-based predictive mapping system being prototyped in Alberta showing various information "windows" overlain on a map display screen.

LITERATURE CITED

1. Burger, D. Forest site classification in Canada. *Mitteilungen des Vereins für Forstliche Standortskunde und Forstpflanzenzuchtung*, 1972, **21**, 20-36.
2. Jones, J.R. Review and comparison of site evaluation methods. USDA Forestry Service Research Paper RM-51. Rocky Mountain Forestry and Range Experiment Station, Fort Collins, Colorado, 1969, 27 p.
3. Williams, T.M. and Gresham, C.A. (Eds.). *Predicting Consequences of Intensive Forest Harvesting on Long Term Productivity by Site Classification.* IEA/BE A3 Report No. 8. Baruch Forest Science Institute of Clemson University, Georgetown, South Carolina, USA, 1988.
4. Bockheim, J.D. (Ed.). *Proceedings: Forest Land Classification: Experiences, Problems, Perspectives.* University of Wisconsin, Department Soil Science., Madison, Wisconsin., 1984, 276 p.
5. Wickware, G.M. and Stevens, W.C. (Ed.). *Proceedings: Site Classification in Relation to Forest Management.* Great Lakes Forestry Centre, Sault Ste. Marie, Ontario, Forestry Canada, Canada Ontario Joint Forest Research Committee publication O-P-14, 1986.
6. Barnes, B.V., Pregitzer, K.S., Spies, T.A., and Spooner, V.H. Ecological forest site classification. *Journal of Forestry*, 1982, **80(8)**, 493-498.
7. Sims, R.A. Forest site classification in Canada: a current perspective. *Forestry Chronicle*, 1992, **68(1)**, 21-22.
8. Jenny, H. *Factors of Soil Formation.* McGraw-Hill, New York, 1941, 281 p.
9. Major, J. A functional, factorial approach to plant ecology. *Ecology*, 1951, **32**, 392-412.
10. Williams, T.M. 1988. Integration of land classification and modelling on lower coastal plain loblolly pine. In: *Predicting Consequences of Intensive Forest Harvesting on Long Term Productivity by Site Classification.* (Eds.) T.M. Williams and C.A. Gresham. IEA/BE A3 Report No. 8. Baruch Forest Science Institute of Clemson University, Georgetown, South Carolina, USA, 1988, pp. 165-173.
11. Pfister, R.D. Site classification and productivity in the western United States. In: *Predicting Consequences of Intensive Forest Harvesting on Long Term Productivity by Site Classification.* (Eds.) T.M. Williams and C.A. Gresham. IEA/BE A3 Report No. 8. Baruch Forest Science Institute of Clemson University, Georgetown, South Carolina, USA, 1988, pp 81-92.
12. Hills, G.A. and Pierpoint, G. Forest site evaluation in Ontario. Ontario Department Lands and Forestry, Research Branch Research Report 42, 1960.
13. Lacate, D.S. Guidelines for bio-physical land classification. Publication No. 1264, Canadian Forestry Service, Ottawa, Ontario, 1969.
14. Moon, D.E. Forest land resources inventory in British Columbia. In: *Proceedings, Symposium on Forest Land Classification: Experiences, Problems, Perspectives.* (Ed.) J.D. Bockheim. University Wisconsin, Department Soil Science., Madison, Wisconsin, 1984, pp. 66-81.

15. Moon, D.E. Interpreting land resource data. In: *Proceedings Approaches to the Evaluation of Forest Soil Productivity in Ontario*. (Ed.) R.K. Jones. April 13-14, 1987 Guelph. Ontario Institute of Pedology Publication No. 88-1, University of Guelph, Guelph, Ontario, 1988, pp. 51-61.

16. Arnold, W.A. A pedological view of forest land classification. In: *Proceedings Symposium Forest Land Classification: Experiences, Problems, Perspectives*. (Ed.) J.D. Bockheim. University of Wisconsin, Department Soil Science., Madison, Wisconsin, 1984, pp. 18-31.

17. Pfister, R.D., Kovalchik, B.L., Arno, S.F., and Presby, R.C. Forest habitat types of Montana. Intermountain Forestry and Range Experiment Station. and Northern Region, U.S. Forest Service, Missoula, Montana, 1974.

18. Jones, R.K. Site classification in Ontario. In: *Proceedings, Symposium on Forest Land Classification: Experiences, Problems, Perspectives*. (Ed.) J.D. Bockheim. University of Wisconsin, Department Soil Science, Madison, Wisconsin, 1984, pp. 82-99.

19. Burger, D. Identification of forest soils on aerial photographs. *Forestry Chronicle*, 1957, **33(1)**, 54-60.

20. Jones, R.K., Pierpoint, G., Wickware, G.M., Jeglum, J.K., Arnup, R.W., and Bowles, J.M. Field guide to forest ecosystem classification for the Clay Belt, Site Region 3E. Ontario Ministry of Natural Resources, Toronto, Ontario, 1983.

21. Corns, I.G.W. and Annas, R.W. Field guide to forest ecosystems of west-central Alberta. Canadian Forestry Service, Edmonton, Alberta, 1986, 251 p.

22. Zelazny, V.F., Ng, T.T.M., Hayter, M.G., Bowling, C.L., and Bewick, D.A. Field guide to forest site classification New Brunswick. N.B. Department of Natural Resources and Energy, Fredericton, New Brunswick Forest Subsidary Agreement Publication, 1989, (unpaginated).

23. Racey, G.D., Whitfield, T.S., and Sims, R.A. *Northwestern Ontario Forest Ecosystem Interpretations*. Ontario Ministry of Natural Resources, Toronto, Ontario, 1989, 160 p.

24. Leary, R.A. Some factors that will affect the next generation of forest growth models. In: *Proceedings, Forest Growth Modelling and Prediction, Volume 1*. (Eds.) A.R. Ek *et al.* IUFRO Conference, August 23-27, 1987, Minneapolis, Minnesota Society American Foresters Publication No. SAF-87.12, 1988, pp. 22-32.

25. Carmean, W.H. Forest site quality evaluation in the United States. *Advances in Agronomy*, 1975, **27**, 209-269.

26. Carmean, W.H. Forest site quality evaluation in northcentral Ontario. In: *Proceedings Approaches to the Evaluation of Forest Soil Productivity in Ontario*. (Ed.) R.K. Jones. April 13-14, 1987 Guelph. Ontario Institute of Pedology Publication No. 88-1, University of Guelph, Guelph, Ontario, 1988, pp. 3-20.

27. Crow, T.R. and Rauscher, H.M. Forest growth models and land classification.

In: *Proceedings, Symposium on Forest Land Classification: Experiences, Problems, Perspectives.* (Ed.) J.D. Bockheim. University Wisconsin, Department Soil Science., Madison, Wisconsin, 1984, pp. 190-204.

28. Stone, E.L. A critique of soil moisture-site productivity relationships. In: *Proceedings, Symposium on Soil Moisture...Site Productivity.* (Ed.) W.E. Balmer. USDA Forestry Service, Southeastern Area, State and Private Forestry, Atlanta, Georgia, 1978.

29. Bouma, J. Land qualities in space and time. In: *Proceedings, Symposium on Land Qualities in Space and Time.* August 22-26, 1988, Wageningen, The Netherlands, 1988, pp. 1-14.

30. Gale, M.R. and Grigal, D.F. Performance of a soil productivity index model used to predict site quality and stand production. In: *Proceedings, Forest Growth Modelling and Prediction, Vol. 1.* (Eds.) A.R. Ek *et al.* IUFRO Conference, August 23-27, 1987, Minneapolis, Minnesota Society of American Foresters Publication No. SAF-87.12, 1987, pp. 403-410.

31. Burrough, P.A. Modelling land qualities in space and time: the role of geographic information systems. In: *Proceedings, Symposium on Land Qualities in Space and Time.* August 22-26, 1988, Wageningen, The Netherlands, 1988, pp. 91-105.

32. Band, L.E., Peterson, D.L., Running, S.W., Coughlan, J., Lammers, R., Dungan, J., and Nemani, R. Forest ecosystem processes at the watershed scale: basis for distributed simulation. *Ecological Modelling*, 1991, (accepted for publication).

33. Moore, D.M., Lees, B.G., and Davey, S.M. A new method for predicting vegetation distributions using decision tree analysis in a geographic information system. *Environmental Management*, 1991, **15(1)**, 59-71.

34. Coughlan, J.C. and Running, S.W. An expert system to aggregate forested landscapes within a geographic information system. *Artificial Intelligence Applications in Natural Resource Management*, 1989, **3(4)**, 35-43.

35. Twery, M.J., Elmes, G.A., and Yuill, C.B. Scientific exploration with intelligent GIS: predicting species composition from topography. *Artificial Intelligence Applications in Natural Resource Management*, 1991, **5(2)**, 45-53.

36. Mackey, B.G. and MacKenney, D.W. Quality control in data base development and spatial analysis. In: *Proceedings, GIS '92.* GIS '92 Symposium, Vancouver, British Columbia, February 10-13, 1992.

37. Mackey, B.G., Nix, H.A., Stein, J.A., and Cork, S.E. Assessing the representativeness of the wet tropics of Queensland World Heritage Property. *Biological Conservation*, 1989, **50**, 279-303.

38. Jones, R.K. and Kansas, J.L. Dynamic ecologically-oriented predictive mapping systems - concept and case study for wildlife. In: *Proceedings Land Reclamation of Resource Developments in West Central Alberta.* (Ed.) T.M. Macyk. Alberta Reclamation Conference 1991, September 24-25, 1991, Olds, Alberta, 1991.

CHAPTER 8

MANAGEMENT SYSTEMS FOR SUSTAINABLE PRODUCTIVITY

P.N. BEETS
New Zealand Forest Research Institute
Private Bag 3020, Rotorua, New Zealand

T.A. TERRY and J. MANZ
Weyerhaeuser Company
Tacoma WA 98531, USA

INTRODUCTION

One of the main tenets of sustainable commercial forest management is to maintain and enhance the capability of forest sites to produce successive rotations of crops at or greater than the base level of productivity that existed prior to management, using cost effective and environmentally sound methods. Wise management of production forests is essential if we are to ensure that future generations will be able to derive the same, if not more, resources from the land than were derived by preceding generations.

In this chapter we define productivity and overview some of the genetic and environmental factors controlling productivity. Emphasis is placed on integrating management practices including tree improvement, cultural treatments, and harvesting operations so that species are better suited to their domesticated roles in intensively managed plantation forests. As part of this integration, it is recognized that appropriate consideration needs to be given to meeting other resource management objectives operating at the local and regional planning levels. We illustrate the important and dynamic role harvesting strategies and tactics can have in maintaining productivity. Management practices will also have to evolve to address environmental changes brought about through anthropogenic causes and natural events. Finally, we summarize the key premises and research recommendations that are considered critical for sustaining productivity on intensively managed sites.

Forest productivity is influenced by many interrelated factors and these need to be incorporated into management information systems to assist managers with their goal of enhancing productivity. Systems for forest management differ from those devised for agronomic crops in one important respect - the planning and management horizon for tree crops typically last several decades. Changes can occur over time in site quality, genotypes, climate, markets, economics, and societal values and concerns. Knowledge also is increasing, and advances continue to be made in forest growing, harvesting, and wood processing technologies. The complication introduced by long time periods is that the pace of change can be so slow and imperceptible that undesirable trends can be easily overlooked from one rotation to the next, yet also rapid enough that forest management considerations made at the beginning of a rotation may no longer be relevant before the end of that rotation. Expected technological advances also sometimes fail to materialize, forcing tactical and sometimes strategic plan changes.

We have taken the view that forest productivity goals can be achieved if: 1) the determinants of sustainable productivity are identified and addressed; 2) appropriate systems are in place to test the underlying concepts and to monitor changes over time; and 3) management regimens are adjusted accordingly. It is therefore essential that memory (of crop and site history), feedback, and adaptability feature prominently in the design of information systems. An important premise is that ecosystems are not static. They will always be in a state of flux regardless of management efforts. This philosophy should be valid whether natural or artificial regeneration systems are being used or whether extensive or intensive forestry is being practised. Throughout this chapter, examples are given of alternative systems for enhancing productivity, and areas are identified which warrant increased research effort.

DETERMINANTS OF FOREST PRODUCTIVITY

Forest productivity, as defined here, is restricted to the ability of a site to produce harvestable crop biomass, so allowing considerations to focus primarily on aspects related to site quality. The productivity of a crop is determined by its genetics and the abiotic and biotic environment; the supply of water, nutrients, sunlight, and interactions with other biota [1]. Integration of all of these factors is essential for sustainable gains to be made through silvicultural management. A possible mechanism for achieving this integration is shown in Table 8.1. The upper right triangle of this interaction matrix contains strategies that could be considered during the operation indicated at the start of each row, followed by an evaluation of operation performance. The lower left triangle contains response strategies that could be considered when planning operations identified at the head of each column, based on feedback from the performance evaluation phase.

FACTORS THAT CAN BE CONTROLLED BY MANAGEMENT

Genetics

Productivity can be increased through the appropriate choice of species and genotypes. The first consideration is to match the appropriate species and seed source to the site, as this consideration yields the greatest and most rapid gains in productivity [2]. Tree improvement through selection and breeding generally focuses on genotypes that grow rapidly on a range of sites, have desirable log and wood quality traits, and are resistant to disease and insect attack. Genetic gains with tree species including *Pseudotsuga menziesii*, *Eucalyptus* spp., and *Pinus radiata* have led to improvements in growth, form, and resistance to disease. A unit area growth gain of 11% was found at rotation age after what was considered to be a low level of genetic improvement relative to that currently available in *P. radiata* [3]. Furthermore, when product quality characteristics were factored in, the estimated gain in value was thought to exceed 20%. At an operational level, growth gains are usually less than obtained in designed tests by 10 to 15% for a number of reasons [2].

TABLE 8.1

Examples of strategies used in sector operations that need to be integrated when developing silvicultural prescriptions that are designed to optimize system economics.

	Regeneration and Growth	Harvesting and Site Preparation	Processing	Environment	Evaluation
Regeneration and Growth	**Growing Technology** - Maintain genetic diversity - Improve breeds - Protect crop - Allocate site resources to crop - Maximizing value through tending	- Consider equipment needs relative to crop characteristics - Optimize log making	- Maximize value - Reduce malformation - Maximize conversion efficiency	- Minimize adverse on-, off-site effects - Enhance aesthetics	- Set target productivity level and yield expectation given genetic, environmental factors
Harvesting and Site Preparation	- Conserve organic matter - Improve conditions for reestablishment - Adapt genotypes - Ameliorate site - Minimize damage to crop element	**Harvesting / Site Prep. Technology** - Develop complementary harvesting/site prep./residue management techniques	- Minimize damage to crop	- Minimize adverse on-, off-site effects - Enhance aesthetics - Improve residue management	- Log making to specifications - Yield acceptable relative to target
Processing	- Tend appropriately - Cycle residues - Produce desired products	- Stand is suitable given specifications - Desired log making feasible	**Processing Technology** - Piece size - Conversion	- Minimize adverse effects	- Products to specification
Environment	- Optimally utilize productive area - Use appropriate genotypes - Reduce chemical use	- Appropriate deployment of harvesting and site prep. equipment	- Health	**Environmental Technology** - Improve human health and safety - Biodiversity	- Practices acceptable and sustainable
Feedback	- Deploy and manage forests appropriately	- Improve harvest planning	- Improve conversion efficiency	- Improve environmental quality	

Genetic potential is thought by some geneticists to be fully expressed only when trees are grown under optimum conditions [2]. Adoption of intensive management practices, entailing increased attention to spacing, weed control, fertilization, and irrigation could be used to fully capture these potential benefits. Some gain is expected without optimal nutritional management, which may not always be feasible or appropriate anyway.

As an example, research is underway in New Zealand to explain the cause of large tree-to-tree differences in health of some *P. radiata* stands. With increasing tree age some dominant trees suffer from excessive needle loss in the upper crown, a condition known as Upper Mid-Crown Yellowing (UMCY), while neighbouring dominant trees can be unaffected. It is known that UMCY is associated with low foliar magnesium [4], and that foliar magnesium concentration differs among clones of *P. radiata* [5]. These same clones were also planted at a site where magnesium supply is not optimum for *P. radiata*. UMCY symptoms were evident only in those clones that Knight [5] had previously characterized as low foliar-magnesium accumulators, while the remaining clones had healthy crowns. Comprehensive investigations are currently underway to validate these observations. Foliar analysis data indicate that the rate of soil magnesium supply is marginal for optimum nutrition of *P. radiata* throughout most of the major plantation areas of New Zealand [6]. The selection and breeding of fast-growing genotypes better adapted to overcome magnesium limitation without the need for large amounts of magnesium fertilizer would be highly desirable.

Tree improvement programs should focus increasingly on identifying fast-growing genotypes that have a low requirement for a range of site resources. For example, nutrient use efficiency differs among genotypes, depending on their ability to take up and utilize nutrients [5, 7, 8]. Similarly, water use efficiency differs among genotypes [9]. Some efficient families can be expected to grow well over a range of sites because genotype x environment interaction has usually been found to be minor, and it would therefore be surprising if having a focus on health and efficiency traits, instead of just growth rate, does not lead to the formation of broadly adapted breeds. However, it still needs to be established if and under what circumstances breeding is a better alternative than fertilization for improving growth on nutrient deficient sites. There is clearly an urgent need to explore the opportunities to improve growth efficiency differences revealed already several decades ago by van Buijtenen and Isbell [10]; a view that has been repeatedly expressed and continues to go largely unheeded by tree breeders [2, 11]. Improved collaboration and integration of research programs is necessary [1, 12].

Precautions should be taken to ensure that suitable seed sources continue to be available in the future, to keep as many breeding options open as possible. Conservation, diversification, and deployment are important breeding tactics irrespective of the regeneration system, as outlined in Ledig and Kitzmiller [13]. Maintenance of genetic diversity is believed to be the best strategy given the certain prospect of a changing future environment. By maintaining collections of characterized "plus trees" and ensuring the conservation of gene pools on sites across the species' natural range, geneticists should be able to recognize and take advantage of the variation that exists in species. Rigorous progeny testing for adaptability and stability over an appropriate range of sites should minimize the chance of maladaptation under future conditions [2].

Diseases and Pest Control

Biotic agencies can limit the productivity of susceptible genotypes and species under stress. Susceptibility to disease depends on the inherent disease-resistance of the crop, on the climate, and on growing conditions. Desirable genotypes should ideally be disease-resistant and preferably different over successive rotations to avoid the development of more virulent pests [2].

Equally important is the need to encourage species and stand conditions that will allow development to maturity without undue mortality. An example of facilitating the development of stands that are not "adapted" to the site include the true firs in the Blue Mountain region of northeast Oregon and southeast Washington, USA. After decades of effective fire suppression and the selective harvest of late-seral tree species, forests were composed of over-stocked, uneven-aged stands of stagnated *Abies grandis* and *Pseudotsuga menziesii*, which occupied sites where the more fire resistant *Pinus ponderosa* once dominated [14]. Epidemic insect infestations, and several years of drought are now causing widespread tree mortality [14]. Early precommercial thinning to reduce density and to alter species composition will have been beneficial in these stands.

The use of chemical sprays requires a well integrated approach, based on a sound ecological understanding. For example, the severity of *Dothistroma pini* infestation of *P. radiata* can be readily controlled with copper oxychloride sprays, but no operationally effective chemical is available for controlling an associated needle cast fungus, *Cyclaneusma minus* [15]. For the latter disease, genetic differences in susceptibility are large, and severe defoliation of the susceptible genotypes can occur irrespective of *D. pini* control. Hence, the use of genotypes that are less susceptible to *C. minus* should increase the cost-effectiveness of *D. pini* control measures.

Vegetation Management
Non-crop vegetation can play several key roles in maintaining long-term productivity by: 1) stabilizing soils to minimize erosion, mass movement, and rainfall impacts; 2) acting as a nutrient sink to reduce leaching; 3) cycling nutrients to crop species; 4) fixing nitrogen; and 5) contributing to a diverse and beneficial soil biota. These essentially beneficial processes must be appropriately balanced against the negative impacts on the allocation of site resources to crop trees owing to the competition for light, nutrients, and water.

Total competition control studies have demonstrated for a range of sites and species that herbaceous and woody vegetation can have a significant negative impact on early plantation growth. For example, on 13 study sites installed in the southern USA, *Pinus taeda* volume after 5 years with total vegetation control averaged about four-fold more than with no control [16]. Likewise, *Pseudotsuga menziesii* on sites in Western Washington and Oregon also responded markedly to total weed control [17]. After six field seasons, weed control treatments had 186% and 260% more total stem volume than controls on sites in Longview, Washington and Springfield, Oregon, respectively.

The importance of nitrogen-fixing plants in providing nitrogen and organic matter inputs into ecosystems and affecting crop growth rates is well documented [18]. Legumes planted with pines can fix 100 kg.ha^{-1}.year^{-1} for a period up to five years [19], *Ceanothus* spp. 0-110 kg.ha^{-1}.yr^{-1} and pure stands of *Alnus rubra* 100-200 kg.ha^{-1}. yr^{-1} [20]. Nitrogen fixation rates in mixed stands of *A. rubra* and *P. menziesii* range from 20-85 kg.ha^{-1}.year^{-1} [21]. The presence of *A. rubra* at

500 stems per hectare on a nitrogen responsive site in southwest Washington, enhanced growth of *P. menziesii* until stand age 8 years [22]. Thereafter the increased competition for light between these species became an overriding factor and the mixed *A. rubra/ P. menziesii* plots lost their growth advantage over the *P. menziesii* only plots. Binkley [21] concluded that the growth of crop trees is generally increased in the presence of N-fixing species only on N-limited sites. N-fixers were generally beneficial if they were of low stature, were removed early to allow site occupancy by the crop, or were established several years after the non-N-fixing trees. Knowing these competitive relationships will allow better decision making.

Nutrient Cycling and Soil Organic Matter

Research strategies and results of long-term productivity studies have previously been summarized [23, 24]. The intensity of biomass and nutrient removal through harvesting or displacement can have a significant impact on long-term productivity [25]. Prudent management regimes must be adopted that use components of the following strategies: 1) conserve nutrients; 2) replace removed nutrients (fertilization amendments); 3) control weeds and crop stocking; 4) encourage processes that can replace removed nutrients (e.g., nitrogen fixation), reduce evaporation, increase water infiltration, and nutrient availability (e.g., adequate soil tilth and mulches); and 5) minimize detrimental soil disturbance.

The importance of developing a harvest strategy concerning the rate of nutrient drain can be illustrated by the work of Compton and Cole [26]. As harvesting intensity increased (bole-only; whole-tree removal; whole-tree plus understorey vegetation, and forest floor removal = complete removal), subsequent plantation performance of *P. menziesii* decreased, with the most pronounced growth reduction evident on the site of lowest quality. Depending on the nutrient, whole-tree harvesting can remove from 1.5-4 times more nutrients than does bole-only harvesting [27].

Consideration must be given to the balance of nutrients and organic matter on all soils. The loss of base cations associated with harvesting and other factors is potentially a threat to site fertility [28, 29, 30]. Marked decreases in base cations, particularly in exchangeable magnesium levels, were evident following harvesting and site preparation in New Zealand [31], and the loss of magnesium has recently been associated with a decline in radiata pine health and vigour [4], and fertilization will be necessary. Nutritional differences occur among species, suggesting fertilizer requirements may be species specific. For example, *Thuja plicata* litter is high in base cations [32], and this species may be particularly efficient at taking up and cycling scarce cations.

The value of organic matter for soil nutrients and water conservation has been well documented for the sandy soils of Australia [33]. Squire [33] described the three alternative methods used in South Australia to meet the nitrogen requirements of second-rotation crops including nitrogen fertilization, nitrogen from legume crops, and litter and logging residue conservation practices. These conservation approaches are applicable to most forest systems. Biomass removal rates should reflect the ability of the site to supply adequate nutrients and organic matter for critical processes (i.e. nutrient cycling and maintenance of soil organic matter levels) and be balanced with natural and supplemental nutrient input rates.

Maximizing the utilization of crop biomass is one objective of forest managers. Merchantability standards, harvest cost, and market prices usually determine the proportion of biomass that can be removed from forest stands economically. The general trend has been for an increasing proportion of stand biomass to be removed from the site during harvesting. Increased utilization or removal of biomass makes it important to balance nutrient inputs and manage organic matter carefully if productivity of short-rotation forest crops is to be maintained.

INTEGRATION WITH HARVESTING TECHNOLOGY

The proposed strategy (Table 8.1) for maintaining long-term site productivity is based on the premise that harvesting operations are an integral part of silvicultural prescriptions. Forest prescriptions or regimens are determined by the inherent nature of the soil and tree species being managed, the associated vegetation, the forest product mix desired, and the harvest strategy. The elements of the harvest strategy that must be reflected in the silvicultural prescriptions include:

1) utilization standards and product sort specifications; 2) amount and degree of soil disturbance that will be tolerated during intermediate and final harvest cuts; 3) level of harvest residuals (foliage/ limbs/ large organic debris – snags) that are desirable to be left in place on the site; 4) distribution pattern desired for the harvest residuals; 5) impacts that are acceptable on residual standing trees (specifications for tolerable levels of stem damage); 6) considerations that must be given to enhance or protect other forest resources including wildlife, fisheries, and water quality; and 7) harvest system "total" economics.

Harvest Tactics

Harvest operations should complement subsequent site preparation and regeneration activities whenever possible. The objective should be to optimize the entire harvesting and regeneration series of treatments rather than evaluating their cost and effectiveness independently. For example, if non-merchantable residual stems, slash debris piles, or high stumps interfere with regeneration treatments, harvesting steps might be modified accordingly at a small incremental cost to off-set subsequent site preparation or regeneration treatment costs. Slash burning in many areas is less acceptable today than in the past owing to the concern over nutrient loss and air pollution. Thus creating large debris piles at landings can be a liability, because they prevent effective regeneration and also represent a source of nutrients that has been removed from the broader area of the site.

Roads and landings must provide for efficient harvesting, yet be optimized to minimize negative on- and off-site impacts. A trade-off must be made between increasing roads and landings to reduce harvest costs and the resultant decrease in the amount of land available for future crop production. Unnecessary loss of productive land should always be avoided. Roads and landings are also the main source of sediments entering streams, and recommended logging practices aim to avoid sediment production [34].

The use of riparian buffers can assist in reducing sediment movement into waterways by trapping sediments from the surrounding landscape and preventing disturbance in and immediately adjacent to waterways. Requirements for riparian protection are likely to vary, depending on the specific nature of each site and on local and regional objectives [35]. There is a need for additional research on buffer design to assist in the development of guidelines for riparian protection [36].

Soil Disturbance Monitoring, Evaluation, and Harvesting

There are three main reasons for having an ongoing monitoring system for measuring harvesting related soil disturbance: 1) to evaluate the relative performance of various equipment configurations; 2) to evaluate ongoing performance against predetermined standards; and 3) to assess the need for harvested units to receive soil amelioration treatments. The diagram in Figure 8.1 provides a concept of how this information can be utilized in the decision making process. Given the levels of soil disturbance that are encountered and the impacts on seedling survival and growth, critical levels of disturbance can be set which prevent a significant net loss of productivity on a unit. Trade-offs can also be made as to the cost of logging and subsequent soil amelioration (e.g., ripping of compacted soil, fertilization) required to obtain full productive potential.

As harvesting systems continue to advance technologically their impacts on forest soils and site productivity need to be re-evaluated. The process outlined should provide the continuum of data to make informed decisions. The database can also be used to set the soil disturbance performance standards which new equipment must meet. To put such a concept into use, a soil resource map that characterizes soil and topographic features, is essential. Such a map can be used for base level forest productivity and equipment operability determinations [37]. The soil survey units can form one of the overlays in the forest inventory system that is referenced to base productivity. Productivity within a given stand or soil survey mapping unit can then be compared through successive rotations at least on a representative sample of the ownership. The survey also can be used to facilitate road construction, harvest planning and setting lay-out and all the other forest activities that are soils/topography dependent [37].

If soil disturbance is to be managed to acceptable levels during harvesting and site preparation operations, a disturbance classification system is very beneficial. Such a system must be relevant to the local field conditions and types of disturbance normally encountered. Simplicity in use is essential. An example (Figure 8.2) for soils in western Washington and Oregon, is given in Miller

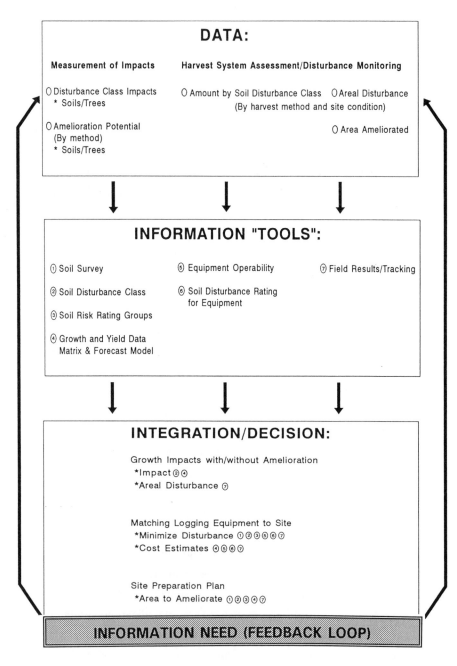

FIGURE 8.1 Soil disturbance management process.

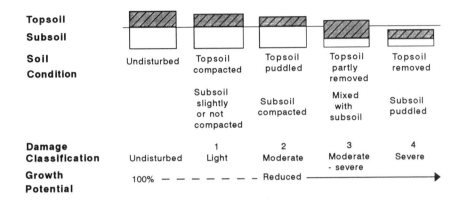

FIGURE 8.2   Site disturbance classification system proposed by Scott (in [38]).

et al. [38]. In this case visual assessment of compaction levels and depth, soil puddling, and soil displacement can be made. The general nature and distribution of disturbance classes across harvested settings will vary depending on felling and forwarding systems (equipment and configurations), topography, soil conditions, traffic intensity, and harvest layout design.

A soil operability rating guide is useful to field personnel and planners. Soils are grouped based on their potential to be impacted detrimentally from trafficking activities. Important characteristics include soil texture, surface soil rock content, soil moisture content, soil drainage, soil organic matter content, and the depth of the A horizon. Information on the time periods that one could expect to operate with ground equipment on the various soils based on average local rainfall patterns and evapotranspiration and experience should be included in the guide.

A measure is required of the impact that the various types of harvesting and site preparation-induced soil disturbance will have on long-term site productivity. In some cases this will be significant [38, 39]. Some level of disturbance will occur with all harvesting operations, and the key questions are: 1) what types and areal extent of disturbance results in long-term productivity impacts; and 2) what levels require amelioration or should be eliminated through better harvest planning or deployment of alternative equipment. The development of operational guidelines pertaining to acceptable levels of soil disturbance depends on this information. Unfortunately, these questions are the most difficult to answer and the background data for many landforms is often lacking.

There have been attempts at defining classification systems for ranking soils as to their susceptibility to harvest-related degradation [37, 40]. Important criteria have included: 1) sites that are already deficient in one or more nutrients; 2) soils that have a relatively shallow A horizon and where a high percent of the available nutrient pool is in the standing biomass; and 3) soils that can compact easily due to fine textured horizons and relatively high moisture contents during periods when trafficking occurs. Once susceptible areas can be identified, it is important to develop a strategy for maintaining both soil physical properties and adequate nutrients through time. In Australia for example two soil nutrient management strategies are being utilized [25]: the Victorian method is to conserve nutrients by residue retention and strip weed control after planting; the South Australian Maximum Growth Sequence places emphasis on complete weed control and nutrient amendment applications. ARACRUZ FORESTAL in South America utilizes a site productivity maintenance strategy that includes both residue management and routine nutritional monitoring to detect and correct nutrient deficiencies through appropriate amendment applications (personal communication, Edgard Campinhos, Jr.).

Another example of a site preparation system that integrates the management of harvest residues, vegetation management, and soil amelioration is a "shovel-scarifier" that has been developed at Longview Washington, (personal communication Greg Jones and Alex Dobkowski, Weyerhaeuser Co.). The machine used is essentially a modified excavation shovel on tracks. The shovel-bucket has been replaced by a "thumb and tine" bucket that has long teeth that can pick up and place residues to create planting sites. It can lift clumps of unwanted vegetation, like vine maple (*Acer circinatum*), directly out of the ground without displacing soil; and can till compacted skid-trails - or replace displaced soil along skid-trails in one pass across a site. The long boom of the shovel, its ability to articulate in a 360-degree circle, and its ability to pick-up and place debris for a traffic-mat minimizes additional unwanted soil disturbance. Small debris piles facilitate planting and conserve nutrients on the site.

The amount of residues to leave on the site is currently in debate, particularly in the Pacific Northwest, USA where large organic matter debris in various stages of decomposition is known to be habitat for many small vertebrates, invertebrates, and microbial populations. The role of this material in maintaining ecological diversity, soil moisture supplies, and nutrient supplies has not been well defined. Nor is it known how stumps, alternative piles of slash, or fertility amendments could accomplish these same objectives. This is an area where long-term studies are critically needed to understand the biological and economic trade-offs.

Each configuration of harvest equipment that is to be used operationally needs to be evaluated to determine its potential impact on the soil. Soil disturbance needs to be evaluated across a range of soil types and under different moisture contents using the setting design layouts that are expected to be most operationally efficient. Disturbance needs to be recorded as to the percent of the area that is in each disturbance class. The resultant data base will provide the basis for allocating equipment to specific settings.

HARVESTING TECHNOLOGY – IMPLEMENTATION STRATEGY AND TACTICS

In this section we summarize key harvesting strategies and associated tactics that can be used to maintain site productivity. We will: 1) demonstrate how both operational forestry and research information can be integrated in the decision-making process; and 2) illustrate that before data or technical information can be useful to operations they have to be organized in a manner that facilitates decision making. By referring to the various informational "tools" managers can then evaluate their alternatives and make sound economic and environmental decisions.

Strategy and Tactics To Minimize Detrimental Soil Disturbances and Nutrient Drain During Harvesting Operations

Specific management recommendations for designing harvest roads, harvest lay-out patterns, equipment configurations, and operational protocol to minimize site damage have been previously summarized [41, 42, 43, 44, 45]. As each land owner has unique considerations that make general recommendations too broad for specific application this section will indicate the key tactics that can be employed to meet previously outlined strategies for maintaining long-term site productivity. A list of equipment design and capability characteristics that should be considered in developing harvesting tactics is presented in Table 8.2. Harvest tactics to minimize impact of soil disturbance are listed below:

- Maximize the integration of harvest and site preparation activities to insure the most efficient overall series of treatments and to reduce traffic on the site.
- Optimize the construction of permanent roads and landings and locate them so that adverse grades for loaded ground transportation are minimized.
- Design harvest patterns so that ground traffic patterns are efficient and so that they avoid areas subject to easy compaction or puddling (e.g. drainages, seeps).
- Prescribe equipment that is matched to the site and can meet soil disturbance guidelines as specified based on monitored field trial results.
- Restrict ground logging systems where Class 2, 3 and 4 will be excessive or cannot be economically ameliorated (Figure 8.2). Avoid using push-blades on skidders or forwarders to create "skid-roads" (Class 4 disturbance).
- Use high-flotation, low ground pressure tires or tracks on equipment with articulated steering instead of skid-steering (i.e. harvesters, skidders and forwarders) to reduce soil compaction levels.
- Designate skid-trails and landing areas to be used when repeated traffic is expected during subsequent harvest operations.
- Ameliorate soils prior to planting with appropriate tillage where compaction is localized and has been demonstrated to be detrimental; e.g., ripping or discing of compacted areas.
- Use equipment and harvest patterns that minimize turning sharp or frequent turns, which causes topsoil displacement and compaction.
- Use harvesting heads that are on extension booms thereby minimizing traffic and equipment turning due to the reach of the booms.
- Where possible utilize the tops and harvest residues as a traffic-mat to displace weight distribution thereby reducing soil compaction.

TABLE 8.2

Examples of equipment design characteristics to consider when developing harvest tactics.

Minimize Soil Disturbance – Ground-based Systems
- Wide tires or tracks
- Independent wheel suspension to reduce shear-forces on the ground surface and to aid climbing over debris and stumps
- Independent wheel articulation to aid turning
- Equipment that steers by articulation versus differentially stopping or slowing tires or tracks on one side of the vehicle (skid-steering)
- Use low aggression tires and tracks by reducing sharp stiff edges and protrusions
- Harvesters and forwarders that have "reach"; e.g., boom extension, to reduce the need to travel to each tree/log
- Processors that can delimb and top trees in place or along corridors to build a travel-mat for flotation as well as distribute other slash as uniformly as possible.
- Forwarders that carry out material instead of skidding it to minimize traffic and soil disturbance
- Forwarders that move material by lifting and extending a boom in repeated cycles; e.g., shovel-logging
- Forwarders that can move large woody debris with a hydraulically articulated boom-head rather than by pushing material out of the way with a blade

Minimize Soil Disturbance – Cable/Aerial Systems
- Cable systems that suspend or skid material without ground-based forwarders
- Balloon or helicopter logging that can lift without any skidding action

Nutrient Conservation
- Single-grip or double-grip harvester that can delimb at the stump or in the traffic path
- Forwarders and skidders that can redistribute slash on return trips rather than creating slash piles at the landing; e.g., grapple-skidders that can remove limbs and tops from the landing and place them back on the site during return trips
- Portable whole-tree flail delimber-debarker – nutrient-rich debris returned to the site by forwarders or skidders
- Shovel with modified bucket with finger-like tines, "shovel-scarifier," that can be used to make small slash piles that do not require burning or take land out of production; competing vegetation can be uprooted with minimal topsoil displacement; slash/debris can be lifted instead of pushed to minimize litter and topsoil displacement; skid-trails can be ripped with bucket-tines if soil moisture is within operating window

- Conserve nutrient drain by utilizing the logging residues for a traffic-mat and subsequently as a mulch for vegetation control and moisture retention, and as a nutrient pool.

An example of a fibre harvesting system where nutrient and soil conservation tactics are utilized is illustrated in Figures 8.3, 8.4, and 8.5.

FIGURE 8.3   Total-tree fibre harvesting operation with a chain-flail delimber-
debarker and mobile chipper.

FACTORS THAT CANNOT BE CONTROLLED BY MANAGEMENT

A number of sometimes insidious environmental factors are of major concern. Managers should be aware of these, to be able to provide good advice for the development of sound national and international policy relating to global change issues. Of particular concern are the anthropogenically induced changes in atmospheric composition. It is important to quantify their impacts, both positive and negative, on growth and to develop appropriate response strategies.

Some types of forest decline have been ascribed to interacting natural and anthropogenic factors, and partitioning of effects is therefore difficult. For

FIGURE 8.4   Grapple-skidder removing foliage, limbs, and bark debris from the landing to be redistributed on the site.

FIGURE 8.5   Small piles of harvest residuals redistributed back on the site along traffic paths after total-tree harvest and chain-flail delimbing and debarking at the landing.

example, the poor health of some forest systems in Europe and Scandinavia has been associated with the direct (acid deposition in the form of sulphur and nitrogen compounds) and indirect effects (soil cation loss and imbalance) of acid rain [46]. While experiments in Germany have demonstrated that the elimination of ionic imbalances through fertilization resulted in revitalization of trees with decline symptoms [46, 47], the decline often seems to involve a complex of factors. Each case needs to be assessed individually. Climatic stress, natural stand dynamics (tree and stand age), soil fertility and buffering capacity, disease, insect damage, and other air pollutants (ozone), have been invoked to explain the nature and severity of the damage.

Changes in climate and atmospheric composition are likely to confound productivity estimates. Atmospheric inputs of nitrogen and sulphur compounds seem to have increased forest growth rates in Europe and Scandinavia, in spite of the sometimes reduced forest health. The physiological responses to the well known increase in atmospheric CO_2 levels are expected to result in greater productivity. These direct effects of CO_2 on growth are still being actively researched, including the influences of temperature, nutrients, and moisture availability [48]. Whether the expected gains in productivity from a CO_2 fertilizer effect will be realized in practice depends on the influence of changed nutrient cycles, and the roles of insect pests and diseases on tree health and vigor. Gains will be site specific.

Catastrophic events such as wind, fire, and large-scale erosion from flooding can result in significant losses in harvestable biomass, and may force a revision of management objectives. The possibility that cyclonic storms may increase in severity should global warming scenarios become a reality is of concern in some regions. More attention will be required to ensure that intensively managed plantation forest is established in areas where the risk of damage is less likely to be catastrophic. Fire risk and severity can also be reduced through appropriate silvicultural regime coupled with education and training of the public and the workforce. Intensive storms can lead to severe soil erosion at sites that have only recently been harvested, and amelioration practices would be necessary if nutrient cycles are adversely disrupted.

AN OPERATIONAL FORESTRY PERSPECTIVE FOR SUSTAINING PRODUCTIVITY IN INTENSIVELY MANAGED SITES

The following are key premises to sustain productivity on intensively managed sites:

1. Strive to enhance productivity

The goal of forest managers should be to enhance forest productivity whenever economic considerations can justify the expense for the additional inputs. The fallacy with the approach of just trying to maintain productivity rather than enhance productivity is that we do not have perfect knowledge and control to precisely prescribe treatments so that productivity is sustained at a given level. The philosophy of "maintaining productivity by minimizing the negative impacts of forest management" is quite different from one where the manager "strives to improve productivity".

2. Develop yield expectations

Forest managers need to determine, for a given regime, base productivity levels and set targets for future yield expectations based on levels of inputs into the system. A difficulty is that base productivity levels generally have not been determined for the various silvicultural regime alternatives that might be considered. This makes it difficult to set targets.

Base productivity levels will need to be defined by species or species mixtures for specific soils, topography, elevation, climatic zones (adjusted for deviations from normals), and cultural inputs. Incremental cultural inputs contributing to yield include genotype, regeneration system, vegetation control, soil amelioration, fertility amendments, amongst others.

Yields are measurable, and can be tracked during stand development through successive rotations. It is recommended that a reliable forest inventory be in place to monitor growth and yield parameters through successive rotations on the same sites and that a cultural treatment matrix be used to monitor the incremental responses to cultural inputs.

3. Inputs must be equal to or greater than outputs

Natural processes can be used to replenish a system's inevitable losses following natural events such as fire or man-induced changes such as harvest removals [49]. Alternatively, cultural inputs can be utilized in conjunction with natural processes to counter losses and more rapidly move the system in the desired direction [26]. The relative efficiency by which natural processes and cultural treatments meet desired crop production levels and timetables primarily determines the silvicultural alternatives that managers will use in their regimes. While biological and environmental considerations are of prime importance, costs (time and capital - site resources as well as dollars), yields (volume and value), risks (probability of meeting short- and long-term yield expectations), and social and political constraints all have a role in determining the preferred alternative.

What the forest manager cannot do is continually remove from the forest more than what is added without eventually noticing a fall-down in productivity. A balance sheet approach will assist in quantifying inputs and losses. Natural inputs are not always sufficient to balance nutrient losses associated with harvesting, in which case enhancements brought about by management practices including residue conservation practices, forest fertilization, use of nitrogen fixers or efficient accumulators and cyclers of nutrients can be considered.

4. Develop resource allocation strategies

Forest management activities are concerned with how to allocate a greater proportion of the available resources to the crop species than would normally occur without management, while maintaining the sustainability of the system. Approaches to achieve this include manipulating genetics, stand density, competing vegetation, and site factors (nutrients, moisture, soil physical properties), with the solution depending on environmental, biological and management constraints.

Process based computer models have been developed to evaluate the effects of forest management on productivity, but a recent review indicates that these models need to be further developed and validated [50, 51]. Few long-term studies exist for this purpose.

A SUMMARY OF KEY RECOMMENDATIONS TO IMPLEMENT RESEARCH
RESULTS IN FOREST OPERATIONS

- Determine database requirements to further refine regional management guidelines to maintain and enhance forest productivity.

- Establish appropriate trials for assessing alternative management options for balancing nutrient inputs and outputs following various levels of harvesting intensity and vegetation control. Undertake process studies to determine the respective contributions of atmospheric, soil, fertilizer, and harvesting residue inputs of nutrients in maintaining long-term productivity, including large organic matter debris. Information on outputs associated with leaching, harvesting, and regeneration will also be required. Selected families or clones should be incorporated in the design, to facilitate physiological and genetic research.

- Research silvicultural and harvesting techniques with the view to reduce unnecessary accumulation of residues at landings, particularly on steep slopes where safety considerations make delimbing at the stump inadvisable. Small branches that break off during felling are desirable, suggesting the need to examine alternative species, breeds, and alternative tending regimes such as delayed thinning.

- Investigate alternative methods to recycle residues that inevitably accumulate at landings and log processing yards. Possible avenues include removal of residues for energy and their subsequent return in the form of pelletized ash. In this regard, the role of organic matter (nutrient vs. organic matter and nutrients) needs to be better defined.

- Evaluate and promote the benefits of setting plantation boundaries with harvesting considerations as the basis. The design of forestry operations from a landscape perspective can improve environmental quality and raise the public image of forestry. Consideration will need to be given to identifying and refining local and regional resource management objectives.

- Develop an understanding of the physiological factors associated with superior performance across the anticipated range of critical environmental conditions (e.g., temperature, moisture, fertility, elevated CO_2), and the heritability of these factors. Progenies can then be appropriately deployed should climate warming scenarios become a reality, management practices alter, or stress-inducing factors change. Both the crop and non-crop elements need to be considered.

- Breed genotypes that are broadly adapted to current and future environmental conditions. Explore the opportunities to develop genotypes that are adapted to regimens that require minimal fertilizer inputs and that can withstand stresses associated with environmental change. Breeds would ideally have improved nutrient-use efficiency, high phosphorus uptake capacity on acid soils having high phosphorus fixation, and have stable performance when nutrient supply is unbalanced.

- Develop and implement information systems that can assist managers to assess risks associated with alternative management options, and thereby enhance productivity and environmental quality. The information and component models often already exist, but need to be refined and integrated into geographic information systems. Topographical, soil, and climate information need to be linked with process based models of potential site productivity and nutrient cycling, determined at least for a representative set of sites. Crop and forest operations monitoring systems, with appropriate feedbacks to research and management, would be essential components in the overall design of a risk assessment system.

LITERATURE CITED

1. Kramer, P.J. The role of physiology in forestry. *Tree Physiology*, 1986, **2**, 1-16.
2. Zobel, B. and Talbert, J. *Applied Forest Tree Improvement*. John Wiley & Sons. New York, 1984.
3. Johnson, G.R., Firth, A., and Brown, P.C. Value gains from using genetically improved radiata pine stock. *New Zealand Forestry*, 1992, **36**, 14-18.
4. Beets, P., Carson, S., Dick, M., Singh, A., Skinner, M., Hunter, I. Mid-Crown Yellowing - On the Increase. What's New in Forest Research No. 206, Forest Research Institute, Rotorua, New Zealand, 1991.

5. Knight, P.J. Foliar concentrations of ten mineral nutrients in nine Pinus radiata clones during a 15-month period. *New Zealand Journal of Forestry Science*, 1978, **8**, 351-368.

6. Hunter, I.R., Rodgers, B.E., Dunningham, A., Prince, J.M., and Thorn, A.J. *An Atlas of Radiata Pine Nutrition in New Zealand.* Forest Research Institute Bulletin No. 165. Forest Research Institute, Rotorua, New Zealand, 1991.

7. Burdon, R.D. Foliar macronutrient concentrations and foliage retention in radiata pine clones on four sites. *New Zealand Journal of Forestry Science*, 1976, **5**, 250-259.

8. Sheppard, L.J. and Cannell, M.G.R. Nutrient use efficiency of clones of *Picea sitchensis* and *Pinus contorta. Silvae Genetica*, 1985, **34**, 126-132.

9. Jackson, D.S., Gifford, H.H., and Hobbs, I.W. Daily transpiration rates of radiata pine. *New Zealand Journal of Forestry Science*, 1973, **3**, 70-81.

10. Van Buijtenen, J.P. and Isbell, R. Differential response of loblolly pine families to a series of nutrient levels. 1st North American Forest Biology Workshop, Michigan State University, East Lansing, 1970.

11. Nambiar, E.K.S. Increasing forest productivity through genetic improvement of nutritional factors. In: *Forest Potentials.* Weyerhaeuser Co., Tacoma, WA, 1984, pp. 191-215.

12. Dixon, R.K. Physiological processes and tree growth. In: *Process Modeling of Forest Growth Response to Environmental Stress.* Timber Press, Portland, Oregon, 1990, pp. 21-32.

13. Ledig, F.T. and Kitzmiller, J.H. Genetic strategies for reforestation in the face of global climate change. *Forest Ecology and Management*, 1992, **50**, 153-169.

14. Clapp, R.A. (Ed.). *Blue Mountain Forest Health Report - "New Perspectives in Forest Health".* USDA Forest Service, Pacific Northwest Region, Malheur, Umatilla and the Wallowa-Whitman National Forests, 1991, 143 p.

15. Hood, I.A. and Vanner, A.L. *Cyclaneusma* (Naemacyclus) needle-caste of *Pinus radiata* in New Zealand. 4: Chemical Control Research. *New Zealand Journal of Forestry Science*, 1984, **14**, 215-222.

16. Miller, J.H., Zutter, B.R., Zedaker, S.M., Edwards, M.B., Haywood, J.D., Newbold, R.A. A regional study on the influence of woody and herbaceous competition on early loblolly pine (*Pinus taeda* L.). *Southern Journal of Applied Forestry*, 1991, **15(4)**, 169-178.

17. Carrier, B.D. and Dobkowski, A. The effects of three different vegetation management regimes on sixth-year Douglas-fir growth in Oregon and Washington. Weyerhaeuser Company, Technical Report 050-3910/31. Weyerhaeuser Company, Centralia, WA, 1991, 32 p.

18. Haines, S.G. and DeBell, D.S. Use of nitrogen-fixing plants to improve and maintain productivity of forest soils. In: *Proceedings: Impact of Intensive Harvesting on Forest Nutrient Cycling.* State University of New York, College

of Environmental Science and Forestry, School of Forestry, Syracuse, NY, 1979, pp. 279-303.

19. Gadgil, R.L. The nutritional role of *Lupinus arboreus* in coastal and sand dune forestry. 3. Nitrogen distribution in the ecosystem before tree planting. *Plant and Soil*, 1971, **35**, 113-126.

20. Binkley, D., Cromack, K., and Fredricksen, R.L. Nitrogen accretion and availability in some snowbrush ecosystem. *Forest Science*, 1982, **28**, 720-724.

21. Binkley, D. Mixtures of nitrogen-fixing tree species. In: *The Ecology of Mixed-Species Stands of Trees*. (Eds.) M.G.R. Cannell, D.C. Malcolm, and P.A. Robertson. Publication No. 11, British Ecological Society. Blackwell Scientific Publications, Oxford, 1992, pp. 99-123.

22. Dobkowski, A. Effects of levels of red alder competition on the growth of young Douglas-fir. Weyerhaeuser Co. Technical Report 050-3910/33. Weyerhaeuser Company, Centralia, WA, 1992, 37 p.

23. Dyck, W.J. and Mees, C.A. *Research Strategies for Long-Term Site Productivity*. Proceedings, IEA/BE/A3 Workshop, Seattle, WA. IEA/BE/A3 Report No. 8. Ministry of Forestry, Forest Research Institute, FRI Bulletin No, 152, 1989, 257 p.

24. Dyck, W.J. and Mees, C.A. *Long-term Field Trials to Assess Environmental Impacts of Harvesting*. Proceedings, IEA/BE T6/A6 Workshop, Amelia Island, FL, USA. IEA/BE T6/A6 Report No. 5. Ministry of Forestry, Forest Research Institute, FRI Bulletin No, 161, 1991, 257 p.

25. Squire, R.O., Finn, D.W., and Campbell, R.G. Silvicultural research for sustained wood production and biosphere conservation in the pine plantations and native eucalypt forests of southeastern Australia. In: *Long-Term Field Trials to Assess Environmental Impacts of Harvesting*. Proceedings, IEA/BE T6/A6 Workshop, Florida, Feb. 1990. (Eds.) W.J. Dyck and C.A. Mees. IEA/BE T6/A6 Report 5. Forest Research Institute, Rotorua, New Zealand, FRI Bulletin 161, 1991, pp. 3-28.

26. Compton, J.E. and Cole, D.W. Impact of harvest intensity on growth and nutrition of successive rotations of Douglas-fir. In: *Long-Term Field Trials to Assess Environmental Impacts of Harvesting*. Proceedings, IEA/BE T6/A6 Workshop, Florida, Feb. 1990. (Eds.) W.J. Dyck and C.A. Mees. IEA/BE T6/A6 Report 5. Forest Research Institute, Rotorua, New Zealand, FRI Bulletin 161, 1991, pp. 151-161.

27. Kimmins, J.P. Evaluation of the consequences for future tree productivity of the loss of nutrients in whole-tree harvesting. *Forest Ecology and Management*, 1977, **1**, 169-183.

28. Hornbeck, J.W. Cumulative effects of intensive harvest, atmospheric deposition, and other land use activities. In: *Impact of Intensive Harvesting on Forest Productivity*. Proceedings, IEA/BE A3 Workshop, South Island, New Zealand, March 1989. (Eds.) W.J. Dyck and C.A. Mees. IEA/BE T6/A6 Report No. 2.

Forest Research Institute, Rotorua, New Zealand, FRI Bulletin No. 159, 1990, pp. 147-154.

29. Johnson, D.W. and Lindberg, S.E. *Atmospheric Deposition and Forest Nutrient Cycling. A synthesis of the integrated forest study.* Springer-Verlag, New York, 1992, 705 p.

30. Kenttamies, K. The effect of acidic deposition on waters. In: *Acidification Research in Finland.* Review of the results of the Finnish acidification research program (HAPRO) 1985-1990. (Ed.) K. Kenttamies. Brochure 39, 1991, pp. 23-35.

31. Ballard, R. Effect of slash and soil removal on the productivity of second rotation radiata pine on pumice soil. *New Zealand Journal of Forestry Science,* 1978, **8**, 248-58.

32. Tarrant, R.F., Isaac, L.A., and Chandler, R.F. Jr. Observation on litterfall and foliage nutrient content of some Pacific Northwest tree species. *Journal of Forestry,* 1951, **49**, 914-915.

33. Squire, R.O. Review of second rotation silviculture of *P. radiata* plantations in southern Australia: establishment practice and expectations. In: *Proceedings: IUFRO Symposium on Forest Site and Continuous Productivity.* (Eds.) R. Ballard and S.P. Gessel. General Technical Report PNW-163, Pacific Northwest Forest and Range Experiment Station, Portland, OR, 1982, pp. 130-137.

34. Vaughan, L. *Logging and the Environment - a review of research findings and management practices.* New Zealand Logging Industry Research Association Report, 1984.

35. Gilliam, J.W., Schipper, L.A., Beets, P.N., McConchie, M. Riparian buffers in New Zealand forestry. *New Zealand Forestry,* 1992, **37**, 21-25.

36. Comerford, N.B., Neary, D.G., and Mansell, R.S. *The Effectiveness of Buffer Strips for Ameliorating Offsite transport of Sediment, Nutrients, and Pesticides from Silvicultural Operations.* NCASI Technical Bulletin No. 631, 1992.

37. Steinbrenner, E.C. Mapping forest soils on Weyerhaeuser lands in the Pacific Northwest. In: Proceedings of the Fourth North American Forest Soils Conference, Laval Univ., Quebec. (Eds.) B. Bernier and C.H. Winget. Les presses De L'Université' Laval, Quebec, 1975, pp. 513-525.

38. Miller, R.E., Stein, W.I., Heninger, R.L., Scott, W., Little, S.N., and Gohean, D.J. Maintaining and improving site productivity in the Douglas-fir region In: *Maintaining the Long-term Productivity of Pacific Northwest Forest Ecosystems.* (Eds.) D.A. Perry *et al.* Timber Press, Portland, OR, 1989, pp. 98-136.

39. Senyk, J.P. and Smith, R.B. Estimating impacts of forest harvesting and mechanical site preparation practices on productivity in British Columbia. In: *Long-Term Field Trials to Assess Environmental Impacts of Harvesting.* Proceedings, IEA/BE T6/A6 Workshop, Florida, USA, Feb. 1990. (Eds.) W.J. Dyck and C.A. Mees. IEA/BE T6/A6 Report No. 5. Forest Research Institute, Rotorua, New Zealand, FRI Bulletin No. 161, 1991, pp. 199-211.

40. Hunter, I.R., Dyck, W.J., Mees, C.A., and Carr, K. Site degradation under intensified forest harvesting: a proposed classification system for New Zealand. IEA/BE Project A3 (CPC-10), Report No. 7. Rotorua, New Zealand, 1988, 22 p.

41. Lantz, R.L. Guidelines for stream protection in logging operations. Oregon State Game Commission, Portland, OR, 1991, 29 p.

42. McKee, W.H. Jr., Hatchell, G.E., and Tiarks, A.E. Managing site damage from logging - a loblolly pine management guide. General Technical Report SE-32, Southeastern Forest Experiment Station, Asheville, NC, 1985, 21 p.

43. Lousier, J.D. and Still, G.W. *Degradation of Forested Land: "Forest Soils at Risk."* Proceedings of the 10th B.C. Soil Science Workshop. Crown Publications Inc., Victoria, B.C., 1988, 331 p.

44. Garland, J.L. Designated skid-trails minimize soil compaction. Extension Circular 1110. In: *"The Woodland Workbook"*. Oregon State University Extension Service, Corvallis, OR, 1986.

45. Froehlich, H.A., Auleerich, D.E., and Curtis, R. Designing skid-trail systems to reduce soil impacts from tractive logging machines. Forest Research Laboratory Research Paper 44, Oregon State University, Corvallis, OR, 1981.

46. Tomlinson, G.H. and Tomlinson, F.L. Cationic nutrients in the soil - inputs, recycling, storage losses and their effects. In: *Effects of Acid Deposition on the Forests of Europe and North America.* CRC Press, Boca Raton, FL, 1990, pp. 85-105.

47. Ke, J. and Skelly, J.M. Foliar symptoms on Norway Spruce and relationship to magnesium deficiencies. *Water, Air and Soil Pollution*, 1990, **54**, 75-90.

48. Eamus, D. and Jarvis, P.G. The direct effects of increase in the global CO_2 concentration on natural and commercial temperate trees and forests. *Advances in Ecological Research*, 1989, **19**, 1-55.

49. Van Lear, D.H., Kapeluck, P.R., and Waide, J.B. Nitrogen pools and processes during natural regeneration of loblolly pine. In: *Sustained Productivity of Forest Soils.* Proceedings of the 7th North American Forest Soils Conference. (Eds.) S.P. Gessel, D.S. Lacate, G.F. Weetman and R.F. Powers. University of British Columbia, Faculty of Forestry Publication, Vancouver, B.C., 1990, pp. 234-252.

50. Yarie, J. Role of computer models in predicting the consequences of management on forest productivity. In: *Research Strategies for Long-Term Site Productivity*. Proceedings, IEA/BE/A3 Workshop, Seattle, WA. (Eds.) W.J. Dyck and C.A. Mees. IEA/BE/A3 Report No. 8. Ministry of Forestry, Forest Research Institute, FRI Bulletin No, 152, 1989, pp. 3-18.

51. Kimmins, J.P. and Scoullar, K.A. *FORCYTE-10: A Users Manual.* 2nd approximation. Contract report to the Canadian Forestry Service, National Forestry Institute, Petawawa, Ontario, 1983, 117 p.

CHAPTER 9

DESIGNING LONG-TERM SITE PRODUCTIVITY EXPERIMENTS

R.F. POWERS
USDA Forest Service
PSW Research Station
Redding, CA, USA

D.J. MEAD
Lincoln University
Canterbury, New Zealand

J.A. BURGER
Virginia Polytechnic Institute and State University
Blacksburg, VA, USA

M.W. RITCHIE
USDA Forest Service
PSW Research Station
Redding, CA, USA

INTRODUCTION

Exploring the long-term sustainability of managed forest ecosystems using the field experimental approach, while recognised as being a research priority, is not a straightforward task. This is true of many long-term forestry trials despite the use of the scientific method, designed experiments, and the advantages of modern equipment. Managers and researchers have frequently underestimated the problems involved in dealing with long-term forest experiments.

The need for long-term forestry experiments has become more apparent in recent years. As forestry has become more sophisticated, as exemplified by the use of a wide range of management models, workers have questioned whether the trends or assumptions based on short-term experiments are necessarily accurate. As we will illustrate later, this is also true of studies involving harvesting practices.

Part of the research problem has to do with the large size of forest trees and their life spans which are greater than the average length of scientists' professional careers. Historically, biological scientists have had difficulty researching and understanding life forms that outlive them. This is often compounded by the complexity of and continual changes to the ecosystems over time.

The following question is a case in point: "how does timber harvesting and stand replacement affect the site's potential to produce the next crop of timber?" Forest scientists asked this decades ago, yet we are still in the process of answering it today.

The relatively slow progress toward answering this question is partly a function of the length of the experimental iteration and the size of the plants, which restrict the experimental approach. Another part of the problem is that scientists have not always clearly defined the problems or the hypotheses that they have sought to test. For example, the question as stated above is poorly defined in that it may not necessarily lead to progress in understanding the underlying mechanisms. If we understand the underlying mechanisms then we should not only be able to answer the question of how harvesting affects site potential, but we can suggest how to prevent any decline.

There are additional institutional constraints to scientific progress. Most forest scientists must conduct their experiments in such a way as to "guarantee" a certain level of confidence in their research conclusions; and this may be very costly. When experiments span decades, continuity is lost by scientist turnover. The cycling of interest, momentum, logistical support, and funding during the course of a long-term experiment can create inefficiencies and sometimes a loss of critical data. Libby *et al.* [1] have estimated that between 50 and 75% of research information from forest genetics research may have been lost by such difficulties, coupled with other disasters, unless special consideration is given to this problem.

Therefore, the greatest constraints to progress in testing hypotheses in long-term site productivity studies have to do with:

1. the cost and difficulty of building scientific rigour into long-term field designs - the need to make the trials "timeless",

2. human limitations, and

3. the institutional constraints to the conduct of long-term research.

THE NEED FOR LONG-TERM DESIGNED EXPERIMENTS

The use of chronosequence, retrospective, or simulation approaches in the study of long-term forest productivity has been discussed in detail in Dyck and Mees [2], in Chapters 2 and 6 of this book, and were also briefly reviewed by Burger and Powers [3].

Chronosequences allow us to study how long-lived systems change with time, but in doing so we assume other factors are constant. Retrospective studies look backward from the present to infer past causes and impacts. Again certain assumptions are made which may be impossible to verify; also there can be design problems and they will be limited by the scope of the original treatments [3]. With simulation models we project what may occur using relationships defined by other studies. All these approaches suffer from being difficult to verify and the results may often be ambiguous.

Designed experiments are another approach to studying long-term phenomena in forestry. Unlike chronosequence, retrospective, and simulation studies that capture in the present events that occur with uncertain influences over long periods, designed studies begin in the present, continue under known conditions, and attain results after the period of interest. Disadvantages of this approach are enormous, not the least of which is that the results, which are often needed immediately, may not be attained within the career of the investigator who installed the study. Furthermore, the hypothesis being tested could become antiquated before the experiment comes to term. Yet another disadvantage is the cost and dedication required to maintain a field trial for periods that may span multiple research administrations whose priorities may be different.

But the advantages of designed experiments are enormous. Unlike other approaches to research:

1. the investigator has complete control over the choice of treatments;

2. systematic errors are controlled so that treatment effects are unbiased;

3. conclusions are valid within the range of the experimental design (results are unambiguous); and

4. the degree of uncertainty or risk as to the validity of the conclusions is quantifiable.

Designed experiments incorporating these advantages are the rule for short-term experiments, where the strengths of designed experiments are taken for granted. In studies exploring long-term changes in forest productivity, chronosequence, retrospective, and simulation approaches have been the rule, but they have been inadequate.

This is illustrated in the pursuit of the persistent question of site productivity declines in the southeastern U.S.A. associated with the removal of organic matter during harvesting and site preparation. There are no rotation-length designed experiments that have explicitly tested the effects of organic matter removal. Burger and Kluender [4] reviewed the literature and developed a rotation-length chronosequence of volume response to the removal of organic residue via windrowing (Figure 9.1). The relative volume of stands on sites prepared by removal versus no removal of organic residue was compared to control stands receiving no site preparation.

This chronosequence suggests that removing organic residue prompts an early

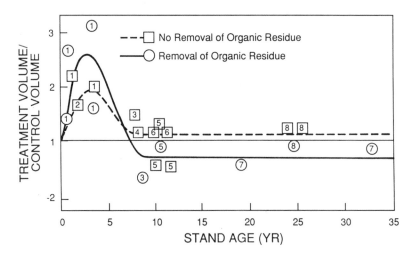

FIGURE 9.1 A chronosequence of relative volume response to organic residue removal. Data are from eight site preparation studies reviewed by Burger and Kluender [4].

response that declines precipitously by age 10 years, thereafter resulting in a volume less than the control. Productivity decline after age 10 may be due to organic matter removal as Fox *et al.* [5] suggested; however, superior early growth probably traces to better weed control because windrowing is usually followed by harrowing. Conclusions are obviously speculative due to the limitations of this chronosequence. The study sites were dissimilar in many ways; the stand character and original levels of organic residue were mostly unknown; and responses were confounded by other site preparation effects such as weed control. Adding to the ambiguity was that some data were based on non-replicated retrospective studies.

Because of the problem's importance and the inadequacies of chronosequence and retrospective approaches, researchers in the Forestry Department at Virginia Polytechnic Institute and State University began a long-term designed experiment to get at the root of this productivity question. The Typic Hapludult great group making up about 80% of the Piedmont physiographic province was chosen as a limited inference space to test the residue-removal hypothesis. Given the known variation in growth responses to site preparation, 12 blocks or replications were installed in a two-state area. Treatment plots were 2 hectares in size so that treatments could be installed at operational levels. This operational scale eliminated the question of experiment-to-field extrapolation. Prior to harvesting, existing stands were characterized to obtain co-variate data. All 12 blocks were installed within a 3-month period. The same operators and equipment prepared all sites for planting. The combination of these design features represents a designed experimental approach that will provide more conclusive results than a combination of studies making up a chronosequence.

Preliminary results from the designed experiment are shown in Figure 9.2. Generally, responses were similar to those hypothesized in the chronosequence. The early volume response to treatment was proportional to the weed control provided by each treatment. At age 9 the study was just at the point of stand closure, the point at which stand nutrient demands approach their maxima. Between ages 6 and 9 treatment responses flattened out after a precipitous drop between ages 3 and 6. By age 9 there was no evidence of productivity decline suggested by the chronosequence studies [4], but it remains to be seen how the removal of organic matter via windrowing ultimately affects volume at rotation age.

The persistence of the question of site productivity as influenced by management, and the inability of reaching sound conclusions as demonstrated above, has prompted forestry researchers to use designed field trials, such as the one described above, even though final conclusions are a rotation length away. For many problems in forestry, such as the question of residue removal, designed experiments are a necessity. The size of the effort required to establish and maintain trials such as this one, means they must be very well planned and supported.

THE NATURE OF STATISTICAL INFERENCE

A major aim of science is to understand the nature of phenomena and to push back the edges of uncertainty. In forestry, we usually deal with populations so large that we cannot examine them fully. Instead, we inspect samples from them in hopes of understanding the large populations better. Often, we compare samples to see if they might come from separate populations, or to see if treating them in a certain way produces a noticeable effect. Always, our aim is to understand

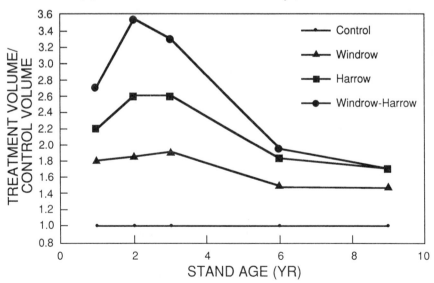

FIGURE 9.2   Relative volume response of loblolly pine stands to windrowing and harrowing.

more about the nature of populations. But because we must deal with samples rather than the whole population, we always are faced with the uncertainty that our treatment effects are simply a product of chance variation or bias, rather than the treatment itself. Modern statistical theory is meant to help us make decisions in light of uncertainty. Much of this is based on probability concepts involving randomness and estimation of errors.

The Normal Distribution and Random Variation
Individuals comprising biological populations are never identical. Instead, there is variability about a central tendency for any particular characteristic. For any characteristic (size, age, etc.), a frequency distribution can be developed for any population, and while these may take virtually any shape, a useful one to consider is the symmetric "normal" distribution. While in biology few population characteristics may follow an idealized normal distribution, the problem is eliminated by the principal of the "Central Limit Theorem" – one of the most remarkable theorems in the whole of mathematics. This theorem shows that the distribution of sample means will tend to become normally distributed as sample size increases. The sample must be random, from a population with a finite standard deviation. This is an immense convenience when the exact distribution of a sampled population is not known.

Much statistical theory of sampling centres on the assumption of randomness; that is, selection is such that the probability of choosing any distinct sample is the same for all possible samples, and that sampling does not affect the population's pattern of variation [6]. Samples taken randomly from a large population give us valid and unbiased estimates of a population's mean and pattern of variation. Both parameters are necessary if one wants to examine samples for possible differences. Non-random samples may yield estimates which are quite precise but wholly inaccurate.

Controlling Variability in Experiments
In any experiment, careful consideration of extraneous factors, replication, and randomness are crucial in avoiding confounding, reducing uncertainty, and producing outcomes free of bias. These points can be illustrated hypothetically with an experiment meant to test the effects of forest fertilization on tree growth.

A site supporting a young pine plantation is selected for study. The site is bisected by a road which the researcher sees as the basis for dividing the land into two experimental units. A fertilizer treatment is assigned to one unit, and the other unit is left as an untreated control. Treatment effect will be defined as the difference in tree growth between fertilized and unfertilized units in the decade following treatment. Tree dimensions are measured at the start of the experiment and 10 years later.

Assume that for some reason (relief, soil change, etc.) one experimental unit (one side of the road) is inherently more productive than the other. If fertilizer is applied to the more productive unit, trees will be considerably larger after 10 years than those on the unfertilized unit. The researcher might attribute this difference to the effects of fertilization, but differences may largely be due to natural differences in site quality (better sites produce larger trees). Alternatively, should the poorer unit be fertilized, the researcher might conclude that fertilization was ineffective (fertilization may have improved growth, but the trees may be no larger than those on the naturally more productive control plot). In both cases the experiment is confounded because the researcher cannot separate the effects of site quality from treatment. Better attention to possible confounding factors such as differences in site quality would have avoided the problem.

No two units of land are exactly alike. Consequently, no two experimental units treated exactly the same way should be expected to respond identically. Experiments in physics and chemistry under controlled conditions produce consistent results with low variability, but biological field experiments incorporate the variability inherent in nature. Because of this any given field experiment contains an element of uncertainty.

As early as 1908 "Student" (W.S. Gosset) said: "Any experiment may be regarded as forming an individual of a population of experiments which might be performed under the same conditions. A series of experiments is a sample drawn from this population." [7]. This means that the results of any experiment that compares a treatment against a control gives us only a single estimate of how the population might respond to the same treatment. Although it may be an unbiased estimate of population response, it should be considered only one of many possible sample estimates of the population. Gaining more certainty about true population response (a better estimate of a mean μ), requires either repeating (replicating) the experiment, or increasing sample size.

Defining the population carefully, selecting multiple experimental units (replicates) from the population, and assigning treatments randomly to those experimental units (randomization) has major advantages over the hypothetical fertilization experiment described above. If treatments and controls are assigned randomly, we can eliminate bias in the sampling procedure and ensure independence of observations. Because treatments are assigned randomly to a variety of sites, the chance is markedly less that treated sites reflect one suite of extraneous factors while controls reflect another. Another advantage of replication and randomization is that it affords an unbiased estimate of the response variability one might expect in the population of interest (such as all pine plantations within a given region). An unreplicated experiment is no more than a case study.

A common failing in the design of field experiments is to confuse subsampling within a treatment plot with sampling from replicated, randomized treatment plots [8]. Assume that the two experimental units in the fertilization experiment above were too large for a complete inventory of tree growth. For practical reasons the researcher installed several small plots for growth measurements. In turn, growth data from these small sampling plots were subjected to an analysis of variance which produced an estimate of overall treatment effect and a measure of experimental error. From this, the researcher concluded that fertilization had a highly significant effect on tree growth. However, the researcher has confused multiple replication of randomized treatments with multiple sampling of an unreplicated treatment. In actuality, the plot data merely produce an estimate of mean tree growth on each experimental unit subject to sampling error. And what was perceived as "experimental error" in the analysis of variance merely is a pooled measure of sampling error from one experimental unit that was fertilized, and one unit that was not. This particular "pseudoreplication" [8] does *not* provide a valid estimate of treatment response or of experimental error. While it may offer some weak insight as to whether growth on the two experimental units are different, it does not tell us why. The experiment remains confounded.

"Blocking" is one way of increasing precision and avoiding confounding if one suspects that extraneous site differences existing before treatment might affect treatment outcome. Blocking is done to help control extraneous variability. It is based on the premise that pretreatment differences among experimental units within a block are less than differences between blocks. In the fertilization experiment above, the two experimental units separated by a road could each be subdivided before treatment into two or more subunits or "plots" – at least two similarly less productive plots, and an equivalent number of plots which were more productive. Fertilizer and control treatments then would be assigned randomly to the plots on the less productive site, and also to the plots on the more productive site. This reveals the true treatment effect by removing the variation due to block (site quality) differences. Blocking also provides insight into the degree to which treatment response is conditioned by extraneous site conditions. While blocking is a powerful tool in experimental field studies, it should not be used indiscriminatly. Blocking without a valid basis reduces the sensitivity of the experiment through the loss of degrees of freedom in experimental error [6].

Inferential Statistics: Types I and II Error

Most forestry research is meant to test some hypothesis concerning one or more population parameters (e.g., mean stand diameter, litter biomass, foliar nitrogen, etc.). The null hypothesis (H_O) states the hypothesis to be tested. For example, one may hypothesize that the mean treatment response is equal to some value: $\mu = \mu_O$, where μ_O is often zero (no treatment response). The decision to reject the null hypothesis is based on a test statistic calculated from sample means (e.g., the standardized difference of the sample means). The entire range of potential sample values is divided into two regions: a rejection region and an acceptance region (Figure 9.3). If the test statistic falls in the rejection region, the null hypothesis is rejected. If it does not fall in this region, the hypothesis is not rejected.

This process is subject to two types of errors: rejection of the null hypothesis when, in fact, it is true (Type I error), or failure to reject the null hypothesis when it is false (some alternative hypothesis is true, a Type II error). The strength of a statistical test of an hypothesis is determined by the probabilities of making a Type I or II error (α and ß, respectively). These probabilities are shown graphically in Figure 9.3. Reducing the size of α will increase ß. If the sample size, n, is increased, and α is the same, both α and ß will decrease, and 1-ß, the complement of ß, or the "power of the test", will increase.

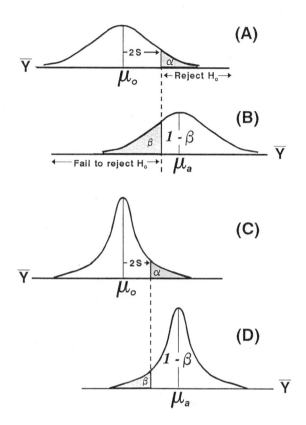

FIGURE 9.3 Distribution of means for H_o (A, C) and H_a (B, D) when sample size is small (A, B) and large (C, D). The consequences of the number of replicates on Type I and Type II error using frequency distributions for sample means $\overline{Y}$ about a true population mean μ are shown. For a given level of alpha, small sample size leads to a large Type II error (B), while larger sample size reduces Type II error (D).

The following is an example of Type I and Type II errors, relative to sample size. Given a true (but unknown) difference between μ_C of a control population and μ_T of a treated population ($\mu_T - \mu_C = \mu$), we wish to see if a treatment has an effect (in this case, positive) on some variable (Y). The researcher decides that for a treatment to have a significant effect, its mean difference must be two or more standard errors larger than the hypothesized difference of zero. Attaining such values will be cause for "rejecting the null hypothesis" that $\mu_T - \mu_C = 0 = \mu$.

An experiment is conducted, means and standard errors are calculated, and a critical value for rejecting the H_O is determined. However, there is an α probability of Type I error - namely, that apparent differences are due merely to chance.

Figure 9.3(A) represents the distribution of sample means ($\overline{Y}$) with mean equal to the hypothesized mean (μ_O) and the standard error of the sample means (s) often noted as $s\overline{y}$ in an experiment with a relatively low number of replicates. Although the control and treated sample means may be identical to the unknown population means, the small sample size has relatively large variance. A treatment effect would have to be relatively large in order to reject the null hypothesis. Figure 9.3(B) illustrates the distribution of sample means under some alternate hypothesis (μ_a) and shows the probability of making a Type II error under this alternative as a shaded area specified as ß.

Figure 9.3(C) represents a second experiment with a greater number of replicates. This reduces the sample standard error, producing a "tighter" arrangement of Y sample means. Because the sample standard error is smaller, the critical value of rejection set by α is drawn closer to the mean, thereby reducing the probability of a Type II error (Figure 9.3(D). Because Type II error is reduced, the power of the test (1-ß) (the ability to detect true differences) is increased. Thus, the power of the test increases with increasing sample size.

One risk of setting an arbitrarily small α level is that, by so doing, the researcher implies that the risk of a Type II error is inconsequential. In some experiments, the "cost" associated with a Type II error may indeed be great - even greater than that of making a Type I error. The legacy of unquestioned application of $\alpha=0.05$ is that the power of tests is often overlooked.

Note that the increase in power associated with any increase in sample size is conditional upon the true unknown value of the difference. In the example above, if the true treatment difference is very large relative to the standard errors, then increasing sample size may do little to increase the power of the test because, for such large differences, a small sample may be adequate to ensure a very powerful test. Suppose, for example, that the treatment response is so large that the value of ß associated with a given α is very small, say 0.01. Then the power of the test is 0.99 already and is effectively as big as it is going to get.

If one is 99% certain of detecting a significant treatment effect, then there is little reason to increase sample size if the only objective is an answer to the question: "Is there a treatment difference?" However another question frequently left unstated is: "Assuming there is a treatment effect, how sizable a difference do we wish to detect?" If one wishes to evaluate the difference in biomass between two treatments within 0.5 cubic meters per hectare, then this should guide determinations of sample size – *not* the question of whether or not there is a treatment difference. An experiment designed to answer the first question may be inadequate for answering the second.

CLARIFYING THE EXPERIMENTAL APPROACH FOR DESIGNED EXPERIMENTS

The trend toward designed experiments for long-term field trials, despite their costs and difficulties, is due to the emergence of forestry research as rigorous science. This trend is analogous to the one identified in ecology by Loehle [9]. Loehle observed that the scientific method, statistical rigour, and hypothesis testing are emphasized as theories and laws in ecology, and are assigned predictive and explanatory power. Users of forestry research results are demanding predictive and explanatory power that can only be provided by a rigorous science and designed experiments.

The experimental approach for designed experiments can be visualized in three phases consisting of set-up, design, and analysis [10]. Applying this to long-term field trials is the same as for any experiment, but the nature of long-term trials makes the set-up and design phases especially important. Unlike short-term laboratory studies that are repeated often to improve confidence in results, only one opportunity exists with a long-term trial. It must be set up and designed correctly the first time. After the data are collected, the analysis phase can be repeated if errors in technique or judgement are made. But errors during set-up or design phases could compromise the study entirely. During the set-up phase, explicit problem definition, identifying testable hypotheses, and over-coming psychological barriers in hypothesis testing are particularly important. During the design phase, treatment interspersion and design sensitivity often are overlooked or overestimated resulting in erroneous specification of Type I and II error probabilities.

Problem Definition

Because forestry research is mostly application of the basic sciences, research tends to seek solutions to problems identified by clients who may be individuals, or groups, or even society as a whole. Information needs of clients define what is relevant in applied science. Working with the client to reduce a problematic situation to an identifiable, researchable problem is a challenging first step in the experimental approach.

According to Andrew and Hildebrand [11], a problem exists when there is a need felt by a client. For the problem to be researchable, the need must be amenable to change as a result of the research process. In our context the problem might be stated as: "Forests must be harvested and regenerated to provide for the fibre needs of society, but the process might degrade long-term site productivity." The "client" (forest landowner) who will be served by its solu-tion has identified this as a problem. If a sustained or an increase in productivity could be a result of researching the problem, then it is reasonable to proceed.

Gordon and Bentley [12] point out that if the clients participate in setting the research goals - in defining the problems - then the clients' use of the research results actually begins at this stage. This is the first crucial step in the research process between scientists and users.

Extension of Results

An adjunct to defining the problem is to understand how broadly one wishes to extend the results. Figures 9.4 and 9.5 help illustrate the significance of this. Figure 9.4 illustrates the relationship between the frequency distribution for a natural population of interest (for example, the frequency distribution of site quality for some particular species) and that for a smaller sample of study sites chosen randomly from this population. The population distribution may be of any shape. But if sites are selected randomly, selection probability is proportional to their abundance in the natural population of site qualities. This means that samples would tend to cluster about the median value. And as the number of randomly sampled sites increases, the shape of their frequency distribution approaches that of the entire population. However, rare values near the extremes of the population are unlikely to be sampled.

What are the consequences of performing experiments on sites chosen this way? As in Figure 9.5, let's assume that treatment response is not uniform but increases with site quality for poor to moderate sites, then declines as site quality improves. Let's also assume that not all sites of a given index respond identically to treatment but that response will vary about any given site index, producing a cloud of treatment responses across the range of site index in the natural population. An experiment established on a single randomly selected site provides an unbiased estimate of the population mean, but no measure of the estimate's standard error. Furthermore, it is unlikely that a single experiment will represent the population mean precisely. A larger number of randomly established experiments will provide unbiased estimates of the population mean and standard error.

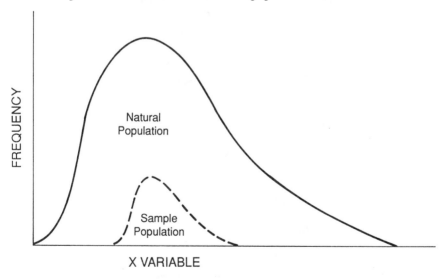

FIGURE 9.4   Relationships between the frequency distributions of large, natural populations and randomly selected samples.

Extremes of the distribution will be poorly represented and, as such, will produce poor (although unbiased) parameter estimates. Furthermore, this procedure may not provide a basis for establishing a response surface over the entire range of interest.

Results obtained from randomly selected sites will be satisfactory for estimating the unconditional expected response to some treatment. That is, the response expected in the absence of knowledge of any concomitant variable which is related to the response. If, on the other hand, site productivity is known, then the researcher may consider fixing various levels of site productivity (e.g., site index) and sampling randomly within each selected level. This would allow a proper description of the response surface across a range of site productivity values. It may be crucial for small samples that an adequate range of predictor values is represented in order to establish a statistically significant relationship between the response and site productivity. By strategically fixing levels of site productivity, one may also be able to evaluate assumptions of model form (see [13] -pages 51-54).

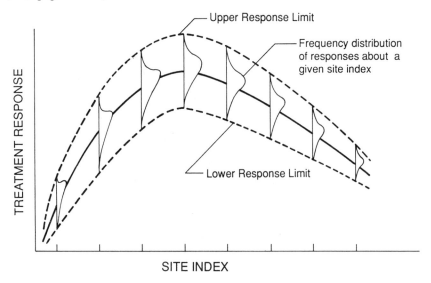

FIGURE 9.5 Sampling scheme for projecting response estimates for any value of site index. Each site index has its own distribution of possible treatment effects.

Figure 9.5 illustrates this concept. First, a variable (such as altitude, or perhaps site index) is chosen as a basis for site stratification. Second, the researcher defines the population of interest (e.g., only commercially important sites). Third, levels of the stratification variable are defined from which samples will be drawn randomly. One aim of this procedure is to adequately define the response surface over this variable of interest. Finally, study sites should be chosen randomly within each of the levels defined. This procedure allows the user to ensure that adequate representation is found across the range of sites in the population.

Mead [14] has described a variant of this approach. He suggested that potential sample sites can be regarded as "existing treatments" (or ecological or environmental treatments) in contrast to applied treatments. A five-step process is used to apply this approach:

1. Identify a large population of observational units.
2. Classify each unit according to its level of each of several existing treatment factors.
3. Further classify each level according to its level of several blocking factors. These factors, unlike the existing treatment factors, are of no direct relevance to the research question.
4. Define the applied treatments.
5. Within the blocks (of similar units), units are randomly selected to represent different existing treatments, and also are allocated levels of the applied treatments.

Note that the design suggests blocking by plot characteristics rather than by plot location on the ground. This presupposes it is possible to identify such significant blocking characteristics as soil type, site quality, or climatic conditions. It also assumes that the past treatments were assigned more or less at random to units now grouped within blocks. However, there is usually a particular reason that one treatment was chosen over another. Namely, that it was considered to be the treatment most appropriate for the site.

Assume for example that the retrospective treatments are three site preparation methods (residue retention, broadcast burning, and windrowing), and that the basis for blocking is soil type. The three observational units within the block are separated geographically. Although it is conceivable that any of the three treatments could have been assigned to any of the three units, chances are good that they were not. For example, residue retention might have been chosen because fuel loading was light on one site. On another site with heavier fuel loading, broadcast burning was the preferred treatment. And on the third, a tractor and brush rake was used to dislodge understorey shrubs and pile them along with logging slash into windrows. While all three sites may have comparable soils (normally, a sound basis for blocking), individual differences at the time of treatment (but undetectable today) determined the choice of treatments and possibly affected the outcome. This was recognized by Burkhart *et al.* [15] who, in a yield study of existing site-prepared plantations, concluded that little could be said about the effect of site preparation on site quality because site preparation methods were not assigned at random in the sample population.

To the extent that potentially confounding factors can be recognized and rationalized on pre-existing treatments within blocks, new applied treatments can be assigned and regarded as a split-plot randomized block experiment or, if several replicates of pre-existing treatments are included in the block, as a factorial randomized block experiment. In some circumstances, covariance analysis could be used instead of blocking. Covariance analysis is a technique combining aspects of regression and analysis of variance. It can be used to adjust treatment means based on some observed continuous variable (a covariate). For example, Burkhart *et al.* [15] used covariance techniques in the analysis of pine plantation yields by adjusting for differences in competition due to natural regeneration from pines and hardwoods. Often, including covariates in the analysis model helps reduce error and provides a more powerful test. A more comprehensive discussion of this technique appears in Steel and Torrie [16]. In the example from Mead [14], covariance analysis involves separating out the effect of blocking factors, the existing treatment factors, the applied treatment factors, and the interaction of existing and applied treatments. In the context of the problem under discussion here, existing treatments could be factors such as numbers of previous rotations, species, or other site characteristics of interest.

Thus, a critical step in planning the experimental approach is in considering how results are to be extended. In the first approach the emphasis was to define responses over a population. In the second approach the emphasis was to study the responses within and between different parts of the population.

Identifying Testable Hypotheses

Andrew and Hildebrand [11] suggest that a problem is best solved by identifying it clearly and posing unambiguous, testable hypotheses. Hypotheses must involve a relationship, at least implicitly, in the form of "if-then" propositions, where "if" describes the relationship between the condition and proposed result. An "if-then" proposition might be: "If forests are harvested and regenerated, then a decrease in long-term productivity will result." Hypotheses derived from the if-then proposition guide the type of data and techniques necessary for analysis.

Hypotheses are traditionally presented as null hypotheses or reference points for comparing alternatives. Null hypotheses assume that a change has not been produced by the condition of interest. For example: "Harvesting and regenerating forests cause no change in long-term site productivity." This approach of falsifying the null hypothesis is rooted in Sir Karl Popper's contention [17] that propositions or theories are scientific only if they are potentially falsifiable. In fact, Chew [18] makes a good case that the null hypothesis always is false in realistic biological experiments. Just as no two snowflakes are alike, there are physical, chemical, and biological reasons *a priori* for assuming that no two treatment means can be identical. Given enough replication, H_o will always be rejected. Chew [18] further contends that less emphasis should be placed on testing hypotheses and more attention given to interval estimation. Point estimation is sufficient for predictive purposes; however, for drawing inferences about true parameter values, interval estimation may be more informative ([19] - pages 113-115).

Hypothesis testing for the purpose of being predictive and to provide explanatory power is the norm in modern forestry research. However, scientific advances in forestry are slow compared to other life sciences. Several explanations for this, unique to forest science and other sciences dealing with long-lived organisms, were mentioned above. Another explanation, perhaps first noted by Chamberlain [20] a century ago, has its roots in human psychological limitations that apply to all scientists, but puts those studying long-lived organisms or systems at a particular disadvantage.

Chamberlain [20] contended that there is a human tendency to concentrate on one idea or hypothesis at a time followed by a subjective affection for the hypothesis that leads to confirmational bias and overall scientific inefficiency. To guard against this he suggested the method of multiple working hypotheses. The purpose was to "..... bring into view every rational explanation .... and every tenable hypothesis".

He observed that when following a single hypothesis, the mind is led towards a single explanation which may provide only a portion or perhaps none of the answer. Presenting a family of hypotheses increases one's chances of properly assigning cause-and-effect and prevents the investigator from unduly fastening his affections upon any one.

Seventy-five years after Chamberlain made his observations about the inefficiencies in conducting investigations, Platt [21] observed that certain fields of science were moving forward faster than others by "an order of magnitude." He cited molecular biology and high-energy physics as two fields making stunning progress. He contended that this rapid progress was simply a function of the use of a systematic, accumulative method of inductive inference that he called "strong inference."

Platt's strong inference consists of applying the following steps to every problem:

1. Devise alternative hypotheses;
2. Devise a crucial experiment with alternative possible outcomes, which will exclude one or more of the hypotheses;
3. Carry out the experiment so as to get a clean result;
4. Recycle the procedure making subhypotheses or sequential hypotheses to refine the possibilities that remain.

Platt's strong inference goes beyond the simple use of the working hypothesis. It involves the use of Chamberlain's multiple working hypotheses to reach out into the unknown. Any conclusion is based on multiple exclusions. Therefore, this is the best method for reaching firm inductive conclusions as rapidly as possible.

Gordon and Bentley [12] discuss how to generate these hypotheses. They note that exploring cause-and-effect relationships is usually based on systematic inquiry into the empirical evidence at hand, rather than relying on paradigms or theory. Hypotheses are therefore derived from a systematic consideration (diagnosis) of the "symptoms." They suggest the following process which will ensure the researcher considers all the possibilities.

First the researcher should ask a sequence of questions to identify possible causes to the "symptom":

1. What is (are) the symptom(s)?
2. What isn't the symptom?

3. Where is the symptom?
4. Where isn't the symptom?

5. When is the symptom observed?
6. When isn't the symptom observed?

7. How frequent, how much, how often is the symptom observed?
8. How frequent, how much, how often is it not observed?

Each pair above leads to two additional questions. What is the difference between the two? When was the difference first observed?

The eight questions are objective and the answers can come from a number of sources. Their purpose is to bring the information together before developing hypotheses and ensure these are considered in depth. Gordon and Bentley [12] suggest this diagnosis be approached methodically so as to separate the objective from the subjective steps. This should lead to better answers.

The initial steps should be to list, discuss, and revise the answers to the questions - avoid leaping to conclusions about possible causes until that part of the process has been completed. Then list the hypotheses - but do not discuss them. The hypotheses should be mechanistic and capture in testable ways the insights gained from collating the answers to the questions. This is the creative and more subjective step. Next, discuss and refine the hypotheses.

Finally, conduct tests (perhaps using other workers) to ensure all questions, etc., are complete. Ask, for example, if the hypothesis explains all of the eight initial questions. If not, it may not be tenable. In some circumstances it may not be possible to answer one or more questions and this in itself can be valuable and suggest preliminary studies may be required. It will also help to clarify the important from the trivial.

Sorting out the causes for the second-rotation decline (2-RD) referred to by a number of investigators [22, 23] provides an example of the need to carefully develop hypotheses. In most cases 2-RD was observed in the second rotation of exotic conifers grown in plantations after the removal of the indigenous forest (i.e., eucalyptus forest converted to radiata pine in Australia; native beech forests converted to pine or spruce in Europe). For a long time, the single working alternate hypothesis used implicitly for investigating the decline was:

H_a: Conifer plantations decrease soil productivity.

Not only did investigations of 2-RD suffer from the use of a single working hypothesis, a "crucial experiment" (Platt's step No. 2) with "a clean result" (step No. 3) was difficult to devise. Evans [23] was one of several investigators who recommended a more detailed look at the 2-RD phenomenon. His discussion suggests that the following alternative multiple working hypothesis might have been used from the outset:

H_a: Land clearing (removal of organic matter) decreases productivity.
H_a: Loss of soil structure and aeration porosity during the process of conversion decrease productivity.
H_a: The first conifer rotation was subsidized by the previous non-conifer vegetative system.

Note these are testable mechanistic hypotheses. A systematic execution 25 years ago of simultaneous crucial experiments testing these and other alternative hypotheses might have propelled progress toward explaining the 2-RD phenomenon.

Psychological Barriers in Hypothesis Testing
In addition to the inherent inefficiency of testing a single hypothesis for inductive exclusion, Chamberlain [20] discussed several psychological factors that reduce efficiency of scientific investigations. In a recent paper, Loehle [9] provided

names and a modern context for Chamberlain's factors. "Confirmational bias is a tendency to try to confirm one's theory, or to not seek out or use disconfirming evidence." When testing a single hypothesis there is a tendency to seek out facts that support or reject the hypothesis. Confirmational bias also occurs when only portions of large data sets are used to confirm ideas while contradictory data are discounted. This is similar to fitting models to data without validation rather than verifying the model with new data.

"Theory tenacity" is another factor that hampers problem solving. Essentially, it is an emotional commitment to basic assumptions. It occurs particularly when a favourite idea is tested as a single operating hypothesis and becomes more tenacious with expended time and energy. Forest scientists are particularly susceptible to theory tenacity because of the unusually long time and large levels of resources needed for long-term problems. Under such circumstances hypotheses that are difficult to test, such as "pine monocultures degrade site productivity," or "clearcutting reduces biodiversity" become "ruling theories" [20] for which "facts are sought" to support the theory.

Based on his experiences in ecology, Loehle [9] concludes that psychological factors such as confirmational bias and theory tenacity constrain scientific progress when time-scales are long and the system is "noisy" and complex. Long-term forestry field trials are usually noisy (highly variable), complex and have many uncontrolled variables. Using Chamberlain's multiple working hypotheses, Platt's strong inference techniques and Gordon and Bentley's structured approach to hypothesis development, reduces these psychological factors and should be standard procedures for all long-term field trials.

Problems of Experimental Execution
After the problem is defined and the alternative hypotheses are constructed, developing an appropriate experimental design is the next step. The design is simply the experimental layout [10] or the logical structure of the experiment [24]. Design, layout, and structure usually include consideration of:

1. number and kinds of treatments,
2. nature of the experimental units,
3. responses to measure,

4. treatment assignment,
5. number of replications of each treatment,
6. arranging experimental units in space and time, and
7. error control.

This discussion then deals with the execution of design elements in field environments "so as to obtain a clean result" [21]. According to Hurlbert [8], successfully executing a designed experiment depends as much on the investigator's artistry, insight, and good judgement as on technical skill.

Treatments Selection

Often, long-term experiments on harvesting impacts will include harvesting or simulated harvesting activities (e.g., removal of organic layers, compaction) and methods of subsequently ameliorating the site. The anticipated size of response should be one guide to selecting treatments. Woollons [25], in considering long-term trials with several rates of fertilizer, demonstrated the utility of restricting treatment choice to those expected to produce differences several times larger than the standard error of the treatment means. Such treatments have major advantages in long-term experiments. First, they are likely to produce sizable, statistically significant, long-lasting responses. Treatments producing smaller responses may have little relevancy in the long run. Second, extreme treatments alter site conditions enough to offer us important insights into major process mechanisms controlling productivity. We recommend that where possible, long-term experiments should include treatments at low, intermediate, and extreme levels so as to encompass the range of conditions possible under present or future management and to define comprehensive response surfaces mathematically. We suggest that the following questions be considered in selecting treatments for long-term site productivity research:

1. What are the most important factors controlling productivity on the study site?
2. What treatments are available for altering these factors?
3. Do treatments merely reflect popular operational practices that may become obsolete before the experiment is completed?

4. Will findings fundamentally advance our knowledge of the processes controlling site productivity?

Treatments that directly and dramatically alter such major factors of productivity as temperature, moisture, and nutrient availability can lead to findings of great significance. For example, in the permafrost-dominated *Picea mariana* region of interior Alaska, Van Cleve *et al.* [26] used heating tapes to raise forest floor temperatures by 8-10°C above ambient. Responses to this extreme treatment included higher rates of organic matter decomposition, nutrient mobility, and photosynthesis which translated to a major gain in net primary productivity. Such extreme treatments are particularly valuable if applied factorially because treatment interactions may be as important as the main treatment effects. Furthermore, factorial experiments minimize the risk of confounding. Such is the rationale for the core experimental design used by the U.S. Forest Service in its national network of long-term soil productivity (LTSP) experiments [27]. In this study, the full factorial arrangement of core treatments separates the influence of soil aeration porosity, site organic matter, competition control, and site differences on biotic diversity and net primary productivity.

Definition of treatments also ties into the choice of the control(s). Are unharvested treatments required? Or should the control be harvested plots with minimal site disturbance? Perhaps "best current practices" is a better standard for comparing the effects of new treatments. Any or all of these definitions of control treatments may be appropriate, depending on the definition of the problem. Including more than one type of control favours strong statistical inference by creating multiple alternative hypotheses and subhypotheses that can be tested systematically to reach unambiguous, progressive conclusions [21]. Should the choice be "more treatments or adequate replication", the choice must be for fewer treatments. However, they must be chosen well.

Plot Size

Plot size has unusual significance in long-term experiments. Small trees become larger. Thinning or natural mortality produces fewer measurement trees over time. Our aim over time must always be to have adequate numbers of individuals responding continually to the specific treatment of interest. The gross area of individual plots required for an experiment is a function of tree spacing, surround width, the nature of the treatments themselves, site, species or species mix, silviculture, and, above all, the length of the trial. The topic was reviewed for plantations by Mead *et al.* [28], who provided the following guidelines:

1. For many trials in reasonably uniform plantations, a minimum size of inner measurement plot (or sub-plot if a split-plot design) size would be 20 x 20 m. Each plot should contain between 30-50 inner measurement trees at time of establishment, which for many species should not lead to problems with mortality until a third of the way through a sawlog rotation. After final thinning to stockings of 300 stems/ha, 12 trees should remain per 0.04 ha plot.

2. For final crop stockings below 300 stems/ha and for high site or crop variability, experimental plots larger than 0.04 ha are needed. For instance a great deal of tree to tree variation can occur in natural stands with irregular spacing and some harvesting treatments may alter the site variability. On the other hand use of clonal material or treatments which result in more stand uniformity (for example ripping) could perhaps allow a smaller plot to be used safely.

3. Surround or buffer width requirements vary with species, site, and treatment and the length of the trial, but should generally be not less than 10 m. Root spread measurements should be assessed to provide a useful guide to the minimum surround width on a particular site. Root spread is often greater on less compact and drier soils; in semi-arid zones they may extend 20 m [13]. Long-term studies of large trees will require a wider buffer than those among studies of shorter duration. Wide surrounds are more important where qualitative growth data are required rather than just testing for significant treatment effects (see [14]).

4. Surrounds should be treated. If there is a need for process studies, the inner plot or surround may need to be increased in size.

5. Experimental layout should utilize soil surveys, aerial photographs and other relevant historical information on the previous crop.

The number of stems to be measured on a given plot can have a substantial influence on variance of the sample mean, as was illustrated by Mead *et al.* [28], who completed a study of three different sites to determine the effect of subsampling on the experimental error (Figure 9.6), and found that halving the number of trees roughly doubled the variance of the sample mean.

Minimal Measurements

Consideration also needs to be given to the base measurements to be taken on the trees. For fast growing crops at least, the diameter at breast height over bark of all trees and heights of a sample of trees in each plot should be measured often enough to describe growth patterns accurately [28]. The close relationship between growth rate and leaf area suggests that direct or indirect measurements of the latter should be taken periodically. Changes in form of pole-sized trees need be assessed only where basal area response to treatment is substantial (say >30%). These and other stand characteristics (e.g., branching habit, wood properties) usually need be collected on only a subsample of trees and at infrequent intervals (perhaps in association with thinning, unless thinning is biased towards a particular class of trees). Periodic assessments should be made of weeds and

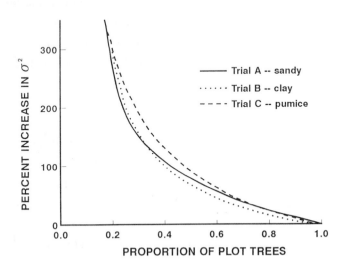

FIGURE 9.6 Effect of subsampling on sample variance ($s\bar{y}^2$) of the sample mean $\bar{y}$, in three trials differing in tree stocking, age, and soil type. (Source: [28]).

damaging agencies. In addition to productivity of trees alone, production of all vegetation will give a more comprehensive measure of total site productivity.

Reducing Errors

The scientist's objective is to make the experiment powerful, avoid systematic error (experimenter bias), and minimize experimental error. Normally, replication and randomization helps to minimize systematic error, but these features are not always enough in field trials. Randomization eliminates investigator bias, but it does not necessarily ensure that assignment of treatments is not in some way (unknown to the researcher) related to some other confounding factor.

Testing hypotheses on responses to silvicultural treatments that ultimately affect rotation-length yields requires large spaces, high costs, and long time periods. These requirements make studies especially prone to errors in testing hypotheses. The "large space" requirement makes field studies especially susceptible to Type I error (rejection of the null hypothesis when in fact it is true). The "high cost" requirement makes them susceptible to Type II error (failure to reject the null hypothesis when it is false). And the "long-term nature" of forestry field experiments makes them susceptible to what we refer to as "Type III error" (correctly rejecting the null hypothesis, but for the wrong reason). Studies of site preparation effects on the rotation-length yield of southern pine plantations illustrate each of these error opportunities.

Type I error

Sometimes plots need to be large in size to overcome high microsite variability or to enable large equipment to be used. With the usual complete block design, block size is then likely to be large. Changes or gradients in soil productivity, vegetation, and microclimate are then likely to occur within the block which would increase the experimental error. This is one reason for trying to keep block size to a minimum, for using incomplete block designs, and for employing covariates to control the error.

Randomization eliminates experimenter bias, but it does not ensure interspersion of treatments across site quality gradients. For example, Figure 9.7 illustrates a hypothetical random arrangement of four treatments in three replicate blocks. By chance, the "harrow" treatment was assigned each time to the highest position along a site quality gradient that may or may not have been apparent to the investigator.

Within large blocks, site quality gradients can be caused by a variety of factors such as water table level, surface erosion, organic matter content, or depth to an impermeable soil layer, to name just a few. Investigators should familiarize themselves with the character of the study blocks and the individual experimental units and ensure that treatment interspersion on site quality gradients occurs. Cox [29] advised applying additional randomization iterations until a layout with reasonable interspersion is obtained. This may help ensure that treatment effects are not confounded.

Type II error

Because of their size and duration, long-term field studies for testing rotation-length silvicultural treatment responses are costly. In planning field

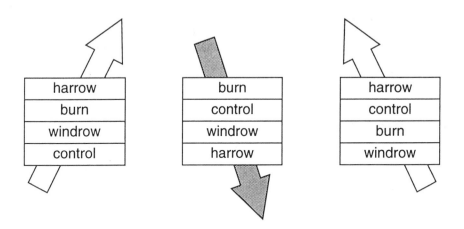

FIGURE 9.7. Hypothetical experimental layout of four site preparation treatments in three replicate blocks. Arrows depict a site quality gradient that could compromise specification of Type I error probability, given that the "harrow" treatment is consistently on the high end of the gradient.

studies, the cost often is the last component calculated after multiple hypotheses, treatments, and response variables have been planned. It seems that the easiest way of reducing the cost of a too-costly experiment is to reduce the number of planned replications. If an experiment has been correctly designed, however, eliminating replications is never reasonable because the number of replications will be based on a pre-determined level of testing power.

The power of a test is the probability of rejecting H_o given H_a is true. Power is denoted statistically as 1-ß. Maximizing the probability of rejecting the H_o when it is false and the H_a is correct is the process of maximizing 1-ß or the power of the test. The larger the sample or the larger the number of replications, the more powerful the test will be. The required testing power, or number of replications needed, depends on the magnitude of response deemed important, the amount of variation in the population being measured, the ability to control the precision of estimated parameters by experimental design and the acceptable level of risk of making Types I and II errors. All of these are calculated design steps that should be made before executing the experiment.

Error variance of the estimate may be reduced by judicious blocking, including using incomplete block designs, adjustment with measurements from neighbouring plots and the use of suitable concomitant information as covariates. These methods may be used in combination if required.

The use of single or multiple covariates which reflect conditions prior to treatment is often preferable to blocking and also may be helpful in explaining treatment effects. Experience has shown that covariance often is more efficient and therefore overrides any block effects in later-age fertilizer trials [30, 31, 32]. Woollons [25, 32] has found that analysis of covariance on a mean tree basis, then converting these means to the equivalent per unit area values, is preferable to analysis of plot totals, where there is a range of stockings.

Therefore, the benefits of blocking should be considered in each individual case and, in addition, it is advisable to measure a range of variables that could be used later as possible covariates. For example, growth of the previous crop and detailed soil or site measures, collected prior to the imposition of treatments, should be assessed on an individual plot basis. These would require the plots to be laid out prior to harvesting. The use of split plot or strip plot layouts also has an influence on the power of the test. Split plot layouts may have practical advantages when looking at harvesting impacts, even though randomisation is restricted. In this context the harvesting "methods" usually would be represented in the main plots with perhaps minor variants or methods of alleviating their impact as sub-plots. The nature of some harvesting treatments would lend themselves to large, long plots. In the LTSP study of the U.S. Forest Service, main effect treatments (compaction, organic matter removal) are applied to 0.4 ha plots, while subtreatments (weed control, no weed control) are applied as split plots of 0.2 ha each [27]. To be successful the anticipated differences between the main-plot treatments should be more marked than the sub-plots as there is usually lower power in the test of main-plot treatments. Care is required to ensure a reasonable number of degrees of freedom for the main-plot error term.

Sometimes it may be desirable to reduce the power of detecting differences among sub-plot treatments and increase the power of studying interactions. The strip-plot (or criss-cross) design would achieve this aim while also allowing large long plots to run in two directions [33].

The number of replications, then, is a direct function of the power of testing a hypothesis that should not be compromised. In most long-term productivity field trials, where large plots make up the experimental unit and their treatment is expensive, the tendency is to treat as few as possible - sometimes at the risk of compromising the entire study. Furthermore, in long-term studies there is always the risk of losing some plots.

Type III error
We coin as "Type III error" one in which a false H_o is correctly rejected, but for the wrong reasons. Often, this occurs in long-term field studies when a new hypothesis or question is superimposed on an existing study that was not designed to adequately test the new hypothesis. Even when an investigator designs a study for multiple testable hypotheses, new questions invariably arise with time. The temptation is to use the study to answer these new questions even though the design may be inadequate.

Type III error is illustrated by several studies on site preparation techniques for establishing southern pine plantations in the USA and maximizing their final yields. These studies began during the 1960's when foresters observed superior tree growth in planted abandoned agricultural fields compared with tree growth on planted clearcuts [34]. Old fields had less woody vegetation, less coarse woody debris, less soil organic matter, and less topsoil. Cut-over sites were prepared by shearing, windrowing, burning, and harrowing in order to simulate old-field conditions. Assessments of plantation establishment success, usually made at about age 3 years, confirmed in many foresters' minds that "creating old-field conditions" greatly enhanced tree growth and overall condition of pine plantations. An example of this early response is shown by data from Burger and Kluender [4] in Figure 9.1. Survival of planted trees usually was better on "created old-fields" (although stocking usually was adequate on cut-over sites) and tree height and volume often were twice that of trees on cut-over sites at this early age. The conclusion was that creating old-field conditions improved early growth.

This general conclusion is a qualitative example of a Type III error - that of correctly rejecting the H_0 that windrowing has no effect on early tree growth, but incorrectly concluding that it was due to creating old-field conditions. In fact, the observed response was mostly a function of herbaceous and woody weed control which reduced competition for soil moisture [34]. Removal of organic matter and topsoil as part of creating the old-field condition ultimately reduces soil fertility and tilth, which may lead to declines in soil productivity as alluded to by the stand response data in Figure 9.1. Attributing growth declines of red spruce in the northeastern U.S. and of loblolly pine in the southern U.S. to causes other than normal senescence or to weed competition also may illustrate Type III errors [35].

In the 1970's and '80's, with the advent of new site preparation machines and herbicides, several long-term field studies were designed and installed to test site preparation treatment effects on the growth and yield of pine plantations (e.g., [36]). These studies were well replicated and provide definitive tests of hypotheses dealing with treatment effects on tree response. In all cases, the investigators went beyond interpreting tree growth treatment effects and ascribed cause and effect by correlating treatment-induced soil changes with tree growth response. These interpretations are attempts to answer questions that the studies were not designed to answer.

For example, data from the study by Stransky *et al.* [36] are presented in Table 9.1 where several site preparation treatments were compared for their effects on tree growth. Coincidentally, a gradient in organic matter was created. Along with the gradient in organic matter, gradients in other soil chemical properties also were created (Table 9.1). Tests of colinearity among these soil properties probably would prove significant, meaning that any or all of the soil properties correlated with tree growth could be causative or deterministic. For that matter, some soil property not measured, but colinear with organic matter, could be the causative agent of the difference in tree growth among treatments. This is an example of how Type III errors are committed. Because of the long duration of some field studies, post facto hypotheses are superimposed on existing designs that are inadequate for a proper test. Although informed speculation of cause and effect processes in such cases are appropriate, investigators must be conscious of the risks of committing Type III errors.

TABLE 9.1
Influence of site preparation treatments on an 8-year-old stand and 5-year soil
properties in a loblolly pine plantation (modified from [36]).

Site treatment	Pine growth				Concentrations in surface soil				
	Surv. (%)	Height (m)	dbh (cm)	Volume (m³/ha)	OM (%)	P	K	Ca	Mg
							(kg/ha)		
Control	57a*	6.1a	6.9a	12.2a	2.6a	1.5a	67a	387a	83a
Slash burning	82b	6.0ab	8.1b	22.5a	2.3ab	1.4a	42a	390a	64a
KG blading & windrowing	95b	6.2b	9.6c	34.7b	2.0b	1.3a	40a	289a	57a

* Means within rows followed by the same letters are not statistically different at $\alpha = 0.05$.

MAKING LONG-TERM STUDIES TIMELESS

Minimizing Hazards and Sources of Confusion

With a clear problem statement, a testable hypothesis, sound design, and flawless execution, an investigator may feel that problem resolution is merely a matter of time. To the contrary, long-term studies require continuous nurturing. Long-term field trials, by their very nature, are prone to all kinds of unwanted intrusions. Vigilance is needed throughout the study to prevent chance events from compromising the experiment.

During the experiment, variation in weather, equipment, and operators can confound treatment responses. Sometimes it is advisable that trials should be installed one complete block at a time so that variation caused by these and many other factors is partitioned as a block effect rather than being confounded with treatment.

After execution, protection must be made against such intrusions as pest infestations, wildfire, vandalism, hurricanes, and other natural and artificial disasters unrelated to treatment. Fences, firelines, and remote locations are usually worth their cost. When planning the design, the consequences of missing values should be considered. Additional replications may be prudent depending upon the risks of unwanted intrusions.

Weed control, thinning, and other tending in long-term trials can also pose problems. Mead *et al.* [28] recommended that weeds and biotic damage should be controlled over the entire trial, unless they are an integral part of the experimental treatments. Powers *et al.* [27, 35] argue that weed and tree growth both are components of productivity and that their effects are not necessarily additive. Consequently, the U.S. Forest Service LTSP study is designed to elucidate weed and non-weed effects on site productivity. Care should also be taken to record the impact of damaging biotic or abiotic agents, or, alternatively, to exclude their influence on some plots if the aim is to evaluate their effects. If thinning is part of the prescription, all replicates for a given treatment should be thinned at the same time. But the timing of thinning should be governed by biological principles, rather than by a rigid schedule applied to all treatments.

Designing Today's Field Experiments for Long-term Relevance

To the extent possible, a useful rule is to concentrate research on tomorrow's conditions and opportunities and not necessarily the current problem as defined by a forest manager. Often this is too restrictive.

However, in the area of sustainability and harvesting effects, many of today's fundamental problems will be problems 30 years from now. But if the hypotheses are poorly stated there is a chance that they may become obsolete; that is, the experiment will answer questions that no one will be asking in the future. It is critical that, as time goes by, the hypotheses remain relevant. One way to ensure this is by defining the hypotheses in mechanistic terms rather than say, by current harvesting technologies. Over this time span it is not wise to rely on management intentions. Rather than compare different logging techniques or equipment, the comparison should be on the basic impacts of harvesting, such as compaction, organic matter changes, effects on weeds, and so forth.

A further safeguard is to use experimental designs, sizes, and treatments that would allow posing additional hypothesis of considerations or problems that arise after a study has started. This would be a case of superimposing a retrospective question on an existing designed experiment.

All long-term field trials should be made as amenable as possible for retrospective use. Field trials should be planned to answer scientific questions not yet envisioned by management. Opportunities for good retrospective studies on designed field trials may be a simple matter of luck, but good trials will be well replicated, have treatment units large enough to accommodate additional treatments in a split plot arrangement, and sites will have been thoroughly characterized so that critical properties can be used for baseline or covariate factors. If possible, soils should be described, sampled, and archived for future comparisons.

Two other related aspects need consideration in making the experiments timeless. These are to always include a broad spectrum, including extremes, in the treatments, and to avoid making early judgements on the answer. By including treatment extremes we will obtain a better prediction of the trends in the treatment effects, and will be in a position to interpolate for new management practices. We will be less likely to be caught out by the unexpected. Furthermore, as we have emphasised before, early judgement of the results also can lead to wrong conclusions.

An example of both these points and their inter-relationship may be seen in the Tikiteri agroforestry trial near Rotorua, New Zealand. This trial is now two-thirds the way through a rotation. When originally conceived, the objective was to look at the opportunities for combining grazing with wood production by planting *Pinus radiata* trees at low stockings onto pasture. At that time, most managers and even legislative constraints would have suggested that planting normal forests on fertile pastures was not a management option. Fortunately, by good design, the trial included a higher stocking, typical of a normal plantation, where there would be a final crop of 400 stems/ha. For many years the researchers suggested that the optimum stocking was 100 to 200 stems/ha. Recently, as trends in growth characteristics have become more apparent, this result has been seriously questioned; the high stocking regime is now seen to have considerable merit over the low stockings [37]. Furthermore, there are now fewer constraints and less reluctance to have plantations or woodlots on many of these sites.

Sustaining Long-Term Studies

The effect of harvesting on long-term productivity is a problem requiring long-term study due to the long life cycle of forest trees. The study would be expected to continue for over a half a rotation and probably for the stand's life cycle. Administrative mechanisms and commitments must be secured for any productivity project lasting more than a few years.

At the very least, long-term field studies must have minimal maintenance to have any hope of achieving their potential. This requires many kinds of support. Recently, Powers and Van Cleve [38] sought the views of an international group of colleagues who had been successful in establishing and maintaining long-term field sites. They concluded that all successful programs were founded on issues of continuing social relevance, such as soil protection, clean and abundant water, or sustained wood production. Such "timeless values" stand as broad, guiding paradigms to enlist the support of both the public and policy-setting politicians. Also, they provide an umbrella to encompass critical studies that address hypotheses central to the guiding paradigms. As emphasised above, hypotheses should be cleanly stated and the experimental design sufficient to judge the validity of multiple alternative outcomes. Preferably, treatments should be bold enough to trigger major changes in response processes, partly because this is how we learn how ecosystems function in respect to change.

Central to all of this is a philosophical commitment of both scientists and administrators to a program's success. Invariably, successful programs of long-term studies have at their core a small nucleus of individuals who are dedicated to the idea of making it work. Guaranteed maintenance support is critical. With this as a baseline "security net," dedicated scientists and administrators are free of concerns for survival. If a worthy study is protected from year to year, bright minds can turn their energies to seeking other sources of funding for advancing the work. Particularly attractive are studies with multidisciplinary elements. Not only do they promise major scientific insights and gains through the synergism of collaboration, but on a practical note they reach into broader areas of interest and support.

SUMMARY

Addressing questions of long-term productivity demands the long-term commitment of individuals, capital, and land. Philosophical, administrative, and physical commitment are vital, but the investments need not be unusually large. If the work is to have lasting value, it must be founded on sound principles of experimental design with a clear understanding of the hypothesis to be tested, the purpose of the experiment, the target population, assumptions and sources of bias, the statistical model, and how the findings will be extended. Long-term study designs should include multiple alternative mechanistic hypotheses that can be tested definitively the first time around. All of these elements are fundamental to the modern scientific method.

LITERATURE CITED

1. Libby, W.J., Stettler, R.F., and Seitz, F.W. Forest genetics and forest tree breeding. *Annual Review of Genetics*, 1969, **3**, 469-494.

2. Dyck, W.J. and Mees, C.A. (Eds.) *Research Strategies for Long-term Site Productivity*. Proceedings, IEA/BE A3 Workshop, Seattle, WA, August 1988. Forest Research Institute, Rotorua, New Zealand, FRI Bulletin No. 152, 1989, 257 p.

3. Burger, J.A. and Powers, R.F. Field designs for testing hypotheses in long-term site productivity studies. In: *Research Strategies for Long-Term Site Productivity*. Proceedings, IEA/BE A3 Workshop, Seattle WA, August 1988. (Ed.) W.J. Dyck and C.A. Mees. IEA/BE A3 Report No. 8. Forest Research Institute, Rotorua, New Zealand, FRI Bulletin No. 152, 1991, pp. 79-105.

4. Burger, J.A. and Kluender, R.A. Site preparation - Piedmont. In: *Symposium on the Loblolly Pine Ecosystem (East Region)*. (Ed.) R.C. Kellison and S.A. Gingrich. USDA Forest Service and North Carolina State University, Raleigh, NC, 1982, pp. 58-74.

5. Fox, T.R., Morris, L.A., and Maimone, R.A. The impact of windrowing on the productivity of a rotation age loblolly pine plantation. In: Proceedings, Fifth Biennial Southern Silviculture Research Conference. USDA Forest Service General Technical Report, 1989, pp. 133-140.

6. Hicks, C.R. *Fundamental Concepts in the Design of Experiments*. Holt, Rinehart, and Winston, New York, 1974.

7. "STUDENT". The probable error of a mean. *Biometrika*, 1908, **7**, 1-26.

8. Hurlbert, S.H. Pseudoreplication and the design of ecological field experiments. *Ecological Monographs*, 1984, **54(2)**, 187-211.

9. Loehle, C. Hypothesis testing in ecology: psychological aspects and the importance of theory maturation. *Quarterly Review of Biology*, 1987, **62(4)**, 397-409.

10. Lentner, M. and Bishop, T. *Experimental Design and Analysis*. Valley Book Company, Blacksburg, VA 24060, 1986, 565 p.

11. Andrew, C.O. and Hildebrand, P.E. *Planning and Conducting Research*. Arno Press, 3 Park Ave., New York, NY 10016, 1976, 116 p.

12. Gordon, J.C. and Bentley, W.R. *A Handbook on the Management of Agroforestry Research*. Winrock International, USA, and Oxford & IBH Publishing Co. Pvt. Ltd., New Delhi, 1990, 72 p.

13. Draper, N.R. and Smith, H. *Applied Regression Analysis*. Wiley, New York, 1980.

14. Mead, R. Designing experiments for agroforestry research. In: *Biophysical Research for Asian Agroforestry*. (Eds.) M.E. Avery, M.G.R. Cannell, and C. Ong. Winrock International, USA, and Oxford & IBH Publishing Co. Pvt. Ltd., New Delhi, 1991, pp. 3-20.

15. Burkhart, H.E., Cloeren, D.C., and Amateis, R.L. Yield relationships in unthinned loblolly pine plantations on cutover, site-prepared lands. *Southern Journal of Applied Forestry*, 1985, **9**, 84-91.

16. Steel, R.G.D. and Torrie, J.H. *Principles and Procedures of Statistics*, (2nd Ed). McGraw-Hill, New York, 1990.

17. Popper, K.R. *The Logic of Scientific Discovery*. Basic Books, New York, 1959.

18. Chew, V. Testing differences among means: Correct interpretation and some alternatives. *Horticultural Science*, 1980, **15**, 467-470.

19. Box, G.E.P., Hunter, W.G., and Hunter, J.S. *Statistics for Experimenters: An Introduction to Design, Data Analysis and Model Building*. Wiley, New York, 1978.

20. Chamberlain, T.C. The method of multiple working hypotheses. *Science*, 1890, **15**, 92. (Reprinted in *Science*, 7 May 1965, **148**, 754-759).

21. Platt, J.R. Strong inference. *Science*, 1964, **146**, 347-353.

22. Keeves, A. Some evidence of loss of productivity with successive rotations of Pinus radiata in the south-east of South Australia. *Australian Forestry*, 1966, **30**, 51-63.

23. Evans, J. Plantations: productivity and prospects. *Australian Forestry*, 1976, **39**, 150-163.

24. Fisher, R.A. *Statistical Methods for Research Workers*. Hafner Publishing Company, New York, 1950.

25. Woollons, R.C. Analysis and Interpretation of Forest Fertilizer Experiments. PhD thesis, University of Canterbury, Christchurch, New Zealand, 1988, 256 p.

26. Van Cleve, K., Oechel, W.C., and Hom, J.L. Response of black spruce (*Picea mariana*) to soil temperature modification in interior Alaska. *Canadian Journal of Forest Research*, 1990, **20**, 1530-1535.

27. Powers, R.F., Alban, D.H., Ruark, G.A., and Tiarks, A.E. A soils research approach to evaluating management impacts on long-term productivity. In: *Impact of Intensive Harvesting on Forest Site Productivity*. Proceedings, IEA/BE T3/A3 Workshop, South Island, New Zealand, March 1989. (Eds.) W.J. Dyck and C.A. Mees. IEA/BE T6/A6 Report No. 2. Forest Research Institute, Rotorua, New Zealand, FRI Bulletin 159, 1990, pp. 127-145.

28. Mead, D.J., Whyte, A.G.D., Woollons, R.C., and Beets, P.N. Designing long-term experiments to study harvesting impacts. In: *Long-Term Field Trials to Assess Environmental Impacts of Harvesting*. Proceedings, IEA/BE T6/A6 Workshop, Florida, February 1990. (Eds.) W.J. Dyck and C.A. Mees. IEA/BE T6/A6 Report No. 5. Forest Research Institute, Rotorua, New Zealand, FRI Bulletin 161, 1991, pp. 107-124.

29. Cox, D.R. *Planning of Experiments*. Wiley, New York, 1958.

30. Whyte, A.G.D. and Mead, D.J. Quantifying responses to fertilizer in the growth of radiata pine. *New Zealand Journal of Forestry Science*, 1976, **6**, 431-442.

31. Woollons, R.C. Importance of experimental design and plot maintenance in forest field experiments in Australasia. *Australian Forest Research*, 1980, **10**, 71-82.

32. Woollons, R.C. Problems associated with analyses of long-term fertilizer X thinning experiments. *Australian Forest Research*, 1985, **15**, 495-507.

33. Cochran, W.G. and Cox, G.M. *Experimental Design*, (2nd Ed). John Wiley & Sons, Inc., NY., 1964, 617 p.

34. Haines, L.W., Maki, T.E., and Sanderford, S.G. The effect of mechanical site preparation treatments on soil productivity and tree (*Pinus taeda* L. and *P. elliottii* Engelm. var. elliottii) growth. In: *Forest Soils and Forest Land Management*. Proceedings, Fourth North American Forest Soils Conference, Montreal, Quebec, August 1973. (Eds.) B. Bernier and C.H. Winget. Laval University Press, Quebec, 1975, pp. 379-395.

35. Powers, R.F., Alban, D.H., Miller, R.E., Tiarks, A.E., Wells, C.G., Avers, P.E., Cline, R.G., Fitzgerald, R.O., and Loftus, N.S., Jr. Sustaining site productivity in North American forests: problems and prospects. In: *Sustained Productivity of Forest Soils*. Proceedings, 7th North American Forest Soils Conference. (Eds.) S.P. Gessel, D.S. Lacate, G.F. Weetman, and R.F. Powers. Faculty of Forestry, University of British Columbia, Vancouver, B.C., 1990, pp. 49-79.

36. Stransky, J.J., Roese, J.J., and Watterston, K.G. Soil properties and pine growth affected by site preparation after clearcutting. *Southern Journal of Applied Forestry*, 1985, **9**, 40-43.

37. Knowles, L. Why we shouldn't grow 100 s/ha. *New Zealand Tree Grower*, 1992, **13(2)**, 17.

38. Powers, R.F. and Van Cleve, K. Long-term ecological research in temperate and boreal perennial ecosystems. *Agronomy Journal*, 1991, **83**, 11-24.

CHAPTER 10

REVIEW OF MEASUREMENT TECHNIQUES IN SITE PRODUCTIVITY STUDIES

H. VAN MIEGROET
Department of Forest Resources, Utah State University
Logan, UT 84322-5215, U.S.A.

D. ZABOWSKI
College of Forest Resources, University of Washington
Seattle, WA 98195, U.S.A.

C.T. SMITH
New Zealand Forest Research Institute, Ltd.
Private Bag 3020, Rotorua, New Zealand

H. LUNDKVIST
Swedish University of Agricultural Sciences
S-750 07 Uppsala, Sweden

INTRODUCTION

Site productivity is a concept that relates (1) to the growth environment created by a combination of physical, chemical, biological, and environmental site characteristics, and (2) to the growth response of the target crop to that particular environment. It is regulated not only by the chemical and biological processes taking place in the soil and forest floor but also by the flow of material between the growth medium and the plants and ultimately by the ability of the trees to convert these inputs into biomass production.

Field studies of the effect of management practices on site productivity should be designed to ascertain (1) the extent to which site properties are changed as a result of the management intervention, and (2) the extent to which these changes affect ecosystem processes that determine tree growth rates. In this chapter we intend to provide a basic measurement framework that can be used in the design and implementation of field studies. This information should prove useful to the applied forest scientists and field managers interested in assessing the nature and extent of management effects on site productivity as well as to basic scientists interested in investigating the underlying mechanisms influencing productivity.

The objective of this chapter is to describe and evaluate techniques that are used to measure static and dynamic ecosystem parameters that may be affected by management practices. They form the basis for explaining and/or predicting the effect of intensive management on ecosystem processes and ultimately on site productivity. In some instances, such measurements may directly quantify a key process affecting site fertility (e.g., nutrient release from mineralization); in other cases, they may characterize a changing environment that indirectly influences site fertility (e.g., the effect of aeration on microbial populations).

Although comprehensive, this review of measurement techniques is not meant to be all-inclusive, and the reader should not expect a detailed description of each measurement protocol, as those can be found in specific reference manuals. Rather, this chapter provides a critical overview of a number of methods that are available for characterizing site quality and forest ecosystem processes, the strengths and limitations of a given technique, and general recommendations for various approaches.

Conceptual Framework

The key ecosystem processes affecting long-term site productivity have been discussed earlier in this volume (Chapter 5) and are touched upon only briefly here. The soil is a source of nutrients and water for the plant, the availability of which is determined by biogeochemical processes and physical characteristics of the soil. Actual uptake of water and nutrients by the target crop depends on climatic factors as well as on the health and condition of the plant, on rhizosphere processes, and on competitive relationships between plants and microorganisms in the soil. Biomass production and growth, the ultimate parameters of interest,

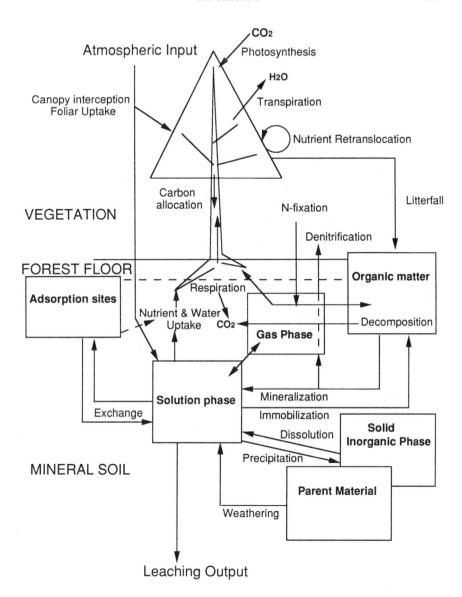

FIGURE 10.1 Schematic representation of the interactions between the solid, solution, and gaseous phases in soils and forest floor and between trees and the growth medium that influence site productivity.

are influenced by the physiological potential of a particular species to respond to above- and belowground growth conditions. The complexity of the interactions between the solid, solution and gaseous phases in the soil and forest floor and between the growth medium and the tree are represented schematically in Figure 10.1. It illustrates that a change in the pool size or the flux rate between eco-system components does not occur in isolation but may trigger a series of other changes within that ecosystem that may ultimately affect site productivity.

Management-induced changes in site productivity occur through changes in nutrient supply and cycling patterns, in water availability and flux rates, and in root and aboveground plant processes. Hence, an evaluation of intensive man-agement impacts on ecosystem processes or the determination of future produc-tivity changes depends on the quantification of changes in these processes or their controlling factors. The discussions in this chapter will focus on methods and measurements to characterize changes in the soil physical environment and to evaluate biogeochemical processes and plant responses. The nutrient cycling processes considered will include: nutrient inputs from external and internal sources; soil biological processes and their effect on nutrient transformations; soil chemical reactions as they affect nutrient availability and as they are in turn affected by physico-chemical soil conditions; and nutrient losses in dissolved, particulate, and gaseous form and associated with biomass removal. The latter part of the chapter reviews techniques to evaluate responses of trees to changes in resource availability following intensive site management practices.

Our discussions will primarily focus on those processes and site parameters that can to some extent be controlled by management practices, and less on those that are global (e.g., climatic change) or stochastic (e.g., hurricane, insect infesta-tion) in nature. Soil hydrology will not be explicitly covered in this chapter, although hydrologic flowpaths and water residence times play a key role in nutri-ent transport through the soil. Issues such as the effect of management practices on wildlife likewise fall outside the scope of this chapter.

Measurement Considerations

The techniques for evaluating ecosystem responses described in this chapter are merely intended to offer guidance to the scientists about to initiate field trials and may suggest more possible measurements than are strictly needed or appropriate for the study in question. The temporal and spatial scale and the intensity of the measurements are largely governed by the specific objectives of the study. Field protocols can be limited to a few key measurements if a broad or cursory characterization of management impacts is intended, whereas site-specific hypothesis testing generally entails more intensive and detailed observations within the confines of a relatively small area. Likewise, the development of general predictive models of harvesting effects across a wide variety of sites will require fewer, more extensive measurements than the development of detailed explanatory models.

Finally, measurements should test hypotheses without being locked too rigidly into current policies or current conceptual models, especially when long-term studies are conducted. Indeed, as the state-of-the-knowledge evolves or agency priorities change, particular research questions may become obsolete over time while others may become more pertinent. Hence, field protocols that are too narrowly focused may lose their utility or fail to record critical changes.

Experimental Design and Statistical Considerations

Research strategies and design of long-term experiments to assess the impact of intensive management have been discussed earlier [1, 2] and are specifically addressed elsewhere in this volume (Chapters 2 and 9). Consequently, they will not be discussed in detail here. Nevertheless, some consideration needs to be given to statistical design when implementing field measurement protocols.

The expensive part of most investigations is the preparation and analysis of samples rather than the actual sampling. This is particularly true for investigations of soil organism communities but also for chemical or physical analyses of soils. Because there often are limitations on the number of samples that can realistically be processed, it is extremely important to optimize sampling procedures with respect to the question that the study aims to address. There are indeed numerous examples of field studies where much effort was put in field work and analyses but where interpretation of data was ultimately obscured by an inadequate initial sampling design.

Statistical considerations in the design of field experiments have been discussed previously ([1], Chapter 9). Within the context of this chapter we will focus on some issues that are particularly relevant to evaluations at the organism, microsite, or landscape levels of resolution. Where possible, money, human resources, and analysis capacity should not be spent on unreplicated experiments. At times, however, replicated plots cannot be established for economic reasons, or true replicates cannot be found [3]. This is, for example, the case for most watershed studies and for a number of field experiments that involve large-scale experimental equipment. If such is the case, pre-treatment measurements to establish baseline characteristics should be the first stage of such time series studies [4, 5]. Replication within a treatment unit, i.e., pseudoreplication, can never be used to compensate for lack of replicates [3, 6]. Although it might be valuable to know the within-treatment unit variation, the approach should be to increase the number of true independent replicates rather than the number of sub-samples per replicate.

Experiments must also have proper controls. This means that field experiments that involve the installation or use of equipment must include controls not only for the actual treatment but also for the equipment used to implement the treatment. For example, if ameliorative practices involve site disturbances (e.g., the installation of irrigation or fertigation equipment, injection of sewage sludge) that by themselves can influence ecosystem processes, then the experimental design of the study should also include a "disturbance alone" treatment such that the role of the disturbance *per se* can be separated from the overall management effect. Although this is conceptually a desirable approach to design field studies, the problem of changes or interference due to equipment is often neglected for economic reasons.

Finally, in order to evaluate experimental treatments in replicated field trials it is also essential to use the statistical design applicable for ALL the components of the study (e.g., soil chemical and biological characterization). Furthermore, when designing laboratory experiments on materials collected in the field, it is important to take advantage of the field design. For example, field experiments are often complemented with laboratory studies including experimental treatments in the laboratory or incubation of soil treated in the field. In such cases, the field experimental design should be used as guidance for the setup and the analyses of the laboratory experiment, and laboratory treatments or analysis should be performed on individual field replicates of the material to the extent possible rather than on subsamples of composited replicates.

ASSESSING NUTRIENT INPUT-OUTPUT BALANCES

Input-output budgets are estimated primarily to assess the net effect of treatments on the ecosystem nutrient capital by contrasting the nutrient fluxes in and out of the system with total and/or available nutrient pools. Nutrient exports during and after logging are evaluated against nutrient inputs from the atmosphere and compared to either pretreatment input-output fluxes (time series approach) or among different treatment regimes (experimental plot approach). If fertilizer application is part of the management regime, then this input flux must also be considered in the calculation, in combination with atmospheric input, nutrient loss via leaching, erosion loss, and nutrient export with biomass removal. Such mass balance calculation takes a "black box" approach to processes and therefore provides limited information as to which and how processes and mechanisms within the ecosystem have changed as a result of site manipulations.

Atmospheric Inputs

Atmospheric inputs critically affect the mass balance calculation and are often the most difficult to measure quantitatively [7]. Nutrients can enter from the atmosphere dissolved in rainwater, as dry deposition of aerosols (fine particulates) and coarse particulates, or in the vapor phase, with the relative contribution of each of these pathways to total atmospheric input varying by nutrient [8]. The instrumentation required to quantify the individual atmospheric input fluxes has varied significantly in level of sophistication and numerous references describe in detail the apparati and techniques used by various researchers (e.g., [9, 10, 11, 12, 13]). Wet precipitation input is the easiest to quantify and is most commonly measured in nutrient cycling studies, either with bulk precipitation gauges or automated wet-only collectors operated continuously or on an event basis and placed in open areas away from forest canopy influences. Separate raingauges are often used to collect samples for chemical analysis and to record hydrologic inputs. Intensive studies have shown that long-term bulk collectors (>2-4 weeks) may become contaminated with particulates leading to higher solution concentrations for some elements than in the event-based wet-only collectors [14], yet fail to capture all components of dry deposition thus significantly underestimating total atmospheric input fluxes [15, 16]. There is also a concern that prolonged (>2 weeks) storage in the field may lead to sample deterioration [8, 14]. Based on these findings, in high-pollution areas automated wet-only collectors are recommended to estimate rainfall inputs, whereas in relatively pristine areas bulk precipitation collectors may prove sufficient, provided that they are sampled regularly.

In areas with frequent high intensity snowfall, conventional gauges may prove insufficient to collect all precipitation input and dedicated snow collectors may need to be installed during winter (e.g., [9, 17]). In sites that experience frequent fog, omission of the cloudwater input can introduce considerable error in total nutrient input estimates (e.g., [10, 15]). There are various types of active and passive fog collectors that can be used to assess this input [13, 16].

Dry deposition varies considerably with vegetation and the type of surface in the ecosystem onto which gases and particles are deposited [10]. It is not measured directly. Micro-meteorological methods and deposition models are used that require more specialized equipment and trained personnel. Air concentrations of small particles and gases are measured with filter packs and multiplied by the appropriate deposition velocities based on meteorological measurements (temperature, windspeed, relative humidity) and features of the deposition surface [11, 18]. For coarse particles, deposition on inert flat surfaces is sometimes utilized and scaled up to the level of the canopy [12, 13]. Because of the instrumentation needs and the variability in dry deposition inputs, and depending on the nature and the structure of the receiving surface, it may not be feasible to derive accurate site-specific dry deposition values. One alternative is to rely on regional or local averages, if such values are available. Another approach is to use broad dry deposition / wet deposition ratios reported in the literature for different elements (e.g., [10, 18, 19]). Finally, it has been found that for some nutrients (e.g., sulfur) regression models using below canopy fluxes (throughfall + stemflow) may provide useful approximations for dry or total atmospheric deposition input (e.g., [15, 18, 20]).

Not included in the routine deposition measurements is N entering the system from the atmosphere through free-living or symbiotic N-fixers. The latter are often associated with pioneer species (e.g., *Alnus* spp, *Ceanothus* spp, lupines) invading newly disturbed sites (see [21]) and plays an important role in the N economy of such sites. The acetylene reduction technique [22] or the N-15 isotope dilution technique [312] are appropriate for small-scale studies of the N-fixation process. However, annual N-fixation rates at the scale of a forest stand have been estimated indirectly from changes in total N over a given period of time (e.g., [21, 23, 24], Summary in [25]).

Leaching Losses

Two basic monitoring approaches can be taken to estimate nutrient exports via leaching: watershed studies and plot- or stand-level lysimeter studies. The watershed approach has been advocated for many years as a relatively simple tool to estimate nutrient input-output budgets for forest ecosystems [26] and has been used in some classical studies on the impact of harvesting on nutrient export [9, 27, 28]. Streamflow and nutrient fluxes out of the watershed are measured utilizing stream gauging weirs or flumes with water level recorders and automatic samplers (e.g., [9, 29, 20]). The watershed is considered as a unit and the stream water chemistry reflects an integration in time and space of all the processes that have influenced the water as it moved through the system.

This approach does not allow the separation between soil, bedrock, and within-stream processes. Also, outputs estimated on the basis of streamwater chemistry may significantly underestimate watershed export of N if denitrification in the riparian zones or within stream channels is significant. Watershed studies are seldom replicated and therefore require extensive pre-treatment measurements to account for between-year variability. The watershed approach works only in systems with the appropriate topography (i.e., some relief is required) and where the watershed is underlain by impermeable bedrock making it a closed system.

The plot- or stand-level approach based on lysimetry is more appropriate for the characterization of changes that have occurred within the rooting zone and allows the evaluation of spatial (vertical and horizontal) and temporal variability in soil water chemistry. Nutrient exports from a particular soil section or horizon are estimated by installing lysimeters at the appropriate soil depth and continuously or periodically collecting solution samples for chemical analysis. Lysimeter types are further discussed later in the chapter. Lysimeters do not always provide reliable water flux information [31], and other methods, such as hydrologic modeling, which requires a more extensive climatic characterization of the site (e.g., [32]), must be used to accurately assess nutrient exports. The lysimeter approach usually entails more replicate sampling and analysis than watershed studies to deal with spatial variability within the stand and to provide representative samples for the area.

Erosion

Although most nutrient cycling studies generally focus on nutrient solute losses, in some climates, erosion may be an equally if not more important form of export for nutrients that are adsorbed to the silt and clay phase or are associated with soil organic matter [30]. It is widely recognized that road construction and skid trails are major causes of water-mediated erosion and particulate export during harvesting [33]. In more arid regions, the removal of a protective vegetation cover may be the critical factor that accelerates wind erosion. There are two basic approaches to evaluate erosion: (1) methods that record particulate loss from a given location by wind, water, or both, and (2) methods that trap or collect displaced sediments. A more detailed description of specific field techniques in each category can be found in Hornung [34] and Reynolds *et al.* [30].

The first set of methods measures the amount of soil loss from a specific source area using exposed tree roots, measuring frames, erosion stakes, or photogrammetric monitoring of surface change [34]. Nutrient losses can only be estimated indirectly based on previously determined or known chemical characteristics of the displaced material. Sediment traps provide samples for chemical analysis, but may require a more careful experimental design to relate measurements to a specific source area, particularly if collection troughs or gutters are used to estimate overland flow in the field [34]. This approach is commonly used in small catchment studies where stream export is the integrator of within-watershed processes (see above). As outlined by Reynolds *et al.* [30] precise estimates of sediment losses in small streams depend on accurate and continuous measurement of streamflow.

Export with Harvesting

Nutrient export with biomass removal can be a major pathway of nutrient loss from harvested forests. Compared to the other fluxes, it is fairly simple to estimate and can account for a major portion of nutrient loss from harvested forests. The total weight of the material harvested is multiplied by the appropriate nutrient concentration, based on methods that have been in use for many decades in nutrient cycling studies (e.g., [35]). Tree biomass weights are derived from forest inventory data and allometric equations of tree biomass *v.* stem diameter (DBH) (e.g., [36]), or directly through destructive subsampling of trees if such data is not yet available (e.g., [37, 38]). Nutrient concentration of the different tree components are determined from tissue subsampling. Nutrient concentrations in wood are generally more stable than in the foliage [39, 40], and various studies have indicated possibly significant variations in foliar concentrations of certain nutrients, e.g., with the age of the tree or foliage, tree and crown position, time of the year, etc., [39, 41, 42]. Where site- or species-specific data are missing, nutrient exports can be roughly estimated with data in the literature for the same or similar species (e.g., [43]). However, because nutrient concentration may vary considerably with site quality, season, stand age, and crown position, using values that are not site-specific may introduce considerable error in these nutrient export estimates.

Total versus Available Nutrient Pools

The mass balance approach provides some insight into the severity of management practices in terms of nutrient removal. However, it cannot assess the impact of these practices on belowground processes, on nutrient source-sink relationships, or on the proportion of available to non-available nutrients. Many soil characterizations routinely include estimates of total nutrient pools as discussed later in the chapter. In most cases, however, total soil nutrient pools are large compared to input/output and internal cycling fluxes and are not significantly affected by management practices. Furthermore, total nutrient capitals are useful to the extent that they identify sites with severe nutrient limitations, but provide limited information on plant-available nutrient pools in any other case. In the following sections we will discuss the biological and chemical processes that play a pivotal role in nutrient availability and form the basis for various nutrient availability indices.

ASSESSING BIOLOGICAL PROCESSES

Different organism groups contribute to site productivity by their effect on: 1) the input of nutrients to the soil (e.g., N-fixation); 2) the release of nutrients via organic matter decomposition; 3) root uptake rates (e.g., mycorrhizae); 4) changes in the chemical environment associated with their activity; or 5) changes in physical soil properties associated with their movement (soil mixing) (Table 10.1). Management activities may alter site productivity by changing the abundance and composition of organism communities, by changing their activity, or by altering environmental conditions that change biological process rates [44].

A major source of nutrient supply within the ecosystem is organic matter decomposition, which occurs in three steps: 1) fragmentation or physical breakdown by soil fauna; 2) leaching of soluble nutrients from exposed fragments; and 3) mineralization or chemical breakdown of the organic constituents mediated primarily by soil microbes [45]. There are several ways of assessing changes in biologically-mediated nutrient release. These involve measurement of: 1) qualitative and quantitative changes in the communities of soil organisms; 2) changes in the activity of soil organisms; 3) changes in decomposition rates; or 4) changes in the indices of nutrient availability. Some of these measurements related to biological processes are summarized in Table 10.1.

Abundance and Composition of Soil Fauna
In most forest ecosystems soil mesofauna and microflora populations have not been studied or are poorly characterized. Assessment of management-induced changes in soil flora and fauna is therefore often problematic. Investigations of

TABLE 10.1

Some important soil biological processes, indicator measurements, and methods used
in site productivity studies.

Measurement	Key Processes	Methods
Abundance of soil fauna	Fragmentation Soil aeration Soil mixing	Soil extraction and counts
Abundance of microflora	Decomposition Nutrient transformation	Dilution Plate Incubation / Fumigation (lab)
Mycorrhizal abundance	Root uptake	Root tip counts
Microbial activity	Decomposition Nutrient transformations Respiration	ATP assays Enzyme assays CO_2 evolution (field)
Litter decomposition	Decomposition Nutrient release	Forest floor thickness Litterbag study Stable isotopes Radioactive tracers
Mineralization potential	Nutrient release Nutrient transformations	Aerobic and anaerobic incubations (lab)
Mineralization	Nutrient release Nutrient transformations	Exchange resins Soil incubations (field)

the effects of forestry practices on soil fauna can be qualitative, quantitative or
both. Basic pre-treatment information on the composition of soil fauna communities and investigations of changes in species diversity following harvesting are,
for example, largely qualitative. In most cases, however, effects of forestry practices are assessed through a quantitative approach that strongly depends on statistical inference. Investigations of food web patterns (see Pimm *et al.* [46] for a
general review of food web theory) in soil organism communities have been
applied in order to assess the impact of the fauna on soil processes (e.g., [47, 48]).

Studies of treatment effects on soil fauna are preferably done in replicated
field experiments that can be complemented by laboratory experiments.
Investigations may concentrate on certain indicator-organisms or organisms that

are quantitatively or qualitatively important for soil processes, and are selected on the basis of specific hypotheses regarding treatment effects that are to be tested in the study. Investigations should focus more broadly on a larger portion of the soil fauna if the objective is to study changes in the composition of the soil organism community and assess resultant changes in the food web structure of the soil. The latter could alter not only the biological but also the physical/chemical properties of the soil causing differences in nutrient dynamics. This is easily demonstrated when earthworms become established in a food web [49, 50, 51, 52].

For an overview of effects of forestry practices on soil fauna, we refer to Shaw *et al.* [53]. A number of soil fauna groups are known to be affected by harvesting as well as by the intensity of harvesting [54, 55]. The most dramatic effect of harvesting on soil fauna abundance have been reported for enchytraeids [56]. Other groups, such as microarthropods [44, 54], nematodes, and rotifers [55] also appear to be affected.

Methods for extracting soil fauna populations are described in a number of textbooks. An extensive review of techniques to assess soil invertebrate populations was recently presented by Edwards [57]. Except for a few semiquantitative methods such as *in situ* formalin extraction of earthworms or trapping of invertebrates in surficial soils, essentially all quantitative techniques to estimate soil fauna involve soil sampling and subsequent treatment of the soil by methods that vary depending on the faunal element to be estimated. These methods are:

1. Mobile fauna in the air space of the soil are extracted in dry funnels using heat from above to force the animals through the soil sample into a closed container with a preservative.

2. Mobile fauna in water films in the soil, such as mesofauna and nematodes, are extracted in wet funnels with a technique built on the same principle as above. Microfauna, such as protozoans, are estimated from dilution samples and with a "most probable number" technique as discussed in the following section.

3. Immobile faunal elements such as eggs and certain larval stages can be separated from the soil via flotation techniques or can be hatched from soil samples and subsequently collected in funnels as above. Other more or less immobile elements of the soil fauna, e.g., molluscs, have to be handsorted.

Abundance and Composition of the Soil Microflora

The procedure most widely accepted to quantify microbial populations is the dilution plate count method. In this method soil or litter samples are dispersed, distributed at different dilutions to a growth medium, and incubated under suitable conditions. Developing colonies are subsequently counted. A detailed description of the different media and isolation techniques can be found in various chapters of Page *et al.* [58] (e.g., [59, 60, 61]). This method does not account for interactions (synergistic or antagonistic) between organisms and is thought to underestimate population abundance [62, 63]. Schmidt and Paul [63] describe alternative methods of enumerating soil bacteria using light microscopy or fluorescence microscopy. The most probable number method estimates population density on the basis of chemical transformations in the medium by particular organism groups [64]. The most common application of this technique is to quantify nitrifiers [65, 66].

Population abundance is also expressed as microbial biomass, which provides a better assessment of the relative importance of the different organism groups than a simple count. Microbial biomass can be calculated based on plate counts, provided information on the size or weight of the individual microorganisms is available [62, 63]. Alternatively, it can be estimated from physiological activity, such as respiration, or chemical characteristics, such as adenosine triphosphate (ATP) content [67], which are further discussed in the next section. Jenkinson and Powlson [68] developed a method to estimate microbial biomass from the CO_2 flush following fumigation and incubation of soil samples [67]. This method was later modified to also assess microbial N and P content of soil microorganisms [69, 70]. The latter provides the first linkage between microbial populations and nutrient release and may be used as one indicator of biologically mediated nutrient availability. Molecular biological methods, e.g. recombinant DNA-technique, PCR (polymerase chain reaction) and DNA-probe techniques, can be applied in investigations of soil microorganism populations [71, 72, 73].

The major drawbacks of most of these microbiological techniques are that they: 1) are laborious; 2) provide static information, i.e., represent a snapshot of a single point in time; and 3) are for the most part organism-specific. They may be more suited for intensive microbiological investigations than for field studies that focus on the broader issue of soil productivity, especially since population abundances *per se* give at best very indirect information on microbial activity and its effect on nutrient availability.

Symbiotic associations between plants and mycorrhizae or N-fixers perhaps constitute an exception to this rule in that the abundance of mycorrhizal root tips and fungal hyphae or the degree of nodulation may have a direct bearing on the nutrient supply to the plant. It may be important to establish the presence of these symbionts and the degree of infection using methods described in the specialized literature (e.g., [22, 60, 74, 75, 76, 77] and various references in Gordon and Wheeler [21]).

Microbial Activity

Because ATP is involved in energy transfer within living organisms, its concentration in the soil is commonly used as an indicator of microbial activity. Several extraction techniques are described in the literature [67, 78, 79, 80]. Others have used the concentration of enzymes involved, in particular microbial transformations or decomposition steps, as a measure of microbial activity (e.g., [81, 82]). Enzymatic assays relevant to carbon, nitrogen, phosphorus, and sulphur cycling are described in Tabatabai [83]. Skujins [84], as cited in Gray and Williams [62], cautioned that a direct link between enzymatic activity, microbial activity, and soil fertility can rarely be made because enzyme assays are not always able to distinguish between the enzymes derived from living cells or dead cells, those adsorbed to soil particles, and free enzymes released by plant roots and soil animals.

Carbon dioxide evolution is thought to establish a more direct link between microbial activity and organic matter decomposition, because CO_2 is the end product of microbial decomposition of organic carbon under aerobic conditions. There are numerous procedures to measure CO_2 evolution from soils in the field, most of which are variations of the inverted chamber method [85]. The amount of CO_2 accumulated in the headspace above the soil from <1 hr to 24 hours is determined by: 1) trapping the gas with alkali (KOH or NaOH) and back-titration with acid after precipitation of CO_3 with $BaCl_2$ (e.g., [86]); 2) trapping the gas with sodalime (e.g., [87, 88]); or 3) passing air through the chamber and determining CO_2 concentrations with an infrared gas analyzer [85, 89]. The latter method requires expensive equipment and electric power. In recent years, modified portable Li-Cor photosynthesis systems have been used to measure soil respiration rates (e.g., [57]).

There are some concerns regarding the efficiency of the trapping agents (e.g., [85, 90, 91], the representativeness of the sampling area, or changes in environmental conditions and soil disturbance caused by the presence of the chambers. These concerns have led to design modifications such as moving chamber systems [89] or larger-scale open chambers [92]. By far the greatest problem with these field measurements is the inability to distinguish between microbial and root respiration [90]. No satisfactory *in situ* methods have been designed to overcome this problem. The most common method to fractionate respiration rates involves removing intact soil cores, soil subsections, or litter material from the field; placing them in sealed respirometer jars in the laboratory; and measuring the CO_2 evolution or O_2 consumption, with the techniques previously described or by collecting gas samples that are analyzed with a gas-chromatograph, while substrates are maintained at approximate field conditions (e.g., [85, 93, 94, 95, 96, 97]. However, it is not known to what extent disturbance during sampling alters the CO_2 evolution rates. Furthermore, the comparison of two different methods introduces error.

Given the inadequacies of field and laboratory methods, and the fact that respiration provides indirect information on nutrient availability through microbial decomposition, we would not recommend these assays for extensive use in soil productivity studies. They have some merit for comparative purposes, e.g., to assess changes in process rates caused by management practices (e.g., [98]), especially when evaluated in conjunction with other physical and chemical parameters.

Organic Matter Decomposition
The research area of decomposition in terrestrial ecosystems, including references to investigation techniques, was extensively covered by Swift *et al.* [99]. Forest floor thickness in forest ecosystems provides a rough indication of organically-bound nutrient pools and decomposition rates. A thick litter mat usually

indicates a large organic nutrient pool that turns over slowly as a result of temperature or water limitations [100]. It may also be the reflection of considerable inputs of fine root biomass in addition to low decomposition rates (e.g., [101]). A thin forest floor, on the other hand, may be due to low litter inputs, rapid decomposition, erosion, or movement of organic matter into the mineral soils. Recording changes in forest floor thickness and morphology as part of a soil profile description (see following section) provides one way of assessing treatment effects on potential nutrient supply from organic matter.

Litter decomposition rates are estimated from litterbag studies that measure nutrient release with time by periodically weighing and analyzing the litter samples as they decompose in the field (e.g., [102, 103, 104, 105]). There are indications that mesh size of the litterbag material affects decomposition rates by excluding particular organism groups [45, 106] and that fine-mesh bags in particular underestimate actual decomposition rates. One-mm bags have been reported to work satisfactorily [103, 107] but litterbag studies using different-size mesh at the top and bottom (2 mm and 10 mm) may provide better estimates of decomposition rates by allowing access for larger invertebrate decomposers (e.g., [105]).

Many harvesting practices result in some woody debris being left on site. Because this substrate influences nutrient dynamics, wood decomposition rates and nutrient release rates should also be assessed from changes in dry weight or specific gravity and changes in chemical composition of the residue. Specific techniques are described by Erickson *et al.* [108], Edmonds *et al.* [109], Harmon *et al.* [110, 111]. Compared to studies of foliar litter decomposition, assessment of nutrient release via wood decomposition involves more spatial variability that is induced by the variable size of the logs and the heterogeneous accumulation of the woody debris on the site. Also, wood decomposes considerably slower than foliage and fine roots, and nutrient releases may consequently be more relevant over the long term.

It is far more difficult to assess fine root turnover. Even though this process may be equally if not more important to nutrient supply than the turnover of above-ground litter [112, 113], root dynamics are seldom included in site productivity studies, because they are so labor-intensive and because good techniques are still lacking [114]. Current methods to measure root production and turnover utilize sequential coring, estimating root dynamics from periodic changes in live and dead fine root biomass (e.g., [107]), or ingrowth cores that estimate colonization by new roots of a homogenized sieved substrate [114]. Considerable controversy still exists about the capability of either method to generate realistic data, and a critical evaluation of both techniques can be found in Persson [114] and Vogt *et al.* [115]. This topic is discussed further later in the chapter.

In recent years, the relative abundance of natural stable isotopes or added radioactive tracers has been used to study organic matter decay. Methods have mainly focussed on the dynamics of carbon [116, 117, 118], nitrogen [116, 119, 120, 121], and sulphur [122, 123], but results are sometimes used to infer the release of other nutrients (e.g., [118]), even though the dynamics of organically bound nutrients may not always coincide [124]. These techniques require specialized laboratory equipment, and usually focus on a specific microbial process. Interpretation of the results is often complicated by isotopic fractionations and dilutions associated with the different transformation processes (e.g., [117, 125]).

Laboratory incubations of soil or forest floor samples, sometimes used in conjunction with fumigation to determine nutrient amounts incorporated in microbial biomass (e.g., [69, 70, 126], provide a measure of mineralization potential. They are mostly used in the context of carbon, nitrogen, sulphur, or phosphorus dynamics [69, 70, 97, 127, 128, 129]. Because samples are removed from the influence of plants in the field and are incubated under controlled temperature and moisture conditions in the laboratory, nutrient release values provide a relative index (usually overestimate) of the supplying power of the soil that may or may not be relevant to actual processes taking place under ambient field conditions (e.g., [130]).

Nutrient Availability Indices
It is difficult to accurately quantify the complexity of biological immobilization and release reactions that occur simultaneously but at different rates depending on the location in the soil profile and the nutrient under consideration. Techniques described in the previous section mostly focus on a particular process or organism group, but seldom provide an integrated picture of all biological reactions. There is nevertheless an acute need for criteria to evaluate site fertility and fertility

changes that has led to the development of chemical and biological availability indices.

Chemical Indices: Chemical indices are based on extraction techniques that characterize different nutrient pools, including labile organic fractions that are thought to degrade readily as a result of microbial activity. Some techniques are further discussed later in the section on soil chemical processes. Specific extraction and fractionation procedures for nutrients that originate from soil organic matter can be found in the literature for phosphorus [116, 131], sulphur [132, 133, 134, 135], nitrogen [116, 136, 137, 138, 139, 140], or a combination of soil nutrients ([141, 142] and various references in Westerman [143]). Most of these indices are empirical and often based on correlations with growth and yield of specific crops or tree species in pot studies. They provide static time-specific information on soil properties but do not characterize rates of nutrient release in the field nor the mechanisms involved.

Biological Indices: Biological indices are intended to characterize microbial conversions of organic into inorganic constituents that are plant-available. They generally involve soil incubations in the laboratory or in the field. Some work has been done on the release of sulphur and phosphorus during mineralization (e.g., [69, 128, 129, 144]), but far more studies have focussed on nitrogen availability indices and their relationship to site productivity and tree growth responses. Various nitrogen availability indices are discussed extensively in Keeney [137], Stanford [138], and Binkley and Hart [140], and will only be described briefly here.

The simplest and easiest laboratory assay measures NH_4-N production in water-saturated soil samples that are incubated anaerobically [137]. The method is quick and a large number of samples can be processed in a short time period, allowing a thorough characterization of spatial variability at a site in intensive studies, or rapid characterization of many different sites during extensive surveys. Because the soils are incubated under waterlogged conditions only potential ammonification can be measured and the method is less suitable for sites where high NO_3 levels are already present in soil ([145], Van Miegroet, unpubl. data).

Powers [146] further points out that the NH_4-N increase during incubation alone is not an adequate measure of potentially available nitrogen and that pre-treatment inorganic nitrogen levels must also be considered.

Two types of aerobic laboratory incubations are in use: the open system technique [147] and the closed container technique [127]. In the first, soil samples, often mixed with sand to improve drainage, are placed in open funnels and leached repeatedly with dilute salts to extract NH_4-N and NO_3-N or SO_4-S; in the second, soils are incubated in sealed containers, maintained at given soil moisture content, and aerated periodically. Subsamples are removed periodically for extraction and analysis. Both methods require longer incubation times and are more laborious than anaerobic incubation techniques. On the other hand, environmental conditions can be chosen to better approach field conditions (e.g., [127, 144, 148]). Samples are often sieved and homogenized before incubation. This disturbance, the absence of plant influences, and the lack of fluctuations in soil environmental conditions that typically occur in the field, tend to over-estimate nutrient release potential.

Field assays are thought to give a more realistic and dynamic picture of microbial transformations, but also entail a more complicated sampling design to deal with spatial and temporal variability. Field incubations utilize one of the following methods: 1) ion exchange resin bags (e.g., [139, 142]); 2) buried bags [149]; 3) closed-top tubes [150]; and 4) resin cores [151]. The review article by Binkley and Hart [140] provides detailed descriptions of these techniques. In the first method, a bag containing a mixture of anion and cation resins is placed in the soil to trap nutrient released during mineralization but not taken up by the surrounding roots. The other three methods measure changes in extractable NH_4-N and NO_3-N content within a soil core that is isolated from living roots and either shielded from precipitation inputs (buried bag, closed-top tubes) or open to the atmosphere (resin cores). Where precipitation is excluded NO_3-N leaching losses should be minimal. In those cases, however, soil samples do not go through the normal wetting and drying cycles that occur in the surrounding soil. The resin core method utilizes exchange resins to capture the inorganic nitrogen that might have otherwise leached and may be most appropriate in frequent-rainfall areas. All coring techniques raise concerns about the disturbance effects

during sampling and the contribution of severed roots to the measured mineralization rates.

Several authors have investigated the relationship between chemical indices,
biological indices, and microbial activity, on the one hand, and plant responses to
fertilization on the other, and results from such comparisons are highly variable
[136, 139, 141, 148, 152, 153, 154, 155]. The most appropriate assay is often
dictated by the objectives of the study.

ASSESSING SOIL CHEMICAL PROCESSES

Soil chemical reactions affect nutrient availability and the chemical environment
of roots, mycorrhizae and soil organisms. The soil operates as a reservoir for
essential plant nutrients. Management practices can alter the availability of nutrients for plant uptake by causing changes in chemical processes occurring in the
soil or alter environmental or physical properties of the soil which subsequently
affect soil chemical processes. The soil solution acts as a solvent for reactions
occurring in the soil and is the primary medium for transfers or transformations of
matter and energy among organisms and various soil components. Chemical
processes not only occur within the soil solution, but interactions also take place
between the solution, solid, and gaseous phases in the soil (Figure 10.1). The
concentration and chemical form in solution of essential nutrients may critically
affect the availability to plants, while the   concentration of other elements
(e.g., pH, Al) may influence site productivity in an indirect way by influencing
biological activity, by regulating chemical reaction rates of other elements, or by
influencing the uptake of essential nutrients. Table 10.2 lists some chemical
processes and soil measurements and indicates those measurements considered
essential in site productivity studies. Long-term studies may need to include
some of the other processes listed. This section will deal with various methodologies for assessing changes in soil chemical changes and their appropriateness
for different objectives.

While the importance of reaction kinetics versus equilibrium conditions will
not specifically be addressed, it is important to consider that rate limitations can

TABLE 10.2

Chemical measurements needed for the characterization of some important soil chemical processes. Measurements considered necessary in site productivity studies are indicated.

Measurement	Important Processes	Necessary Measurement
Soil Solution Analysis	Nutrient Uptake / Availability	Yes
	Mineral Stability / Weathering	
pH	H and Al exchange	Yes
Partition Coefficient	Nutrient Buffering / Supply	Yes*
Total Soil Analysis	Translocation	
Soil Organic Matter /		
Carbon	Energy / Food Supply	Yes
	Sorption Reactions	
	Decomposition	
Soil Nitrogen	Nutrient Availability	
	Substrate Palatability	
Mineralogy	Weathering / Surface Reactions	
Ion Exchange / Sorption	Nutrient Availability / Ion Buffering	
Soil Extractions	Nutrient Availability	
CO_2	Respiration / Decomposition	
	Carbonate Speciation	
O_2	Aeration / Oxidation / Reduction	
Profile Description	Soil Development	Yes

* If partition coefficients are not measured, then nutrient availability should be measured by some other method.

be very important in many soil processes. Rates of some solution-solid reactions and solution-gas reactions can be much lower than rates of reactions occurring within the solution phase only; two-phase processes may be rate limiting, and it is possible that equilibrium conditions are not often reached. In addition, some soil reactions are considered to be irreversible (for a discussion of solution and soil kinetics, see [156, 157, 158]).

Soil Solutions

The soil solution is central to the transfer and transformation of nutrients between microorganisms, plants, and the solid and gaseous phases of the soil. Next to the soil gaseous phase, it is the most dynamic and easily affected part of the soil. Changes in the quantity of water entering the soil, the rate at which it moves, its temperature, and the interaction time between the solution and soil solids or gases affect the composition of the solution, and ultimately the amount and form of nutrients available for plant uptake. Harvesting impacts are readily detected by changes in solution composition, as has been shown by many studies, due to the dynamic nature of the soil solution. Collection and analysis of soil solutions is recommended for site productivity studies, as much information about rapid reactions and subtle long-term changes in site chemistry can be detected in the soil solution that are not easily detectable in the soil solid phase.

Collection: Numerous methods have been used to collect the solution fraction of the soil system. Many of these methods add water or other solutions to the soil prior to extraction; because these methods are not collecting the *in vivo* soil water, they will not be considered here. Techniques which collect *in vivo* water include: gravitational lysimeters, tension lysimeters, centrifugation with double-bottom cups, and centrifugation with immiscible displacement.

Gravitational lysimeters use only the force of gravity to collect water flowing down in the soil above the collector at zero matric potential. These devices often collect water intermittently when the soil moisture content is greater than field capacity. Both trough and plate type lysimeters have been used [159, 160, 161].

Tension lysimeters collect water from both saturated flow and unsaturated flow which is held at tensions less than that of the vacuum system [162, 163, 164, 165, 166]. Suction may be applied to tension lysimeters with handpumps, vacuum pumps [167], or hanging columns where no power is available [168]. Lysimeter plates are constructed from various materials such as ceramic, teflon, alundum, fritted glass, polyethylene, and polyacrylic cloth (see [31]). Because some of these materials may release contaminants or adsorb solution constituents [13], the appropriate lysimeter type is dictated by the solution acidity and the chemical analysis to be performed in the study. Field installation of both gravity and tension lysimeters can be quite extensive, depending on the number of horizons sampled and the number of replicate collectors per horizon. Maintenance of the collection system can also involve considerable effort at some sites, as problems such as damage by bears or mice, freezing of collection or vacuum lines, or other unanticipated events can cost considerable time and money.

Soil water held at field capacity or greater tension can also be collected through soil centrifugation using double-bottom cups [169, 170, 171, 172] or the immiscible displacement technique [173, 174]. Few field materials are needed, but centrifuging should be done as quickly as possible after sampling and samples should be kept refrigerated prior to and during centrifugation. Centrifugation with immiscible displacement is somewhat more complex, as most immiscible solvents used with this technique are toxic. Care must be taken to avoid using a solvent which will partition elements of interest between the solvent and soil water [175].

Any of these soil solution collection methods are suitable for process studies, but some methods may be preferred for specific objectives, as different techniques collect different water fractions [176]. For example, if the aim is to examine the flow of nutrients from the soil, then lysimeters should be used (see discussion above). If an examination of nutrient availability and its seasonality are the objectives, then centrifugation would be more appropriate as this method collects soil water that is more representative of the water available for plant uptake.

Solution Analyses: Because the soil solution also acts as a medium for biological processes, solution/solid interactions and solution/gas interactions, the chemical reactions and solution composition can become quite complex. Within the solution, interactions also occur between both organic and inorganic solutes.

An analysis of acidity (pH) and inorganic constituents of the soil solution should be done for any productivity study which includes an examination of soil processes. Total element concentrations in the soil solution can readily be determined using atomic adsorption spectrophotometry, inductively coupled plasma emission spectrophotometry or neutron activation [177, 178, 179, 180]. If this equipment is unavailable, wet bench chemistry techniques can be used to determine major components (see [58]). Ion chromatography is also an increasingly popular method for measuring both solution anions and cations. Alkalinity can be measured by titration [181].

Many of the organic components found in the soil solutions do not have exact chemical compositions, and can not be readily identified. For example, humic and fulvic acids are identified by their reactions to a series of solution extractions and this method does not provide information on their exact composition or equilibrium constants [182, 183]. Many low molecular weight organics can be identified using chromatographic, nuclear magnetic resonance or spectrographic techniques [182, 183, 184].

A speciation of the solution ions can be accomplished and solution activities calculated once a comprehensive analysis of the soil solution components has been completed. Knowledge of activities of ions in the form suitable for plant uptake is much more informative for understanding nutrient availability or toxicity than are total concentrations. Several computer speciation programs are available, such as GEOCHEMPC [185], MINEQL+ [186], and MINTEQA2 [187]. GEOCHEMPC speciates solutions at 25 °C only. It requires extensive input to the data base to run at another temperature, but has data base tailored to soil studies. Both MINTEQA2 and MINEQL+ will speciate at any given temperature, but enthalpy data are not available for all reactions. In addition, all of these programs can also provide information regarding solid/solution and solution/gas reactions.

Solution - Solid Interactions

Chemical reactions between the solid and the solution phase are numerous. Dissolution, precipitation, adsorption, desorption and exchange reactions are the most common (Figure 10.1). Harvesting or management practices may have impacts on weathering, exchange reactions or sorption reactions individually, or in combination. Measuring changes in these processes over longer time periods may be critical in highly weathered or low exchange capacity soils. Many of the processes involving essential nutrients are rapid reactions which are readily detectable and may be highly susceptible to harvesting impacts. Others constitute long-term sources of nutrients whose depletion may not be immediately detectable, but would appear only in longer-term studies. The separation between individual chemical processes may be difficult without extensive field and laboratory experiments, but may, in some cases, be unnecessary. For example, a distribution coefficient may be adequate to characterize both exchange and sorption reactions (Table 10.2). In studies where more long-term process information is desired, it may be necessary to examine solution-solid reactions in more detail, or individually. In some cases, it may also be necessary to examine total quantities of an element in the solid phase. In this section, methods for examining solid/solution properties and processes will be considered separately.

Weathering: Four approaches have been used to examine mineral weathering: 1) rock alteration; 2) laboratory simulations; 3) ecosystem mass balance; and 4) modelling [188]. The rock alteration technique has two different approaches, mineral grain etching and rock weathering rinds [189, 190, 191]. These methods provide gross weathering rates over extremely long time periods, and are of limited use for studies of current or recent site productivity. Laboratory simulations are most suited for examinations of the effect of specific processes on weathering rates [192, 193], but are difficult to extrapolate to actual field conditions. The mass balance technique has been the most commonly used method for examining weathering inputs to ecosystems. With this method, weathering is calculated as the difference between atmospheric inputs to the soil and soil leaching losses plus changes in soil storage. Both watershed/ weir and plot/lysimeter designs have been used to measure leaching losses [9, 194, 195, 196] (see earlier discussion). Because weathering rates are estimated by difference between fluxes that each have their own error term, the degree of uncertainty associated with this approach is high. The use of modelling to determine mineral weathering inputs to ecosystems has remained limited [197, 198, 199]. While this technique is promising, it requires extensive information on soil properties that are difficult and costly to determine, such as water flow rates, primary and secondary mineral information, and water-surface contact.

Exchange and Sorption Reactions: While weathering reaction may be critical to the long-term supply of nutrients to a forest, it is the sorption and exchange reactions which will largely control the short-term or immediate availability of nutrients. The measurement of sorption reactions is frequently done with adsorption/desorption isotherms. Often, a partition or distribution coefficient (K_d) is determined that indicates the relationship between the quantity of a ion in solution to the amount adsorbed to the soil particles. This value is an empirical one, and must be determined for each soil horizon separately; it can be used to estimate the nutrient buffering power of the soil [200].

Soil organics, amorphous minerals and oxides are important to sorption reactions in soils. Soil exchange reactions can be measured using both anion exchange capacity (AEC) and cation exchange capacity (CEC) techniques. In agricultural soils, CEC has typically been measured in pH 7 buffered ammonium acetate solution, but several other unbuffered solutions have been used for forest soils [13, 201]. In forest soils the use of an unbuffered solution is recommended because of the large effect pH has on CEC. Anion exchange capacity is of most importance in andic and tropical forest soils, and can be measured using the method of Uehara and Gillman [202]. Accurate measurement of soil pH is also important, because pH is an indicator of many other soil properties (e.g., exchangeable aluminum or base saturation) and because it affects many chemical reactions in the soils. A water saturated paste, a fixed quantity of soil to solution (e.g., 1:1 ratio of soil to water) and a $CaCl_2$ solution have been used to measure pH [181, 203].

Total Analyses: While total element analysis is of little use for examining nutrient availability, it can offer insight into longer term soil processes and indicate nutrient cycling changes. Soil carbon is particularly relevant as this value often correlates with good soil fertility and physical properties, except in cold and very wet ecosystems. Soil carbon can be reported as soil organic matter, total carbon, organic carbon, or oxidizable carbon. Soil organic carbon is generally determined by converting loss on ignition values using a correction factor of 0.5 or 0.58 [204]. Oxidizable carbon is measured using dichromate oxidation [204, 205], and total carbon is determined by combustion. If no carbonates are present in the soil, total carbon equals organic carbon, otherwise, inorganic carbon must be measured and subtracted.

Total soil nitrogen can be determined by wet oxidation (Kjeldahl digestion), by combustion [206], or by sulfuric acid digestion [207]. For total elemental content of other elements in soils, a perchloric acid–hydrofluoric acid digestion or neutron activation is recommended [178, 208, 209].

Extractions: Chemical extractions are used to quantify soluble nutrients and to compare and index the amount of nutrients or metals that are accessible to plant roots from the solid phase. There are four different categories of extractions based on the extractants used: 1) water; 2) salts; 3) acids; and 4) organic chelators. Comparisons between extraction methods do not accurately assess differences in nutrient availability, as each method will not necessarily extract nutrients from the same solid phase binding sites. However, extractions are one way to evaluate changes in nutrient availability among treatments or with time. In general, salt and water extractions are used for macronutrients, while dilute acid and organic chelator extractions are used for some micronutrients and metals. A comprehensive list of the numerous extractants used to assess soil nutrient availability is not possible here, but a few examples will be mentioned. Some widely used methods for forest soils include, 2 M KCl extractions to measure inorganic soil nitrogen [210], saturated paste extracts for soil cations or salts [181], Bray-extractions for phosphorus [211], hot water extractions for boron [212], and DTPA extractions for some micronutrients and some trace metals [213]. Soil acid neutralizing capacity can be measured by titration of water extracts [181].

Soil Air

The composition of the soil air is influenced by many factors the most important being the abundance and activity of organisms and roots in the soil, decomposition rates, and exchange reactions with the soil solution (Figure 10.1). At the same time, it is an indicator of and a factor in the control of biological and chemical process rates. For example, microbial and root respiration consume soil O_2 and release CO_2 into the soil air; optimum plant growth conditions require good gas exchange between the soil and atmospheric air to prevent high CO_2 buildup or low O_2 levels in the soil. Harvesting can affect both respiration and decomposition rates (see section III). Management practices which alter the soil structure (see discussion in section V) can also change the composition of the soil air by changing gas diffusion rates or soil water content, which in turn influence gas diffusion reactions. Many of these properties and processes are critical to maintaining a healthy rooting environment or to creating conditions that are beneficial to microbial processes.

Collecting uncontaminated soil air samples can be difficult. Methods used to collect samples include tubes inserted into the soil to the desired depth, or open chambers inserted into horizons with tubing to the surface for access. Most soil air collection methods have the problem that the associated soil disturbance affects the accuracy of the measurement. Because no collection method appears free of this problem, it is recommended that consistency in sampling method and handling be the highest priority. In addition, sample chambers or access tubes should be covered between collections to reduce artificial exchange with atmospheric air through the sampler. The chamber should also be evacuated and the air discarded immediately before collecting the sample. If vacutainers are to be used for sample holders, the vials should be overfilled to prevent air seepage into the container with time. Smith and Arah [214] describe some soil gas collectors and gas chromatographic techniques used to analyze them. Analysis of the samples can be done readily by gas chromatography for many components such as O_2, N_2, and CO_2 [214, 215]. Field analysis of soil CO_2 can also be done using Dreager CO_2 indicator tubes [216]. This method is extremely quick and cheap, but is not as accurate as gas chromatography. Membrane covered O_2 electrodes can also be used to determine soil O_2 [217].

Soil Development

The influence of different types of forest vegetation is a major determinant in the genesis of a soil, and removal or alteration of the forest cover can be a direct cause of changes in soil development. For example, Fisher [218] reported loss of the Oe and Oa horizons and the replacement of an E and B horizon by an Ah within 18 years of the conversion of a white pine stand to a hardwood stand. Griffith *et al.* [219] also found that white pine growing in an abandoned field resulted in the formation of an E horizon within 70 years, but this horizon disappeared after conversion of the pine to hardwoods. While harvesting impacts can alter soil development, changes in soil development may in turn affect the long-term productivity and nutrient cycling of the site. Nutrient leaching rates, number of sorption sites in the soil, movement of dissolved carbon and chelators, mineral stability, pH, and numerous other factors which are a function of soil development processes could be altered as the result of forest harvesting. Some of the changes in soil processes or soil properties caused by harvesting are reversible and may return to preharvest conditions; harvesting may also cause changes that are more prolonged or irreversible.

Pedogenic Processes: Most of the major pedogenic processes such as eluviation, leaching, melanization, podzolization and others can be influenced by changes in soil physical and chemical properties caused by harvesting. Complete soil profile descriptions should therefore be an integral part of any field study (see following section). Although soil forming processes have major cumulative effects on the soil, changes caused by harvesting can be extremely difficult to detect in the bulk soil except in long-term field studies where chemical changes become physically evident in the soil profile. Therefore, the examination of soil solutions to assess rapid changes in soil processes resulting from harvesting impacts is recommended [161, 165]. Alternatively, soil extractions can be used to compare changes in chemical form or movement of soil carbon, iron, aluminium, and other prevalent soil elements in studies of intermediate length.

Soil Morphology and Structure: As pointed out previously, a complete soil profile description before and after harvesting can provide much information towards evaluating the severity of management disturbances and should therefore be a standard procedure in any field study. A profile description is an excellent source of information about the general physical properties of a soil such as the physical structure of aggregates, particle size, color, and porosity, but will also provide information about horizon development. Soil morphology is changed by harvesting to the extent that mixing or removal of the upper horizons occurs (either by erosion or in association with the harvesting equipment and method used) and soil structure is altered. These changes can subsequently affect hydrologic and thermal characteristics of the surface soil and the horizons below. The soil should be classified according to a standard system to allow comparison with soil information collected at other sites. Many soil classification schemes are used throughout the world, of which the U.S. Soil Conservation Service [220, 221], and the FAO system [222] are among the most encompassing due to the large variety of soil types covered.

ASSESSING SOIL PHYSICAL PROPERTIES

Changes in soil physical properties can be a major factor in altering long-term site productivity. Manipulations of forest sites can alter soil structure, thermal properties, or water relations that result in increased or decreased plant growth. Harvesting activities, for example, can affect gas exchange, root penetration, water movement and water availability, soil temperature, and influence chemical and biological reaction rates. Several important soil physical properties and measurements considered essential for long-term process studies are summarized in Table 10.3.

Soil Solid Phase

One of the most widely recognized and common impacts of timber harvesting on the physical properties of the soil is a change in bulk density, or soil porosity. Tree growth reductions as a result of high soil bulk density are not uncommon. Increases in bulk density can be very detrimental to root extension. Such compaction may also affect the function roots and the activity of soil micro-organisms through changes in gas diffusion rates, aeration, soil water potential, and water flow. Measurements of soil physical properties may thus largely contribute to an understanding of productivity change or lack of change at a site following harvesting.

Bulk Density and Soil Porosity: Bulk density (ρ) is the ratio of soil mass to soil volume. It provides information about the total porosity of a soil horizon. It can be determined using core, clod, excavation, or radiation methods [223, 224]. Essentially, the core, clod and excavation methods measure the dry weight of a

TABLE 10.3

Physical measurements needed for the characterization of some important soil processes. Measurements considered necessary in site productivity studies are indicated.

Measurement	Important Processes	Necessary Measurement
Hydraulic Conductivity	Water Flow Rates	Yes
Soil Infiltration	Water Input Rates	
Soil Water Characteristic Curves	Water Availability / Flow	Yes
Soil Water Content	Water Availability	
Solar Radiation / Heat Input	Heat Supply to Soil	Yes
Thermal Conductivity	Heat Transfer within Soil	Yes*
	Root Growth / Activity	
Soil Temperature	Root Growth	
	Microbial Activity	
Bulk Density	Root Penetration	Yes
	Compaction	
Porosity	Root Impedance	
	Gas Diffusion	
	Water Diffusion	
Soil Structure	Root Penetration / Impedance	
	Water Flow	
Soil Profile Description	Soil Development /	Yes
	Soil Structure / Morphology	

* Either thermal conductivity or periodic soil temperature measurements should be made.

known volume of soil, differing only in how the soil volume is obtained and measured. With the core method, a small metal cylinder is used to remove the field sample at known volume. This method is quick, allows a large number of replicate samples, but is unsatisfactory in soils with high stone content. It has the least accuracy due to artifacts from the coring (e.g., compaction during sampling of organic-rich soils). With the excavation method, the soil volume of consecutive soil layers is determined by filling the hole with sand or water. This method is generally more accurate and better suited for gravelly soils. It is more labour-intensive and may not allow for many replicates. The ped or clod method uses peds removed intact from the soil and coated with liquid plastic or paraffin to preserve the volume which is then measured by displacement in water. The radiation method [225] indirectly determines bulk density by the quantity of gamma radiation received from a source lowered into the soil.

Shifts in pore size distribution can result from compaction during logging; this frequently results in a decline in air-filled pores (and aeration) although not always in a reduction of water-filled porosity [226]. Total porosity can be calculated from ρ measurements and average particle density, which in turn may be deduced from soil mineralogy or measured directly [225]. The pore size distribution is determined in the laboratory by applying variable suction to an intact soil core placed on a porous plate, and measuring water release [227]. Because this method requires intact soil samples, which are often difficult to obtain, and because of the time involved in processing each soil core, this procedure may only be practical for small-scale intensive field studies.

The degree of soil compaction can be evaluated indirectly from the soil resistance against the vertical penetration using a penetrometer. Several types and sizes of penetrometers are available [228]. Penetrometers are easy to use and enable an extensive characterization of spatial variability. Furthermore, soil resistance has some relationship to the degree of difficulty with which root elongation and penetration is likely to take place. The major drawbacks of this method are: 1) that no direct ρ or porosity measurements are obtained; and 2) soil resistance changes with moisture content, which requires prior calibration of the results and determination of soil moisture content.

Particle Size: Soil texture can be assessed in the field as part of a profile description.   If a more accurate determination of soil particle distribution is needed, laboratory methods can be used. Sieves can be used to determine larger size fractions, while the finer sand, silt, and clay fractions can be measured using a hydrometer and a settling tube, or using the pipette method [229]. For most field studies, repeated measures of soil texture are not necessary, as harvesting usually has little impact on this soil property in the short term. Longer-term studies may need to consider changes in particle size distribution. In those cases, field determination of such changes coupled with laboratory analyses of a few selected samples to verify the field assessment should be sufficient.

Soil Water

Water constitutes one of the basic requirements for plant growth, and soil water relations should be established before and after harvesting (and for long-term studies also periodically thereafter) to fully characterize potential site productivity changes. Two general types of measurements can be made with respect to soil water: 1) moisture content, and 2) water potential. The former provides information on the total amount of water per unit soil mass or volume, but does not provide any information on the extent to which that water is available. Soil water potential, or soil moisture tension, is an expression of the tension with which the water is held in the pore spaces or, conversely, the ease with which it becomes available for plant uptake. Both soil moisture content and soil water potential fluctuate considerably throughout the year depending on specific weather conditions and the amount and condition of the vegetative cover. Therefore, if only a limited number of measurements can be taken, then they should preferentially be taken at the same time across sites or from one year to the next. However, multiple measurements throughout the year are generally recommended.

Soil Water Content: Soil moisture content is the easiest to determine among soil water properties, but also the least meaningful characteristic in and of itself, because the same water content may represent different amounts of available water in differently textured soils. At any particular site or in a given soil, it can nevertheless be useful for comparative purposes, for example, to broadly characterize seasonal differences in soil water availability or to evaluate treatment effects on soil water among sites.

Soil water content can be determined directly, by removing the water from a given soil sample, or indirectly, by using soil properties that change with moisture content as an indicator of soil water content [230, 231]. The gravimetric method is the most common and easiest of the direct methods. A known amount of wet soil taken from field samples is ovendried (100-110 °C) and weighed again. The weight loss represents soil water and is expressed per unit of dry soil weight. This method has minimal equipment requirements. If a known volume of soil is used, values can also be converted to volumetric moisture content. Alternative procedures for soil drying are described in Gardner [230].

Several methods have been developed to infer soil moisture content from changes in: 1) the electrical or thermal conductivity of materials (resistance bridges, porous blocks); 2) the scattering pattern of emitted neutrons (neutron probe); and 3) the behaviour of electromagnetic waves in the soil (time-domain reflectometry (TDR)). The cost and maintenance of the neutron probes or TDR instrumentation may make these approaches prohibitive to some projects. Furthermore, the procedures involve permanent or semi-permanent installations in the field, and with the exception of the TDR, require calibration of the readings for the specific site/soil in which the instrument will be used [230, 232, 233]. The TDR method does not appear very sensitive to soil temperature, bulk density, soil texture, salt content, and mineral composition of the soil. Neutron probes have been used successfully for forest soil water measurements, but can also be expensive to maintain due to storage and handling of radioactive materials.

Soil Water Potential – Moisture Release Curves: Soil moisture potentials can be measured directly in the field with tensiometers [231, 234], thermocouple psychrometers [235], electrical resistance blocks [236], or piezometers [237]. Tensiometers are based on the principle that the water inside the porous endcup is in equilibrium with the soil water, and changes in suction within the tensiometer are used as a measure of soil moisture tension. Both tensiometers and resistance blocks are relatively inexpensive and easy to install, allowing sufficient replication to adequately characterize spatial patterns in soil water tension within a study site. The useful range of tensiometers is limited between 0 and 80 KPa (0.8 bar) tension, so tensiometers are ineffective in drier soils. Soil moisture tension can also be derived from TDR readings, but prior calibration for each individual soil type is needed. Piezometers measure hydraulic head of soil water or depth to the water table. An open-ended pipe is inserted into the soil to within the water table, and depth of water within the pipe relative to a reference level indicates the hydraulic head [237]. With several piezometers installed across an area, they can be used to examine changes in the depth of the water table or water flow directions and gradients over larger areas and over time.

Soil moisture release curves indicate the relationship between water tension and water content in a given soil, and must be determined in the laboratory using a tension plate apparatus or a pressure membrane apparatus [231, 238]. Intact cores are placed on a ceramic plate to which either a suction (0-100 KPa) or pressure (up to 10,000 KPa) is applied to extract the water from the soil sample. At the different water potentials, gravimetric soil moisture contents are determined to establish the correlation between both variables. This relationship is soil specific, varies with soil structure and pore size distribution, and can thus be altered significantly with soil compaction (see discussion above). Once generated, the curves can be used to convert soil water content measurements into water potential values.

Hydraulic Conductivity: Hydraulic conductivity is a measure of the rate of flow of water in the soil, and can be determined for either saturated or unsaturated conditions [239, 240, 241]. This measurement is important both for calculating the rate of infiltration of precipitation or throughfall water into the soil, the water flow throughout the soil profile, and the movement of water to plant roots.

Thermal Properties
Soil temperature in conjunction with soil water relations will reflect the degree of change in soil microclimate resulting from management practices (particularly the removal of vegetative cover), which in turn may influence chemical and biological reaction rates or alter site conditions beyond the range that are optimal for microbial and root processes. This type of field information should be collected if possible, because it may play a critical role in evaluating site quality changes or in explaining changes in belowground nutrient dynamics.

There are numerous kinds of thermometers to measure soil temperature that are based on expansion-contraction properties of particular substances (Hg or liquid thermometers, bimetal strips) or changes in electrical properties of metals (thermistors, thermocouples). They can be either installed permanently within a given soil horizon, which reduces subsequent soil disturbance and allows them to be hooked-up to a data recorder, or may be inserted each time to a given soil depth. The former method might be especially appropriate in conjunction with soil mineralization studies [242]. The measurement frequency will depend on the specific data needs of the project, as well as the field infrastructure (presence of datalogger, power availability, etc.), but should be no less than once every month.

Solar radiation or heat input to the soil combined with soil thermal conductivity can provide information about soil temperature for predicting root growth, chemical reactions and microbial activity. For most long-term site productivity studies, thermal conductivity can be calculated from solar radiation inputs and changes in soil temperature (over a shorter specified period of time) in conjunction with moisture content and bulk density (assuming an average specific heat for mineral soils) [243]; measurements should be made both before and after any site manipulation. Alternatively, laboratory methods can be used to measure thermal conductivity [244]. If thermal conductivity and heat input to the soil cannot or will not be determined, then periodic measurements of soil temperature should be made.

ASSESSING PLANT RESPONSES

In research trials established to determine the impacts of intensive harvesting on site productivity, plant responses can be viewed as biological indicators of changed site processes and site quality. Plant responses to beneficial or harmful alterations in site quality are expected to be manifested at levels ranging from biochemical and physiological processes to whole plant and plant community dynamics. We hypothesize that if intensive harvesting practices alter site nutrient availability, trees will respond by showing: 1) changes in nutrient uptake rates; 2) altered physiological function; 3) changed foliar surface area and mass; 4) altered carbon allocation patterns within plants which will affect above-ground development, fine root standing crops, and shoot-to-fine root ratios; and 5) altered above- and below-ground productivity per plant and unit area. In this section, we will evaluate the relevant literature from studies in forest nutrition, and physiological and ecosystem ecology for protocols and techniques that appear to be useful for detecting plant responses to alterations in site quality following intensive harvesting.

Tree Growth and Development

In a production forestry context, the most important response variable to measure in intensive harvesting trials is tree growth and development. In this section we will evaluate several techniques for assessing plant responses on physiological and whole organism levels and will begin by evaluating techniques for estimating growth responses of the above-ground parts of trees.

Dimensional measurements of trees should be taken to assess tree growth responses and to provide information that will be useful for conventional mensurational analysis of stand volume response to treatments. Measurements should be taken that permit one to determine the amount of growth that is allocated to stem wood versus other components of trees. Market demand and merchantability standards will influence the priorities for dimensional analysis; however, rapidly changing utilization standards and uncertainty about the future suggest that dimensional analysis not be limited to current merchantability standards for the main stem of the tree. Researchers should consult mensuration texts by Husch *et al.* [245] or Avery and Burkhart [246], and a recent review by Telewski and Lynch [247] for a basic consideration of forest mensuration techniques.

The most common tree measurements collected in basic forestry work are tree diameter and height, since those measurements can be used directly in estimating merchantable stand volume and value. Tree diameters are typically measured at breast-height (DBH, 4.5 ft or 1.4 m above ground) for trees that are sapling size and larger. This height has been traditionally used for stem diameter measurement since it is closely related to total and merchantable stem volume, and can be done accurately. In young trees that have not grown above breast height, stem diameters are usually measured at the root collar. Root collar diameters are usually sensitive indicators of differences in seedling growth rates and vigour, since the foliar surface area above that height is closely related to stem diameter increment. Tree DBH measurements are generally converted to cross sectional area, referred to as stem basal area, and summed to provide an estimate of stand basal area. Basal area estimates have been shown to be useful for determining vegetative cover, while sapwood basal area, more specifically, has been shown to be closely related to foliar surface area across a wide range of ecosystems. Stand basal area estimates are also used in conjunction with estimates of stand density (e.g., stems per hectare) to determine whether stands have adequate amounts of growing stock, given specific silvicultural goals and management strategies. Average stand diameter is typically expressed as the diameter of the tree with average basal area (quadratic diameter) since this term can be used with stand density estimates to estimate total stand basal area, and hence, total stand volume (see [245] for more detail).

Dimensional analysis of trees in fertilizer trials and studies of site productivity suggest that stem diameter be determined along the entire length of the bole to permit analysis of taper and form, and to ensure that measurements are sensitive to tree response. For example, base-green-crown diameter growth is more closely linked to photosynthate production than diameter at breast height on trees where these points are several meters apart. Thorough discussions of mensurational considerations in assessing tree response to environmental stresses can be found in Solomon and Brann [248].

Tree diameter growth rates are not independent of stand density, since diameter is closely related to crown mass above that point, and crown mass is related to the degree of crowding in a stand. Therefore, evaluations of stand growth should also include height measurements. Height growth rates are typically used to assess differences in site quality, as seen in the use of site index curves. However, height has not been found to be changed as much as diameter by fertilizer applications to *Pinus radiata* [249], largely because fertilizer strongly affects crown mass. The greatest utility of height measurements can be found in estimating total stem volume.

Tree diameter and height measurements are converted to estimates of total and merchantable stem volume, using assumptions about stem geometric form, since stems rarely are perfect cylinders due to taper (see [245] for details on volume equations). Tree dimensional measurements should be taken frequently since tree growth rates are generally curvilinear. Optimum frequency will be influenced by tree growth rate, but measurements should be made at least every 5 years.

Many ecologists find that solely measuring tree diameter and height is not adequate to evaluate stand growth response to environmental factors because it is not a direct measure of net primary productivity. Volume estimates do not account for differences in wood density and can not provide information regarding mass or carbon accumulation and allocation patterns by trees. As a result, most studies of tree response to environmental conditions conducted in the last 30 years have measured the weight of various tree components, expressed as oven-dry weight, or biomass. Oven-dry weight is usually determined at 70 °C for ecological studies, especially where nutrient analyses are to be conducted on plant tissue samples. Oven-dry weight is the preferred unit of measure because of differences in moisture content within and between trees that may otherwise confound comparisons.

Tree component moist weights are usually determined by direct measurement in the field, then converted to oven-dry weight following drying of subsamples at 70° C. Estimates of tree nutrient contents can be determined following analysis of tissue samples dried for moisture determinations. Detailed procedures for tissue analysis were evaluated by Walker [250]. Tree subsampling considerations for biomass and nutrient content have been discussed by Valentine *et al.* [251]. Regression analysis is used to determine the relationship between tree size and component weight and nutrient content. Various equation forms are used to predict weight from tree dimensions, depending on the statistical assumptions associated with regression analysis (see [252] or [310]). The resulting weight equations, referred to as biomass or allometric equations, can then be used to estimate mass on a stand level. An efficient technique for determining above-ground stand biomass following appropriate allometric analysis is the basal area ratio method [253]. Estimates are based on the following relationship:

$$\text{plot weight} = (\Sigma \text{ sample tree weight})/(\Sigma \text{ sample tree basal area})$$
$$* \text{ (plot basal area)}.$$

Detailed discussion of statistical considerations associated with stand level estimates of biomass can be found in Madgwick and Satoo [254].

Site Nutrient Availability

Intensive harvesting operations cause significant removal and redistribution of organic matter and nutrients as a result of product removals and delimbing and topping operations. In addition, harvesting may cause significant soil distur-bance, which may mix or remove soil organic or surface mineral horizons. These disturbances, and the temporary elimination of plant uptake on an ecosystem scale shortly after harvesting, may increase levels of plant available nutrients [311] or reduce nutrient availability. Altered nutrient availability can be expected to result in positive or negative changes in plant uptake and tissue nutrient concentrations. However, changes in foliar nutrient concentrations can be caused by both altered nutrient availability and nutrient antagonisms, as well as by rapid biomass growth that results in nutrient dilution within plant tissues. Common questions addressed by intensive harvesting research should include the relation-ship between harvesting impacts and tree nutritional status.

Monitoring and interpreting changes in foliar nutrient concentrations is required for evaluating and testing hypotheses concerning the mechanisms that relate intensive harvesting impacts to tree nutrition. Reviews written by Mead [255] and Timmer [256] provide thorough evaluations of various approaches for diagnosing nutrient deficiencies and interpreting nutritional analyses. While the context for these papers is primarily plantation forests and conifer seedlings, two suggested approaches, evaluation of "critical" nutrient concentrations and vector analysis, have direct application in intensive harvesting research.

The "critical level" of a nutrient may be defined as the concentration below which one might expect a rapid decline in productivity, and a positive response to additions of that nutrient. Evaluation of tissue nutrient status in research trials may be useful in regions where past research has established protocols for tissue collections, and the relationship between species productivity and nutrient concentration or supply is known [257]. In such regions, it may be possible to determine whether intensive harvesting treatments cause significant declines in tissue nutrients, depending on the rigor of the experimental design. However, critical levels established for one region may not be accurate in other regions where factors affecting nutrient availability (e.g., climate and soils) differ [255]. For example, boron availability is affected by soil moisture levels and critical foliar levels of boron established in wet climates may not be adequate for drier regions, where boron uptake may be limited by soil mass flow rates. Lambert [258] discusses additional limitations of foliar analysis for specific objectives under Australian conditions.

The vector analysis approach suggested by Timmer [256] has been demonstrated to be a useful method of interpreting the reasons for changes in foliar nutrient concentrations. This is done by analyzing directional shifts in nutrient concentration, nutrient content, and average foliar dry weight from a reference population of plants to populations affected by various experimental treatments. Vector analysis is best utilized where there is adequate experimental rigor to manipulate plant nutrient status. Perhaps the most significant advantage of vector analysis over the critical level approach is that interpretations are not solely based on nutrient concentration. In addition, it permits one to distinguish between nutrient dilution, deficiency, luxury consumption, and antagonistic relationships between nutrients. This approach, for example, has been shown to be useful in elucidating the reason for reduced foliar boron concentrations in *Pinus radiata* following high urea-N additions in trials where boron uptake was reduced by nitrogen antagonisms following forest floor removal during intensive harvesting [259] and drought [260]. An alternative graphical representation of the effect of interacting nutrients is described in Valentine and Allen [261].

Research conducted in Australia indicates that litter nitrogen concentrations may be a better indication of site nitrogen availability than live foliage. For example, Raison *et al.* [262] reported that annual weighted mean nitrogen concentration in *Pinus radiata* needle litter was positively correlated with nitrogen uptake from soil, and more sensitive to differences in site nitrogen availability than weighted foliar concentrations. Birk and Vitousek [263] similarly found litterfall nitrogen in several *Pinus taeda* plantations to be a good indicator of nitrogen availability. Monitoring litter nitrogen concentrations should thus be an efficient alternative to costly *in situ* nitrogen availability studies in intensive harvesting trials; and is a good example of a technique that links key ecosystem processes in an attempt to measure differences in site quality.

Physiological Function and Leaf Surface Area

Most field trials designed to determine the impacts of intensive harvesting or availability of water and nutrients on site productivity measure crop tree growth rates and biomass production as dependent variables of the experimental treatments. If increased growth or net primary productivity (NPP) is observed in such a study, it is tempting to conclude that the response was due to an underlying increase in net photosynthesis rates and foliar surface area. However, many field studies do not quantify whether basic plant physiological functions, such as photosynthesis, are altered by experimental independent variables. In this sub-section, we review current techniques to determine photosynthetic rates in trees and discuss some evidence indicating whether changes in nutrient availability affect net photosynthesis rates and/or leaf surface area.

Altered nutrient uptake rates may directly affect plant metabolic rates and physiological processes. There is evidence, for example, that nitrogen deficiencies in conifers and hardwoods may reduce the rate of photosynthesis per unit of leaf area [264, 265, 266, 267, 268]. However, there may not be consistent evidence that seedlings and mature trees respond similarly, due to such effects as mutual shading of foliage in large crowns, and several age classes of foliage with differing nutrient status and vigour. Research with mature trees suggests that nutrient deficiencies may result in greater reductions in the development of leaf surface area than on the rate of photosynthesis per unit of leaf area [269, 270, 271]. Macronutrients other than nitrogen have not been shown to have such an effect on photosynthetic rates, although balanced nutrition is essential [267].

Techniques for determining photosynthesis rates in tree foliage have been recently reviewed by Leverenz and Hallgren [272]. Several methods for estimating photosynthetic rates are available for intact leaves, including a photoacoustic method; chlorophyll fluorescence and luminescence measurement; and measurement of CO_2 and O_2 gas exchange rates using several techniques. Measurement of CO_2 gas exchange rates by infrared gas analyzer is currently the most commonly used method; and considered the most precise and accurate under a wide range of conditions by Leverenz and Hallgren [272]. While it may be possible to accurately measure tree photosynthetic response to changing environmental conditions for individual leaves or shoots, it may be difficult to scale up from individual leaf or shoot photosynthetic rates to stand level estimates. These difficulties may be overcome through the development of models to estimate the distribution of sun and shade leaves in a stand (e.g., [273]); however, this work is not practical for many forest stands that have mixed species and age classes.

Variability in field conditions associated with intensive harvesting trials may make it difficult and very expensive to detect significant changes in plant physiological processes. These difficulties will hinder much progress in determining whether increased foliar mass is due to increased rates of C capture, decreased respiration, or decreased allocation of C to below-ground growth, as claimed by Gower *et al.* [274]. However, in the case of N and P, the net effect of increased plant nutrient uptake should be manifested in increased leaf surface area and foliar mass per plant, as has been observed for *Pseudotsuga menziesii* [275, 276], *Pinus nigra* [277], *Pinus elliottii* [278], and *Pinus sylvestris* [279]. Altered foliar area, or leaf area index, can be expected to directly affect net carbon fixation and whole-plant productivity.

Since leaf area is such a basic determinant of forest productivity, intensive harvesting trials should include techniques to destructively sample foliage in order to periodically estimate leaf surface area and foliar mass on an individual plant and stand basis. An efficient technique for estimating foliar surface area in conifers utilizes the product of Specific Leaf Area (SLA, ratio of fresh needle area to dry mass) and foliar biomass (see [274]). Estimates of SLA are obtained for sample needles collected in representative crown positions, and foliar biomass from destructive sampling and allometric equations. Gower *et al.* [280] have developed a technique for estimating leaf area with a video camera linked to an image processing system. Similar approaches can be used efficiently for broadleaved species. Detailed discussion of leaf area measurements can be found in Larsen and Kershaw [313].

Carbon Allocation

At a fundamental level, forest managers must determine how to efficiently maximize carbon storage in the merchantable portions of trees. From this perspective, silvicultural practices are designed for various stages of stand development in order to economically increase the rate of carbon storage by crop species. For example, site preparation, planting, and weed control practices increase the speed of carbon storage by the desired tree species; thinning increases the rate of carbon storage on an individual crop tree basis; and fertilizer and irrigation treatments increase carbon storage rates by overcoming limiting site factors. However, recent research indicates that most edaphic and environmental factors that define site productivity have significant impacts on all aspects of plant carbon storage. More specifically, light, temperature, atmospheric composition, moisture, nutrients, and other biota affect rates of carbon capture, loss via respiration, and within-plant allocation. Understanding the impacts of silvicultural practices on carbon capture and storage in various plant organs should be a basic objective of intensive harvesting research. In this section we discuss the effect of site quality on allocation of carbon to above- and below-ground parts of plants.

Research with conifers indicates that nitrogen fertilizer alters carbon allocation patterns in the above-ground parts of the tree. These allocation shifts can have positive and negative consequences for production forestry depending on the relative growth of foliage, branches, and stemwood. Nitrogen fertilization stimulates greater production of branch biomass relative to stems in *Pseudotsuga menziesii* [275] and *Pinus radiata* [249, 259, 281, 282]; and increases stemwood production per unit of foliage in *Pinus sylvestris* and *Picea abies* [283]. Similar dynamics in above-ground components have not been demonstrated for other nutrients. Above-ground dry matter partitioning may be affected by intensive harvesting if nitrogen availability is affected for an extended period of time. Gains in productivity caused by increased nutrient availability may be offset economically by declines in value if terminal (main stem) dominance is reduced, and branch diameters (knot sizes) are increased. Declines in tree form have been demonstrated in *Pinus radiata* following nitrogen fertilization; and on old-field sites with high residual phosphorus availability [284]. Intensive harvesting trials should be designed to evaluate carbon allocation patterns, since a change favoring large branches at the expense of stemwood has significant economic implications.

Below-ground carbon dynamics in forests are relatively poorly understood, yet recent studies estimate that fine root systems (often <2 mm diameter) are a major sink for carbon, perhaps accounting for 8 to 67% of NPP [285]. Below-ground carbon allocation patterns are affected by site nutrient status. Nutrient-poor sites have generally been found to have greater fine root biomass and reduced shoot-to-root ratios than nutrient-rich sites [286, 287]. Experimental increases in nitrogen and water availability have resulted in reduced carbon allocation to fine roots [274, 288]. Other nutrients may have differential effects. For example, magnesium-deficient trees have been shown to develop less fine root length and to increase shoot-to-root ratios [289]. The results indicate the need to clarify the effects of individual nutrient availability and nutrient inter-actions on carbon allocation patterns. Coarse root biomass (>2 mm diameter) has been shown to change proportionally to stem diameter [290], and should be pos-itively correlated with nutrient availability.

Relatively basic questions regarding fine root dynamics are very difficult to answer because of the difficulty in directly measuring fine root physiological function, growth, and mortality without altering the rhizosphere. Methods to estimate fine root dynamics in response to site factors include the use of repeated core sampling [291]; combined repeated coring and budget estimation [292]; ingrowth cores [114]; combined repeated coring and root observation windows [287]; mini-rhizotron video observation tubes (Bartz Technology Co., Santa Barbara, CA, USA) [293]; and a variety of elemental budgeting approaches [91, 285].

Estimates derived from any one of these techniques may be biased by both sampling methodology and conceptual assumptions [294]. For example, inten-sive harvesting or normal seasonal variations in ecosystem processes may affect the fine root standing crop either through changes in carbon allocation to fine roots, or through changes in fine-root decomposition; but repeated determination of the standing crop biomass of fine roots can not distinguish between these alternatives. Researchers are cautioned to adapt sampling methods to specific sites and types of vegetation [115], and to seek new root production estimation techniques that do not rely solely on biomass data [295]. Some combination of estimates may be desired until additional information is available. For example, Santantonio [296] suggested that an examination of the correlation between fine-root production and foliar mass will facilitate understanding within-tree functional relationships, especially under rigorous experimental conditions where site variables are manipulated. We recommend that researchers evaluate methods discussed by Vogt and Persson [297] in their review of root sampling methods.

Nutrient and Water Availability

Intensive field studies involving experimental manipulations of both nutrient and water availability have been conducted for only three forest types: *Pinus radiata* in Australia [298], *Pinus sylvestris* in Sweden [279], and *Pseudotsuga menziesii* var. glauca in New Mexico [274]. The results of these studies have indicated that responses in canopy development and peak needle mass to both increased water availability and nutrient status must be understood in the context of an interaction between these environmental factors [266, 274, 298]. Generally, fertilizer increases leaf area by greater amounts than irrigation alone; although increased availability of both water and nutrients results in the greatest increase. Similar interactions have been observed in 24-year-old Pacific coastal *Pseudotsuga menziesii* after fertilization and increased water availability due to thinning [299]. In the latter study, the mechanism behind the interaction can be partially explained by an increase in water use efficiency, attributed to changes in needle stomatal control over water loss, that was caused by nitrogen fertilizer.

Research experience in the Australian Capital Territory has shown that trees will respond to strong climatic stresses, such as drought, by reducing their foliar surface area [300]. The changes in foliar mass after the onset of such events will be a function of crown size prior to the new stress, and the severity of stress which forces a new equilibrium crown carrying capacity. Thus, stand crown mass for a species at a given age is a direct function of climatic conditions and site fertility; and every combination of these factors will result in some maximum crown carrying capacity.

Research in southwest Victoria, Australia has indicated that intensive harvesting may have significant impact on soil moisture availability through changes in forest floor and harvesting residues [301]. In this semi-arid region, conservation of organic residues can help to maintain forest productivity by conserving nutrients contained in the organic matter, and through a mulching effect, which conserves soil moisture. These results highlight the importance of considering interactions between nutrients and water in experimental designs for intensive harvesting trials, even where experimental manipulations of irrigation or fertilizer are not planned.

The studies which have experimentally manipulated both water and nutrients have demonstrated that it is essential to verify changes in tree water status with direct measurements. From a practical perspective, plant water status has a major effect on plant physiological processes [302] and primary production. Techniques for measuring tree water status were reviewed recently by Pallardy *et al.* [303]. Although early studies measured plant water content (water mass per mass plant tissue dry weight) to quantify plant water status, studies during the past 30 years have used plant water potential (ψ, expressed in terms of pressure or tension, MPa) because of a closer relationship between plant physiological processes and water potential than water content. This is also true for quantifying soil moisture status, as described earlier in this chapter. The most reliable method for determining plant water potential in the field is a pressure chamber system, also referred to as a Scholander bomb [304]. Ritchie and Hinckley [305] have reviewed operating considerations for pressure chamber systems.

Instantaneous measures of ψ in the field may not be adequate for predicting tree productivity if care is not taken to relate these measures to some more cumulative estimate of plant water stress [303]. Recent work suggests that predawn leaf water potential may be a reasonable measure of plant water status, since $\psi leaf = \psi root = \psi soil$ may be expected after overnight equilibration of the soil-plant water system. Raison and Myers [298] highlight the usefulness of an index, developed by Myers [306] for *Pinus radiata* in Australia, for determining tree water status. This index, termed Water Stress Integral, integrates pre-dawn needle water potential over time as an index of cumulative plant water stress. The Water Stress Integral may have application in intensive harvesting research where site disturbance may affect both nutrient and water availability.

Basic Experimental Design and Mensurational Considerations
Field designs for intensive harvesting trials must be selected carefully to provide rigorous tests of research hypotheses. In some cases, this may suggest planting densities and weed control strategies not conforming to conventional silvicultural practices for a given region. For example, research trial plot stocking may be

higher than in conventional plantations in order to achieve crown closure and full site occupancy at an early age. Intensive harvesting trials established in New Zealand have utilized *Pinus radiata* at high planting density (2m x 2m spacing) to achieve early site occupancy and high rates of nutrient uptake demand. In these trials, the trees are essentially a bio-assay of the nutrient supplying capacity of the site. Stand density will be reduced from 2500 stems/ha to 1250 stems/ha soon after initial crown closure; and reduced with a second, final thinning to 625 stems/ha after crown closure is achieved a second time. Thinning will be systematic, rather than random or to select for crop trees, in order to maintain uniform spacing. Foliar surface area will not be reduced by live-crown pruning to maintain treatment related differences in crown mass. These silvicultural practices should place maximum potential demand on site nutrient supplying capacity; will permit evaluation of treatment effects on individual tree growth and form; and will maintain adequate sample size for the duration of the experiment (one rotation). These practices may differ from those used throughout New Zealand; however, they are efficient for testing hypotheses concerning tree response to harvesting and nutrient supply manipulations.

We recommend that field trials be designed after suggestions of Burger and Powers [1]. Researchers should consider eliminating weed competition from trial plots in order to provide a more rigorous test of the effects of harvesting distur-bances on site productivity, and avoid confounding the effects of weeds and harvest [307]. The results of Squire *et al.* [301] indicate that stocking differences between trials can directly affect estimates of site productivity. They demonstrated that volume production would vary directly with stocking until the onset of competition between trees, which would be expected to be appreciable by 2 years of age, and intense at age 5 years in *Pinus radiata*. Basically, increased stocking would result in increased total volume production in the early years; and reduced average volume production per tree after the onset of competition. These results suggest that trial stocking be standardized. Additional discussion of research design considerations can be found in Dyck and Mees [308, 309].

DISCUSSION

Site productivity is the unifying term that identifies the growth environment created by a combination of physical, chemical, and biologic site characteristics and incorporates the response by the target production crops to this particular environment. Some site productivity factors are beyond the control of the manager (e.g., climate, natural disturbance), while others are clearly affected by management practices. This chapter has attempted to review a variety of tools and measurement techniques that may be used in process studies to qualify and quantify changes in site productivity brought about by intensive harvesting. Site manipulations can affect site productivity by altering soil physical properties; by affecting chemical interactions between the solid, gaseous, and solution phases in the soil; by influencing the abundance and activity of the soil flora and fauna; by influencing processes at the plant level; or more likely by a combination of any of the above.

For practical reasons, the site productivity factors were separated into different categories and discussed separately. In actual field situations, however, there is a constant interplay between these different site characteristics and key processes, with site productivity ultimately determined by that interplay. Also, not all the factors / processes that have been discussed here are relevant to all sites at all times. While this chapter provides some guidance, the investigator is still faced with a double challenge: 1) to merge programmatic, financial, and logistical considerations with the need or desire for an integrated and comprehensive study approach; and 2) to identify, out of the list measurement protocols, those measurements / processes that are most likely to be crucial to the issues at hand. For those reasons, a clear statement of objectives and a well-conceived experimental design *at the onset* of the investigation are key to the success of any process study, as they ultimately determine the level of intensity of the measurements and the site productivity components the study should focus on.

ACKNOWLEDGMENTS

Helga Van Miegroet was supported through funding by the Biofuels Feedstock Program, Biofuels Systems Division of the Department of Energy under contract DE-AC05-84OR21400 with Martin Marietta Energy Systems, Inc. during the preparation of this chapter while employed at the Environmental Sciences Division of Oak Ridge National Laboratory. The contribution of Darlene Zabowski was supported by the United States Department of Agriculture - Forest Service, Pacific Northwest Research Station, Forestry Sciences Laboratory in Wenatchee, Washington. Tat Smith acknowledges the support of the College of Life Sciences and Agriculture of the University of New Hamsphire during the preparation of this chapter.

LITERATURE CITED

1. Burger, J.A. and Powers, R.F. Field designs for testing hypotheses in long-term site productivity studies. In: *Long-term Field Trials to Assess Environmental Impacts of Harvesting*. Proceedings, IEA/BE T6/A6 Workshop, Florida, USA. February 1990. (Eds.) W.J. Dyck and C.A. Mees. IEA/BE T6/A6 Report No. 5. Forest Research Institute, Rotorua, NZ, FRI Bulletin No. 161, 1991, 79-105.

2. Mead, D.J., Whyte, A.G.D., and Woollons, R.C. Designing long-term experiments to study harvesting impacts. In: *Long-term Field Trials to Assess Environmental Impacts of Harvesting*. Proceedings, IEA/BE T6/A6 Workshop, Florida, USA. (Eds.) W.J. Dyck and C.A. Mees. IEA/BE T6/A6 Report No. 5. Forest Research Institute, Rotorua, NZ, FRI Bulletin No. 161, 1991, pp. 107-124.

3. Hurlbert, S.H. Pseudoreplication and the design of ecological field experiments. *Ecological Monographs*, 1984, **54**, 187-211.

4. Stewart-Oaten, A., Murdoch, W.W., and Parker, K.R. Environmental impact assessment: pseudoreplication in time? *Ecology*, 1986, **67**, 929-940.

5. Carpenter, S.R., Frost, T.M., Heisey, D., and Kratz, T. Randomized intervention and the interpretation of whole-ecosystem experiments. *Ecology*, 1989, **70**, 1141-1152.

6. Eberhardt, L.L. and Thomas, J.M. Designing environmental field studies. *Ecological Monographs*, 1991, 61:53-73.

7. Hicks, B.B., Weseley, M.L., and Durham, J.L. Critique of methods to measure dry deposition: Workshop Summary. EPA 600/9-80-050. U.S. Environmental Protection Agency, Environmental Sciences Research Laboratory, Research Triangle Park, North Carolina, 1980.

8. Rosen, K. Measuring nutrient inputs to terrestrial ecosystems. In: *Nutrient Cycling in Terrestrial Ecosystems: Field Methods, Application, and Interpretation.* (Eds). A.F. Harrison, P. Ineson, and O.W. Heal. Elsevier Applied Science, London and New York, 1990, pp. 1-10.

9. Likens, G.E., Bormann, F.H., Pierce, R.S., Eaton, J.S., and Johnson, N.M. *Biogeochemistry of a Forested Ecosystem.* Springer-Verlag, New York, 1977, 146p.

10. Ulrich, B. and Pankrath, J. (Eds.). Effects of Accumulation of Air Pollutants in Forest Ecosystems. Proceedings of a workshop held at Göttingen, West Germany, May 16-18, 1982. D. Reidel Publ. Co., Dordrecht, The Netherlands, 1983.

11. Fowler, D., Cape, J.N., and Leith, I.D. Field methods for determining the atmospheric inputs of major plant nutrients. In: *Nutrient Cycling in Terrestrial Ecosystems: Field Methods, Application, and Interpretation.* (Eds.) A.F. Harrison, P. Ineson, and O.W. Heal. Elsevier Applied Science, London and New York, 1990, pp. 36-45.

12. Lindberg, S.E., Harris, R.C., Hoffman, W.A. Jr., Lovett, G.M., and Turner, R.R. Atmospheric chemistry, deposition, and canopy interactions. In: *Analysis of Biogeochemical Cycling Processes in Walker Branch Watershed.* (Eds.) D.W. Johnson and R.I. Van Hook. Springer-Verlag, New York, 1989, pp. 96-163.

13. Lindberg, S.E., Johnson, D.W., Van Miegroet, H., Taylor, G.E. Jr., and Owens, J.G. Sampling and analysis protocols and project description for the Integrated Forest Study. ORNL/TM 11214. Oak Ridge National Laboratory, Oak Ridge, Tennessee, 1989.

14. Richter, D.D. and Lindberg, S.E. Wet deposition estimates for long-term bulk and event wet-only samples of incident precipitation and throughfall. *Journal of Environmental Quality,* 1988, **17**, 619-622.

15. Lindberg, S.E. Atmospheric deposition and canopy interactions of sulfur. In: *Atmospheric Deposition and Nutrient Cycling in Forest Ecosystems.* (Eds.) D.W. Johnson and S.E. Lindberg. Springer-Verlag, New York & Berlin, 1992, pp. 74-90.

16. Lovett. G.M. Rates and mechanisms of cloud water deposition to a balsam fir forest. *Atmospheric Environment,* 1984, **18**, 361-367.

17. Ineson, P. Field methods for estimation of nutrient inputs to terrestrial ecosystems. In: *Nutrient cycling in terrestrial ecosystems: Field methods, Application, and Interpretation.* (Eds.) A.F. Harrison, P. Ineson, and O.W. Heal. Elsevier Applied Science, London and New York, 1990, pp. 69-74.

18. Lindberg. S.E., Bredemeier, M., Schaefer, D.E., and Qi, L. Atmospheric concentrations and deposition of nitrogen and major ions in conifer forests in the United States and Federal Republic of Germany. *Atmospheric Environment*, 1990, **24A**, 2207-2220.

19. Lindberg, S.E., Lovett, G.M., Richter, D.D., and Johnson, D.W. Atmospheric deposition and canopy interactions of major ions in a forest. *Science*, 1986, **231**, 141-145.

20. Lovett, G.M. Atmospheric deposition and canopy interactions of nitrogen. In: *Atmospheric Deposition and Nutrient Cycling in Forest Ecosystems.* (Eds.) D.W. Johnson and S. Lindberg. Springer-Verlag, New York & Berlin, 1992, pp. 152-166.

21. Gordon, J.C. and Wheeler, C.T. *Biological Nitrogen Fixation in Forest Ecosystems: Foundations and Applications.* Martinus Nijhoff/Dr W. Junk Publishers, The Hague, Netherlands, 1983.

22. Weaver, R.W. and Frederick, L.R. Rhizobium. In: *Methods of Soil Analysis. Part 2. Chemical and Microbiological Properties*, Second Edition. (Eds.) A.L. Page, R.H. Miller, D.R. Keeney. ASA-SSSA, Madison, Wisconsin, 1982, pp. 1043-1070.

23. Cole, D.W., Gessel, S.P., and Turner, J. Comparative mineral cycling in red alder and Douglas-fir. In: *Utilization and Management of Alder.* (Eds.) D.G. Briggs, D.S. BeBell, and W.A. Atkinson. USDA Forest Service General Technical Report PNW-70, Pacific Northwest Forest Range Experiment Station, Portland, Oregon, 1978, pp. 327-336.

24. Bormann, B.T. and DeBell, D.S. Nitrogen content and other soil properties related to age of red alder stands. *Soil Science Society of America Journal*, 1981, **45**, 428-432.

25. Binkley, D. Nodule biomass and acetylene reduction rates of red alder and Sitka alder on Vancouver Island, B.C. *Canadian Journal of Forest Research*, 1981, **11**, 281-286.

26. Bormann, F.H. and Likens, G.E. Nutrient cycling – Small watersheds can provide invaluable information about terrestrial ecosystems. *Science*, 1967, **155**, 424-429.

27. Bormann, F.H. and Likens, G.E. *Pattern and Process in a Forested Ecosystem.* Springer-Verlag, New York, 1979.

28. Swank, W.T. and Crossley, D.A. Jr. (Eds.) *Forest Hydrology and Ecology at Coweeta.* Ecological Series 66. Springer-Verlag, New York, 1987, 469 p.

29. Johnson, D.W. and Van Hook, R.I. (Eds.). *Analysis of Biogeochemical Cycling Processes in Walker Branch Watershed.* Springer-Verlag, New York, 1989, 401 p.

30. Reynolds, B., Hudson, J.A., and Leeks, G. Field methods for estimating solute and sediment losses in small upland streams. In: *Nutrient Cycling in*

Terrestrial Ecosystems: Field Methods, Application, and Interpretation. (Eds.) A.F. Harrison, P. Ineson, and O.W. Heal. Elsevier Applied Science, London and New York, 1990, pp. 103-129.

31. Haines, B.L., Waide, J.B., and Todd, R.L. Soil solution nutrient concentrations sampled with tension and zero-tension lysimeters: Report of discrepancies. *Soil Science Society of America Journal*, 1982, **46**, 658-661.

32. Vose, J.M. and Swank, W.T. Water balances. In: *Atmospheric Deposition and Nutrient Cycling in Forest Ecosystems.* (Eds) D.W. Johnson and S.E. Lindberg. Springer-Verlag, New York and Berlin, 1992, pp. 27-49.

33. Swift, L.W. Jr. Forest access roads: Design, maintenance, and soil loss. In: *Forest Hydrology and Ecology at Coweeta.* (Eds.) W.T. Swank and D.A. Crossley Jr. Ecological Series 66. Springer-Verlag, New York, 1987, pp. 313-337.

34. Hornung, M. Measurement of nutrient losses resulting from soil erosion. In: *Nutrient Cycling in Terrestrial Ecosystems: Field Methods, Application, and Interpretation.* (Eds.) A.F. Harrison, P. Ineson, and O.W. Heal. Elsevier Applied Science, London and New York, 1990, pp. 80-102.

35. Reichle, D.E. *Analysis of Temperate Forest Ecosystems.* Springer-Verlag, New York, 1970.

36. Tritton, L.M. and Hornbeck, J.W. Biomass equations for major tree species of the Northeast. USDA Forest Service General Technical Report NE-69, 1982, 46 p.

37. Johnson, D.W., West, D.C., Todd, D.E., and Mann, L.K. Effects of sawlog vs. whole-tree harvesting on nitrogen, phosphorus, potassium, and calcium budgets of an upland mixed oak forest. *Soil Science Society of America Journal*, 1982, **46**, 1304-1309.

38. Smith, C.T. Jr., McCormack, M.L. Jr., Hornbeck, J.W., and Martin, C.W. Nutrient and biomass removals from red spruce-balsam fir whole-tree harvest. *Canadian Journal of Forest Research*, 1986, **16**, 381-388.

39. Comerford, N.B. and Leaf, A.L. An evaluation of techniques for sampling forest tree nutrient content. I. Sampling the crown for total nutrient content. *Forest Science*, 1982, **28**, 471-480.

40. Comerford, N.B. and Leaf, A.L. An evaluation of techniques for sampling forest tree nutrient content. II. Sampling for stem nutrient content. *Forest Science*, 1982, **28**, 481-487.

41. Morrison, I.K. Within-tree variation in mineral content of leaves of young balsam fir. *Forest Science*, 1974, **20**, 276-278.

42. Turner, J., Dice, S.F., Cole, D.W., and Gessel, S.P. Variation of nutrients in forest tree foliage – A Review. University of Washington, Institute of Forest Products Contr. 35. College of Forest Resources, University of Washington, Seattle, WA, 1978.

43. Cole, D.W. and Rapp, M. Elemental cycling in forest ecosystems. In: *Dynamic Properties of Forest Ecosystems*. (Ed.) D.E. Reichle. Cambridge University Press, Cambridge, United Kingdom.. 1980, pp. 341-409.

44. Blair, J.M. and Crossley, D.A. Jr. Litter decomposition, nitrogen dynamics and litter microarthropods in a southern Appalachian hardwood forest 8 years following clearcutting. *Journal of Applied Ecology*, 1988, **25**, 683-698.

45. Richards, B.N. *Introduction to the Soil Ecosystem*. Second Printing. Longman Group Ltd, London, United Kingdom, 1976.

46. Pimm, S.L., Lawton, J.H., and Cohen, J.E. Food web patterns and their consequences. **Nature**, 1991, **350**, 669-674.

47. Moore, J.C. The influence of microarthropods on symbiotic and non-symbiotic mutualism in detrital-based below-ground food webs. *Agriculture, Ecosystem and Environment*, 1988, **24**, 147-159.

48. Moore, J.C. and de Ruiter, P.C. Temporal and spatial heterogeneity of trophic interactions within below-ground food webs. *Agriculture, Ecosystem, and Environment*, 1991, **34**, 371-397.

49. Satchell, J.E. Soil and vegetation changes in experimental birch plots on a Caluna podzol. *Soil Biology and Biochemistry*, 1980, **12**, 303-310.

50. Satchell, J.E. Earthworm populations of experimental birch plots on a Caluna podzol. *Soil Biology and Biochemistry*, 1980, **12**, 311-316.

51. Springett, J.A. Effect of introducing *Allobophora longa* (Ude) on root distribution and some soil properties in New Zealand pasture. In: *Ecological Interactions in Soils: Plants, Microbes and Animals*. (Ed.) A.H. Fitter. Blackwell, Oxford, 1985, pp. 399-405.

52. Edwards, C.A. and Bater, J.E. The use of earthworms in environmental management. *Soil Biology and Biochemistry*, 1992, **24**, 1683-1689.

53. Shaw, C.H., Lundkvist, H., Moldenke, A., and Boyle, J. The relationship of soil fauna to long-term forest productivity in temperate and boreal eco-systems: Processes and research strategies. In: *Long-term Field Trials to Assess Environmental Impacts of Harvesting*. (Eds.) W.J. Dyck and C.A. Mees. IEA/BE T6/A6 Report No. 5. Forest Research Institute, Rotorua, New Zealand, FRI Bulletin No. 161, 1991, pp. 39-78.

54. Huhta, V., Nurminen, M., and Valpas, A. Further notes on the effect of silvi-cultural practices upon the fauna of coniferous forest soil. *Annales Zoologici Fennici*, 1969, **6**, 327-334.

55. Sohlenius, B. Short-term influence of clear-cutting on abundance of soil microfauna (Nematoda, Rotatoria and Tardigrada) in a Swedish pine forest soils. *Journal of Applied Ecology*, 1982, **19**, 349-359.

56. Lundkvist, H. Effects of clearcutting on the enchytraeids in a Scots pine forest soil in central Sweden. *Journal of Applied Ecology*, 1983, **20(3)**, 873-885.

57. Edwards, C.A. The assessment of populations of soil-inhabiting invertebrates. In: *Modern Techniques in Soil Ecology*. (Eds.) D.A. Crossley Jr., D.C. Coleman, P.F. Hendrix, W. Cheng, D.W. Wright, M.H. Beare, and C.A. Edwards. Elsevier, Amsterdam, 1991, pp. 145-176.

58. Page, A.L., Miller, R.H., and Keeney, D.R. (Eds.) *Methods of Soil Analysis, Part 2. Chemical and Microbiological Properties*, Second Edition, ASA-SSSA, Madison, Wisconsin, 1982.

59. Wollum, A.G. II. Cultural methods for soil microorganisms. In: *Methods of Soil Analysis. Part 2. Chemical and Microbiological Properties*, Second Edition. (Eds.) A.L. Page, R.H. Miller, and D.R. Keeney. ASA-SSSA, Madison, Wisconsin, 1982, pp. 781-802.

60. Parkinson, D. Filamentous fungi. In: *Methods of Soil Analysis. Part 2. Chemical and Microbiological Properties*, Second Edition. (Eds.) A.L. Page, R.H. Miller, and D.R. Keeney. ASA-SSSA, Madison, Wisconsin, 1982, pp. 949-968.

61. Williams, S.T. and Wellington, E.M.H. Actinomycetes. In: *Methods of Soil Analysis. Part 2. Chemical and Microbiological Properties*, Second Edition. (Eds.) A.L. Page, R.H. Miller, and D.R. Keeney. ASA-SSSA, Madison, Wisconsin, 1982, pp. 969-988.

62. Gray, T.R.G. and Williams, S.T. *Soil Micro-organisms*. Longman Group Ltd., London, United Kingdom, 1979.

63. Schmidt E.L. and Paul, E.A. Microscopic methods for soil microorganisms. In: *Methods of Soil Analysis. Part 2. Chemical and Microbiological Properties*, Second Edition. (Eds.) A.L. Page, R.H. Miller, and D.R. Keeney. ASA-SSSA, Madison, Wisconsin, 1982, pp. 803-820.

64. Alexander, M. Most probable number method for microbial populations. In: *Methods of Soil Analysis. Part 2. Chemical and Microbiological Properties*, Second Edition. (Eds.) A.L. Page, R.H. Miller, and D.R. Keeney. ASA-SSSA, Madison, Wisconsin, 1982, pp. 815-820.

65. Schmidt, E.L. and Belser, L.W. Nitrifying bacteria. In: *Methods of Soil Analysis. Part 2. Chemical and Microbiological Properties*, Second Edition. (Eds.) A.L. Page, R.H. Miller, and D.R. Keeney. ASA-SSSA, Madison, Wisconsin, 1982, pp. 1027-1042.

66. Hankinson, T.R. and Schmidt, E.L. Examination of an acid forest soil for ammonia- and nitrite-oxidizing autotrophic bacteria. *Canadian Journal of Microbiology*, 1984, **30**, 1125-1132.

67. Parkinson, D. and Paul, E.A. Microbial biomass In: *Methods of Soil Analysis. Part 2. Chemical and Microbiological Properties*, Second Edition. (Eds.) A.L. Page, R.H. Miller, and D.R. Keeney. ASA-SSSA, Madison, Wisconsin, 1982, pp. 820-831.

68. Jenkinson, D.S. and Powlson, D.S. The effects of biocidal treatments on the metabolism in soil. V. A method for measuring soil biomass. *Soil Biology and Biochemistry*, 1976, **8**, 209-213.

69. Brookes, P.C., Powlson, D.S., and Jenkinson, D.S. Measurement of microbial phosphorus in soil. *Soil Biology and Biochemistry*, 1982, **14**, 319-329.

70. Brookes, P.C., Landman, A., Pruden, G., and Jenkinson, D.S. Chloroform fumigation and the release of soil nitrogen : a rapid extraction method to measure microbial nitrogen in soil. *Soil Biology and Biochemistry*, 1985, **17**, 837-842.

71. van Elsas, J.D. and Waalwijk, C. Methods for the detection of specific bacteria and their genes in the soil. *Agriculture Ecosystems and Environment*, 1991, **34**, 97-105.

72. Torsvik, V., Salte, K., Sørhem, R., and Goksøyr, J. Comparison of phenotypic diversity and DNA heterogeneity in a population of soil bacteria. *Applied and Environmental Microbiology*, 1990, **56**, 776-781.

73. Torsvik, V., Goksøyr, J., and Daae, F.L. High diversity in DNA of soil bacteria. *Applied and Environmental Microbiology*, 1990, **56**, 782-787.

74. Giovanetti, M. and Moss, B. An evaluation of techniques for measuring vesicular arbuscular mycorrhizal infection in roots. *New Phytologist*, 1980, **84**, 489-500.

75. Schenk, N.C. (Ed.) *Methods and Principles of Mycorrhizal Research.* American Phytopathological Society, St. Paul, Minnesota, 1982.

76. Reid, C.P.P. Mycorrhizae. In: *The Rhizosphere.* (Ed.) J.M. Lynch. John Wiley and Sons, Chichester, United Kingdom, 1990, pp. 317-353.

77. Zuberer, D.A. Soil and rhizosphere aspects of N_2-fixing plant-microbe associations. In: *The Rhizosphere.* (Ed.) J.M. Lynch. John Wiley and Sons, Chichester, United Kingdom, 1990, pp. 281-315.

78. Paul, E.A. and Johnson, R.L. Microscopic counting and adenosine 5'-triphosphate measurement in determining microbial growth in soils. *Applied and Environmental Microbiology*, 1977, **34**, 263-269.

79. Eiland, F. An improved method for determination of adenosine triphosphate (ATP) in soil. *Soil Biology and Biochemistry*, 1979, **11**, 31-35.

80. Jenkinson, D.S. and Oades, J. A method for measuring adenosine triphosphate in soil. *Soil Biology and Biochemistry*, 1979, **11**, 193-199.

81. Spalding, B. Enzymatic activities related to the decomposition of coniferous leaf litter. *Soil Science Society of America Journal*, 1981, **41**, 622-627.

82. Nannipieri, P., Johnson, R.L., and Paul, E.A. Criteria for microbial growth and activity in soil. *Soil Biology and Biochemistry*, 1978, **10**, 223-229.

83. Tabatabai, M.A. Soil enzymes. In: *Methods of Soil Analysis. Part 2. Chemical and Microbiological Properties*, Second Edition. (Eds). A.L. Page, R.H. Miller, D.R. Keeney. ASA-SSSA, Madison, Wisconsin, 1982, pp. 903-947.

84. Skujins, J.J. Enzymes in soil. In: *Soil Biochemistry.* (Eds.) A.D. MacLaren and G.H. Peterson. Arnold, London, 1967, pp. 371-414.

85. Anderson, J.P.E. Soil respiration. In: *Methods of Soil Analysis. Part 2. Chemical and microbiological properties*. Second Edition. (Eds.) A.L. Page, R.H. Miller, and D.R. Keeney. ASA-SSSA, Madison, Wisconsin, 1982, pp. 831-871.

86. Witkamp, M. Decomposition of leaf litter in relation to environment, microflora, and microbial respiration. *Ecology*, 1966, **47**, 194-201.

87. Howard, P.J.A. A method for the estimation of carbon dioxide evolved from the surface of soil in the field. *Oikos*, 1966, **17**, 267-271.

88. Edwards, N.T. The use of soda-lime for measuring respiration rates in terrestrial ecosystems. *Pedobiologia*, 1982, **23**, 321-330.

89. Edwards, N.T. A moving chamber design for measuring soil respiration rates. *Oikos*, 1974, **25**, 97-101.

90. Anderson, J.M. Carbon dioxide evolution from two temperate, deciduous woodland soils. *Journal of Applied Ecology*, 1973, **10**, 361-378.

91. Raich, J.W. and Nadelhoffer, K.J. Belowground carbon allocation in forest ecosystems: global trends. *Ecology*, 1989, **70**, 1346-1354.

92. Schwartzkopf, S.H. An open chamber technique for the measurement of carbon dioxide evolution from soils. *Ecology*, 1978, **59**, 1062-1068.

93. Klein, D.A., Mayeux, P.A., and Seaman, S.L. A simplified unit for evaluation of soil core respirometric activity. *Plant and Soil*, 1972, **36**, 177-183.

94. Jorgensen, J.R. and Wells, C.G. The relationship of respiration in organic and mineral soil layers to soil chemical properties. *Plant and Soil*, 1973, **39**, 373-387.

95. Edwards, N.T. and Harris, W.F. Carbon cycling in a mixed deciduous forest floor. *Ecology*, 1977, **58**, 431-437.

96. Gloser, J. and Tesarova, M. Litter, soil, and root respiration measurement. An improved compartmental analysis method. *Pedobiologia*, 1978, **18**, 76-81.

97. Persson, H., Lundkvist, H., Wiren, A., Hyvonen, R., and Wessen, B. Effects of acidification and liming on carbon and nitrogen mineralization and soil organisms in mor humus. *Water, Air, and Soil Pollution*, 1989, **45**, 77-96.

98. Edwards, N.T. and Ross-Todd, M. Soil carbon dynamics in a mixed deciduous forest following clear-cutting with and without residue removal. *Soil Science Society of America Journal*, 1983, **47**, 1014-1021.

99. Swift, M.J., Heal, O.H., and Anderson, J.M. *Studies in Ecology, Volume 5: Decomposition in Terrestrial Ecosystems*. Blackwell, Oxford, 1979.

100. Meentemeyer, V. Macroclimate and lignin control of litter decomposition rates. *Ecology*, 1978, **59**, 465-472.

101. Vogt, K.A., Grier, C.C., Meier, C.E., and Keyes, M.R. Organic matter and nutrient dynamics in forest floors of young and mature *Abies amabilis* stands in western Washington, as affected by fine-root input. *Ecological Monographs*, 1983, **53**, 139-157.

102. Edmonds, R.L. Decomposition and nutrient release in Douglas-fir needle litter in relation to stand development. *Canadian Journal of Forest Research*, 1979, **9**, 132-140.

103. Berg, B. and Staaf, H. Decomposition rate and chemical changes of Scots pine needle litter. I. Influence of stand age. *Ecological Bulletin* (Stockholm), 1980, **32**, 363-372.

104. Berg, B. and Staaf, H. Leaching, accumulation and release of nitrogen in decomposing forest litter. *Ecological Bulletin* (Stockholm), 1981, **33**, 163-178.

105. Kelly, J.M., and Beauchamp, J.J. Mass loss and nutrient changes in decomposing upland oak and mesic mixed-hardwood leaf litter. *Soil Science Society of America Journal*, 1987, **51**, 1616-1622.

106. Witkamp. M. and Olson, J.S. Breakdown of confined and non-confined oak litter. *Oikos*, 1963, **14**, 138-147.

107. Vogt, K.A., Edmonds, R.L., and Grier, C.C. Seasonal changes in biomass and vertical distribution of mycorrhizal and fibrous-textured conifer fine roots in 23-year and 180-year-old subalpine *Abies amabilis* stands. *Canadian Journal of Forest Research*, 1980, **11**, 223-229.

108. Erickson, H.E., Edmonds, R.L., and Peterson, C.E. Decomposition of logging residues in Douglas-fir, western hemlock, Pacific silver fir, and ponderosa pine ecosystems. *Canadian Journal of Forest Research*, 1985, **15**, 914-921.

109. Edmonds, R.L., Vogt, D.J., Sandberg, D.H., and Driver, C.H. Decomposition of Douglas-fir and red alder wood in clearcuttings. *Canadian Journal of Forestry Research*, 1986, **16**, 822-831.

110. Harmon, M.E., Franklin, J.F., Swanson, F.J., Sollins, P., Gregory, S.V., Lattin, J.D., Anderson, N.H., Cline, S.P., Aumen, N.G., Sedell, J.R., Liemkaemper, G.W., Cromack, K. Jr., and Cummins, K.W. Ecology of coarse woody debris in temperate ecosystems. In: *Advances in Ecological Research, Volume 15*. (Eds.) A. MacFayden and E.D. Ford. Academic Press, London, 1986, pp. 133-302.

111. Harmon, M.E., Cromack, K. Jr., and Smith, B.G. Coarse woody debris in mixed conifer forests, Sequoia National Park, California. *Canadian Journal of Forest Research*, 1987, **17**, 1265-1272.

112. Staaf, H. and Berg, B. Plant litter input to soil. *Ecological Bulletin* (Stockholm), 1981, **33**, 147-162.

113. Vogt, K.A., Grier, C.C., and Vogt, D.J. Production, turnover, and nutrient dynamics of above- and belowground detritus of world forests. *Advances in Ecological Research*, 1986, **15**, 303-377.

114. Persson, H. Methods of studying root dynamics in relation to nutrient cycling. In: *Nutrient Cycling in Terrestrial Ecosystems: Field Methods,*

Application, and Interpretation. (Eds.) A.F. Harrison, P. Ineson, and O.W. Heal. Elsevier Applied Science, London and New York, 1990, pp. 198-217.

115. Vogt, K.A., Grier, C.C., Gower, S.T., Sprugel, D.S., and Vogt, D.J. Overestimation of net root production: a real or imaginary problem? *Ecology*, 1986, **67**, 577-579.

116. Stevenson, F.J. *Cycles of Soil Carbon, Nitrogen, Phosphorus, Sulfur, and Micronutrients.* John Wiley and Sons, Inc., New York, 1986.

117. Andreux, F., Cerri, C., Vose, P.B., and Vitello, V.A. Potential stable isotope ^{15}N and ^{13}C methods for determining input and turnover in soils. In: *Nutrient Cycling in Terrestrial Ecosystems: Field Methods, Application, and Interpretation.* (Eds.) A.F. Harrison, P. Ineson, and O.W. Heal. Elsevier Applied Science, London and New York, 1990, pp. 259-275.

118. Harrison, A.F., Harkness, D.D., and Bacon, P.J. The use of bomb-^{14}C for studying organic matter and N and P dynamics in a woodland soil. In: *Nutrient Cycling in Terrestrial Ecosystems: Field Methods, Application, and Interpretation.* (Eds.) A.F. Harrison, P. Ineson, and O.W. Heal. Elsevier Applied Science, London and New York, 1990, pp. 246-258.

119. Bremner, J.M. and Hauck, R.D. Advances in methodology for research on nitrogen transformations. In: *Nitrogen in Agricultural Soils.* (Ed.) F.J. Stevenson. American Society of Agronomy, Madison, WI, 1982, pp. 467-502.

120. Davidson, E.A., Stark, J.M., and Firestone, M.K. Microbial production and consumption of nitrate in an annual grassland. *Ecology*, 1990, **71**, 1968-1975.

121. Nommik, H. Application of ^{15}N as a tracer in studying fertilizer nitrogen transformations and recovery in coniferous ecosystems. In: *Nutrient Cycling in Terrestrial Ecosystems: Field Methods, Application, and Interpretation.* (Eds.) A.F. Harrison, P. Ineson, and O.W. Heal. Elsevier Applied Science, London and New York, 1990, pp. 276-290.

122. Strickland, T.C. and Fitzgerald, J.W. *In situ* mobilization of ^{35}S-labelled organic sulphur in litter and soil from hardwood forests. *Soil Biology and Biochemistry*, 1986, **18**, 463-468.

123. McLaren, R.G., Keer, J.I., and Swift, R.S. Sulphur transformations in soils using sulphur-35 labelling. *Soil Biology and Biochemistry*, 1985, **17**, 73-79.

124. McGill, W.B. and Cole, C.V. Comparative aspects of cycling of organic C, N, S, and P through soil organic matter. *Geoderma*, 1981, **26**, 267-286.

125. Shearer, G., Duffy, J., Kohl, D.H., and Commoner, B. A steady-state model of isotopic fractionation accompanying nitrogen transformations in soil. *Soil Science Society of America Proceedings*, 1974, **38**, 315-322.

126. Kapoor, K.K. and Haider, K. Mineralization and plant availability of phosphorus from biomass of hyaline and melanic fungi. *Soil Science Society of America Journal*, 1982, **46**, 953-957.

127. Sabey, B.R., Frederick, L.R., and Bartholomew, W.V. The formation of nitrate from ammonium nitrogen in soils: III. Influence of temperature and initial population of nitrifying organisms on the maximum rate and delay period. *Soil Science Society of America Proceedings*, 1959, **23**, 462-465.

128. Kowalenko, C.G. and Lowe, L.E. Evaluation of several extraction methods and of closed incubation method for soil sulfur mineralization. *Canadian Journal of Soil Science*, 1975, **55**, 1-8.

129. Maynard, D.G., Stewart, J.W.B., and Bettany, J.R. Sulfur and nitrogen mineralization in soil compared using two incubation techniques. *Soil Biology and Biochemistry*, 1983, **15**, 251-256.

130. Johnson, D.W., Edwards, N.T., and Todd, D.E. Nitrogen mineralization, immobilization, and nitrification following urea fertilization of a forest soil under field and laboratory conditions. *Soil Science Society of America Journal*, 1980, **44**, 610-616.

131. Hedley, M.J., Stewart, J.W.B., and Chauhan, B.S. Changes in inorganic and organic soil phosphorus fractions induced by cultivation practices and by laboratory incubations. *Soil Science Society of America Journal*, 1982, **46**, 970-976.

132. Bettany, J.R., Saggar, S., and Stewart, J.W.B. Comparison of the amounts and forms of sulfur in soil organic matter fractions after 65 years of cultivation. *Soil Science Society of America Journal*, 1980, **44**, 70-75.

133. Tabatabai, M.A. Sulfur. In: *Methods of Soil Analysis. Part 2. Chemical and Microbiological Properties*, Second Edition. (Eds.) A.L. Page, R.H. Miller, and D.R. Keeney. ASA-SSSA, Madison, Wisconsin, 1982, pp. 501-537.

134. David, M.B., Schindler, S.C., Mitchell, M.J., and Strick, J.E. Importance of organic and inorganic sulfur to mineralization processes in a forest soil. *Soil Biology and Biochemistry*, 1983, **15**, 671-677.

135. Wainwright, M. Field methods used in the determination of sulphur transformations in soils. In: *Nutrient Cycling in Terrestrial Ecosystems: Field Methods, Application, and Interpretation*. (Eds.) A.F. Harrison, P. Ineson, and O.W. Heal. Elsevier Applied Science, London and New York, 1990, pp. 218-232.

136. Keeney, D.R. and Bremner, J.M. Comparison and evaluation of laboratory methods of obtaining an index of soil nitrogen availability. *Agronomy Journal*, 1966, **58**, 498-503.

137. Keeney, D.R. Nitrogen availability indices. In: *Methods of Soil Analysis. Part 2. Chemical and microbiological properties*, Second Edition. (Eds.) A.L. Page, R.H. Miller, and D.R. Keeney. ASA-SSSA, Madison, Wisconsin, 1982, pp. 711-733.

138. Stanford, G. Assessment of nitrogen availability. In: *Nitrogen in Agricultural Soils*. (Ed.) F.J. Stevenson. American Society of Agronomy, Madison, WI, 1982, pp. 651-688.

139. Binkley, D. and Matson, P. Ion exchange resin bag method for assessing soil nitrogen availability. *Soil Science Society of America Journal*, 1983, **47**, 1050-1052.

140. Binkley, D. and Hart, S.C. The components of nitrogen availability in forest soils. *Advances in Soil Science*, 1989, **10**, 57-112.

141. Hart, S.C. and Binkley, D. Correlation among indices of forest soil nutrient availability in fertilized and unfertilized loblolly pine plantations. *Plant and Soil*, 1985, **85**, 11-21.

142. Lundell, Y. *In situ* exchange resin bags to estimate forest site quality. *Plant and Soil*, 1989, **119**, 186-190.

143. Westerman, R.L. (Ed.). *Soil testing and plant analysis. Third Edition.* Soil Science Society of America, Madison, WI, 1990.

144. Strickland, T.C. and Fitzgerald, J.W. Formation and mineralization of organic sulfur in forest soils. *Biogeochemistry*, 1984, **1**, 79-95.

145. Robinson, J.B.D. Anaerobic incubation of soil and the production of ammonium. *Nature*, 1967, **214**, 534.

146. Powers, R.F. Mineralizable soil nitrogen as an index of nitrogen availability to forest trees. *Soil Science Society of America Journal*, 1980, **44**, 1314-1320.

147. Stanford, G. and Smith, S.J. Nitrogen mineralization potentials of soils. *Soil Science Society America Proceedings*, 1972, **36**, 465-472.

148. Mahendrappa, M.K. Relationship between different estimates of mineralizable N in the organic materials under black spruce stands. *Canadian Journal of Forest Research*, 1980, **10**, 517-522.

149. Eno, C. Nitrate production in the field by incubating soil in polyethylene bags. *Soil Science Society of America Proceedings*, 1960, **24**, 277-279.

150. Raison, R.J., Connell, M., and Khanna, P. Methodology for studying fluxes of soil mineral-N *in situ*. *Soil Biology and Biochemistry*, 1987, **19**, 521-530.

151. DiStefano, J.F. and Gholz, H.L. A proposed use of ion exchange resins to measure nitrogen mineralization and nitrification in intact soil cores. *Communications in Soil Science and Plant Analysis*, 1986, **17**, 989-998.

152. David, M.B., Mitchell, M.J., and Nakas, J.P. Organic and inorganic constituents of a forest soil and their relationship to microbial activity. *Soil Science Society of America Journal*, 1982, **46**, 847-852.

153. Lea, R. and Ballard, R. Predicting loblolly pine growth response from N fertilizer, using soil-N availability indices. *Soil Science Society of America Journal*, 1982, **46**, 1096-1099.

154. Powers, R.F. Estimating soil nitrogen availability through soil and foliar analysis. In: *Forest Soil and Treatment Impacts.* Proceedings, 6th North American Forest Soils Conference. (Ed.) E.L. Stone. Dept. Forestry, Wildlife, and Fisheries, University of Tennessee, Knoxville, TN, 1984, pp. 353-379.

155. Myrold, D.D. Relationship between microbial biomass nitrogen and a nitrogen availability index. *Soil Science Society of America Journal*, 1987, **51**, 1047-1049.

156. Stumm, W. and Morgan, J.J. *Aquatic Chemistry. An Introduction Emphasizing Chemical Equilibria in Natural Waters*. John Wiley & Sons, New York, 1981, 780 p.

157. Sparks, D.L. *Kinetics of Soil Chemical Processes*. Academic Press, San Diego, California, 1989, 210 p.

158. Sparks, D.L. and Suarez, D.L. *Rates of Soil Chemical Processes*. Soil Science Society of America, Special Publication No. 27. Madison, WI, 1991, 302 p.

159. Shilova, Ye.I. A method for obtaining soil solution under natural conditions. *Pochvovendeniye*, 1955, No. 11.

160. Jordan, C.F. A simple tension-free lysimeter. *Soil Science*, 1968, **105**, 81-86.

161. Vedy, J.C. and Bruckert, S. Soil solution: composition and pedogenic significance. In: *Constituents and Properties of Soils*. (Eds.) M. Bonneau and B. Souchier. Academic Press, New York, 1982, pp 184-213.

162. Cole, D.W., Gessel, S.P., and Held, E.E. Tension lysimeter studies of ion and moisture movement in glacial till and coral atoll soils. *Soil Science Society of America Proceedings*, 1961, **25**, 321-324.

163. Johnson, D.W., Cole, D.W., Gessel, S.P., Singer, M.J., and Minden, R.V. Carbonic acid leaching in a tropical, temperate, subalpine and northern forest soil. *Arctic and Alpine Research*, 1977, **9**, 329-343.

164. Joslin, J.D., Mays, P.A., Wolfe, M.H., Kelly, J.M., Garber, R.W., and Brewer, P.F. Chemistry of tension lysimeter water and lateral flow in spruce and hardwood stands. *Journal of Environmental Quality*, 1987, **16**, 152-160.

165. Ugolini, F.C., Stoner, M.G., and Marrett, D.J. Arctic pedogenesis: 1. Evidence for contemporary podzolization. *Soil Science*, 1987, **144**, 90-100.

166. van Grinsven, J.J.M., Booltink, H.W.G., Dirksen, C., van Breemen, N., Bongers, N., and Waringa, N. Automated *in situ* measurement of unsaturated soil water flux. *Soil Science Society of America Journal*, 1988, **52**, 1215-1218.

167. Cole, D.W. A system for measuring conductivity, acidity, and rate of water flow in a forest soil. *Water Resources Research*, 1968, **4**, 1127-1136.

168. Riekerk, H. and Morris, L.A. A constant-potential soil water sampler. *Soil Science Society of America Journal*, 1983, **47**, 606-608.

169. Davies, B. and Davies, R. A simple centrifugation method for obtaining small samples of soil solution. *Nature*, 1963, **198**, 216-217.

170. Gillman, G.P. A centrifuge method for obtaining soil solution. Division of Soils, Divisional Report 16, CSIRO, Melbourne, Australia, 1976.

171. Adams, F., Burmester, C., Hue, N., and Long, F. A comparison of column-displacement and centrifuge methods for obtaining soil solutions. *Soil Science Society of America Journal*, 1980, **44**, 733-735.

172. Zabowski, D. and Ugolini, F.C. Lysimeter and centrifuge soil solutions: Seasonal differences between methods. *Soil Science Society of America Journal*, 1990, **54**, 1130-1135.

173. Murbarak, A. and Olsen, R.A. Immiscible displacement of the soil solution by centrifugation. *Soil Science Society of America Journal*, 1976, **40**, 329-331.

174. Whelan, B.R. and Barrow, N.J. A study of a method for displacing soil solution by centrifuging with an immiscible liquid. *Journal of Environmental Quality*, 1980, **9**, 315-319.

175. Kittrick, J.A. Accuracy of several immiscible displacement liquids. *Soil Science Society of America Journal*, 1983, **47**, 1045-1047.

176. Zabowski, D. Lysimeter and centrifuge soil solutions: a comparison of methods and objectives. In: *Research Strategies for Long-term Site Productivity*. (Eds.) W.J. Dyck and C.A. Mees. IEA/BE Report No. 8., Forest Research Institute, Rotorua, New Zealand, Bulletin 152, 1989, pp. 139-148.

177. Baker, D.E. and Suhr, N.H. Atomic absorption and flame emission spectrophotometry. In: *Methods of Soil Analysis. Part 2. Chemical and Microbiological Properties*, Second Edition. (Eds.) A.L. Page, R.H. Miller, and D.R. Keeney. ASA-SSSA, Madison, Wisconsin, 1982, pp. 13-27.

178. Helmke, P.A. Neutron activation analysis. In: *Methods of Soil Analysis. Part 2. Chemical and Microbiological Properties*, Second Edition. (Eds.) A.L. Page, R.H. Miller, and D.R. Keeney. ASA-SSSA, Madison, WI, 1982, pp. 67-84.

179. Soltanpour, P.N., Jones, J.B. Jr., and Workman, S.M. Optical emission spectrometry. In: *Methods of Soil Analysis. Part 2. Chemical and Microbiological Properties*, Second Edition. (Eds.) A.L. Page, R.H. Miller, and D.R. Keeney. ASA-SSSA, Madison, Wisconsin, 1982, pp. 29-65.

180. Sharp, B.L. Inductively coupled plasma spectrometry. In: *Soil Analysis, Modern Instrumental Techniques*. (Ed.) K.A. Smith. Marcel-Dekker, New York, 1981, pp. 63-109.

181. Rhoades, J.D. Soluble salts. In: *Methods of Soil Analysis. Part 2. Chemical and Microbiological Properties*, Second Edition. (Eds.) A.L. Page, R.H. Miller, and D.R. Keeney. ASA-SSSA, Madison, WI, 1982, pp. 167-179.

182. Schnitzer, M. Organic matter characterization. In: *Methods of Soil Analysis. Part 2. Chemical and Microbiological Properties*, Second Edition. (Eds.) A.L. Page, R.H. Miller, and D.R. Keeney. ASA-SSSA, Wisconsin, 1982, pp. 581-594.

183. Stevenson, F.J. *Humus Chemistry. Genesis, Composition, Reactions.* John Wiley and Sons, New York, 1982, 443 p.

184. Wilson, M.A. Analysis of functional groups in soil by Nuclear magnetic resonance spectroscopy. In: *Soil Analysis, Modern Instrumental Techniques.* (Ed.) K.A. Smith. Marcel-Dekker, New York, 1991, pp. 601-645.

185. Parker, D.R., Norvell, W.A., and Chaney, R.L. GEOCHEM-PC: A chemical speciation program for IBM and compatible personal computers. In: *Soil Chemical Equilibrium Models.* (Eds.) R.H. Loeppert *et al.* SSSA Special Publication. ASA, Madison, WI, (In Press).

186. Schecher, W.D. and McAvoy, D.C. *MINEQL+: A Chemical Equilibrium Program for Personal Computers.* Environmental Research Software, Edgewater, MD, 1991.

187. Allison, J.D., Brown, D.S., and Novo-Gradac, K.J. MINTEQA2/PRODEFA2, A geochemical assessment model for environmental systems: Version 3.0. Center for Exposure Assessment Modeling, U.S. Environmental Protection Agency, Athens, GA., 1990.

188. Zabowski, D. Role of mineral weathering in long-term productivity. In: *Impact of Intensive Harvesting on Forest Site Productivity.* (Eds.) W.J. Dyck and C.A. Mees. IEA/BE T6/A6 Report No. 2. Forest Research Institute, Rotorua, New Zealand, Bulletin 159, 1990, pp. 55-71.

189. Clayton, J.L. Nutrient supply to soil by rock weathering. In: *Proceedings, Impact of Intensive Harvesting on Forest Nutrient Cycling.* (Ed.) A.L. Leaf. Syracuse, New York, 1979, pp. 75-96.

190. Birkeland, P.W. *Soils and Geomorphology.* Oxford University Press, New York, 1984.

191. Cremeens, D.L., Norton, L.D., Darmody, R.G., and Jansen, I.J. Etch-pit measurements on scanning electron micrographs of weathered grain surfaces. *Soil Science Society of America Journal*, 1988, **52**, 883-885.

192. Boyle, J.R. and Voigt, G.K. Biological weathering of silicate minerals. *Plant and Soil*, 1973, **38**, 191-201.

193. Cronan, C.S. Chemical weathering and solution chemistry in acid forest soils: differential influence of soil type, biotic processes and H^+ deposition. In: *The Chemistry of Weathering.* (Ed.) J.I. Drever. Reidel Publishing, Dordrecht, 1985.

194. Cole, D.W., Gessel, S.P., and Dice, S.F. Distribution and cycling of nitrogen, phosphorus, potassium and calcium in a second-growth Douglas-fir ecosystem. In: *Primary Productivity and Mineral Cycling in Natural Ecosystems.* (Ed.) H.E. Young. University of Maine Press, Orono, Maine, 1967, pp. 197-232.

195. Velbel, M.A. Geochemical mass balances and weathering rates in forested watersheds of the southern Blue Ridge. *American J. Sci.*, 1985, **285**, 904-930.

196. Clayton, J.L. Some observations on the stoichiometry of feldspar hydrolysis in granitic soil. *Journal of Environmental Quality*, 1988, **17**, 153-157.

197. Paces, T. Rate constants of dissolution derived from the measurements of mass balance in hydrological catchments. *Geochim. Cosmo. Acta.*, 1983, **47**, 1855-1863.

198. Lasaga, A.C. Chemical kinetics of water-rock interactions. *Journal of Geophysical Research*, 1984, **89**, 4009-4025.

199. Sverdrup, H. and Warfinge, P. Weathering of primary silicate minerals in the natural soil environment in relation to a chemical weathering model. *Water, Air, and Soil Pollution*, 1988, **38**, 387-408.

200. Van Rees, K.C.J., Comerford, N.B., and Rao, P.S.C. Defining soil buffer power: implications for ion diffusion and nutrient uptake modeling. *Soil Science Society of America Journal*, 1990, **54**, 1505-1507.

201. Gillman, G.P. and Sumpter, E.A. Modification to the compulsive exchange method for measuring exchange characteristics of soils. *Australian Journal of Soil Research*, 1986, **24**, 61-66.

202. Uehara, G. and Gillman, G.P. *The Mineralogy, Chemistry and Physics of Tropical Soils with Variable Charge Clays*. Westview Tropical Series #4, Westview Press, Boulder, Colorado, 1981.

203. McLean, E.O. Soil pH and lime requirement. In: *Methods of Soil Analysis. Part 2. Chemical and Microbiological Properties*, Second Edition. (Eds.) A.L. Page, R.H. Miller, and D.R. Keeney. ASA-SSSA, Madison, Wisconsin, 1982, pp. 199-224.

204. Nelson, D.W. and Sommers, L.E. Total carbon, organic carbon, and organic matter. In: *Methods of Soil Analysis. Part 2. Chemical and Microbiological Properties*, Second Edition. (Eds.) A.L. Page, R.H. Miller, and D.R. Keeney. ASA-SSSA, Madison, Wisconsin, 1982, pp. 539-579.

205. Walkley, A. and Black, I.A. An examination of the Degtjareff method for determining soil organic matter and a proposed modification of the chromic acid titration method. *Soil Science*, 1934, **37**, 29-38.

206. Bremner, J.M. and Mulvaney, C.S. Nitrogen-total. In: *Methods of Soil Analysis. Part 2. Chemical and Microbiological Properties*, Second Edition. (Eds.) A.L. Page, R.H. Miller, and D.R. Keeney. ASA-SSSA, Madison, Wisconsin, 1982, pp. 595-624.

207. Parkinson, J.A. and Allen, S.E. A wet oxidation procedure suitable for the determination of nitrogen and mineral nutrients in biological material. *Communications in Soil Science and Plant Analysis*, 1975, **6**, 1-11.

208. Lim, C.H. and Jackson, M.L. Dissolution for total elemental analysis. In: *Methods of Soil Analysis. Part 2. Chemical and Microbiological Properties*, Second Edition. (Eds.) A.L. Page, R.H. Miller, and D.R. Keeney. ASA-SSSA, Madison, Wisconsin, 1982, pp. 1-12.

209. Salmon, L. and Cawse, P.A. Instrumental neutron activation analysis. In: *Soil Analysis, Modern Instrumental Techniques*. (Ed.) K.A. Smith. Marcel-Dekker, New York, 1991, pp. 377-432.

210. Keeney, D.R. and Nelson, D.W. Nitrogen – inorganic forms. In: *Methods of Soil Analysis. Part 2. Chemical and Microbiological Properties*, Second Edition. (Eds.) A.L. Page, R.H. Miller, and D.R. Keeney. ASA-SSSA, Madison, Wisconsin, 1982, pp. 643-698.

211. Olsen, S.R. and Sommers, L.E. Phosphorus. In: *Methods of Soil Analysis. Part 2. Chemical and Microbiological Properties*, Second Edition. (Eds.) A.L. Page, R.H. Miller, and D.R. Keeney. ASA-SSSA, Madison, Wisconsin, 1982, pp. 403-430.

212. Bingham, F.T. Boron. In: *Methods of Soil Analysis. Part 2. Chemical and Microbiological Properties*, Second Edition. (Eds.) A.L. Page, R.H. Miller, and D.R. Keeney. ASA-SSSA, Madison, Wisconsin, 1982, pp. 431-447.

213. Lindsay, W.L. and Norvell, W.A. Development of a DTPA soil test for zinc, iron, manganese, and copper. *Soil Science Society of America Journal*, 1978, **42**, 421-428.

214. Smith, K.A. and Arah, J.R.M. Gas chromatographic analysis of the soil atmosphere. In: *Soil Analysis, Modern Instrumental Techniques*. (Ed.) K.A. Smith. Marcel-Dekker, New York, 1991, pp. 505-546.

215. Bremner, J.M. and Blackmer, A.M. Composition of soil atmospheres. In: *Methods of Soil Analysis. Part 2. Chemical and Microbiological Properties*, Second Edition. (Eds.) A.L. Page, R.H. Miller, and D.R. Keeney. ASA-SSSA, Madison, Wisconsin, 1982, pp. 873-901.

216. Miotke, F.D. Carbon dioxide and the soil atmosphere. *Abh. Karst Hohlenkunde Ser. A*, 1974, **9**, 1-49.

217. Phene, C.J. Oxygen electrode measurement. In: *Methods of Soil Analysis, Part 1. Physical and Mineralogical Methods*, Second Edition. (Ed.) A. Klute. ASA-SSSA, Madison, Wisconsin, 1986, pp. 1137-1159.

218. Fisher, R.T. Soil changes and silviculture on the Harvard forest. *Ecology*, 1928, **9**, 6-11.

219. Griffith, B.G., Hartwell, E.W., and Shaw, T.E. The evolution of soils as affected by the old field white pine-mixed hardwood succession in central New England. Harvard Forest Bulletin 15, 1930, 82 p.

220. Soil Survey Staff. *Soil Taxonomy*. USDA Soil Conservation Service, Washington, D.C., 1975.

221. Soil Survey Staff. *Keys to Soil Taxonomy*. USDA Soil Management Support Services Tech. Monograph No. 19., Virginia Polytechnic Institute, Blacksburg, VA, 1990.

222. Food and Agricultural Organization of the UN Educational, Scientific and Cultural Organization. Soil Map of the World, Revised Legend. UNESCO-FAO World Soil Resources Report 60, 1988.

223. Blake, G.R. and Hartge, K.H. Bulk density. In: *Methods of Soil Analysis. Part 2. Chemical and Microbiological Properties*, Second Edition. (Eds.) A.L. Page, R.H. Miller, and D.R. Keeney. ASA-SSSA, Madison, WI, 1986, pp. 363-375.

224. Soil Conservation Service. *Procedures for Collecting Soil Samples and Methods of Analysis for Soil Survey*. USDA Soil Survey Investigations Report No. 1, 1984.

225. Blake, G.R. Bulk density. In: *Methods of Soil Analysis. Part 1. Physical and Mineralogical Properties, Including Statistics of Measurement and Sampling*. (Eds.) C.A. Black *et al.* ASA-SSSA, Madison, Wisconsin, 1965, pp. 374-390.

226. Froelich, H.A. and McNabb, D.H. Minimizing soil compaction in Pacific Northwest soils. In: *Forest Soils and Treatment Impacts*. (Ed.) E.L. Stone. Proceedings of the Sixth North American Forest Soils Conference. Dept. Forestry, Wildlife and Fisheries, University of Tennessee, Knoxville, TN, 1984, pp. 159-192.

227. Danielson, R.E. and Sutherland, P.L. Porosity. In: *Methods of Soil Analysis, Part 1. Physical and Mineralogical Methods*, Second Edition. (Ed.) A. Klute. ASA-SSSA, Madison, WI, 1986, pp. 443-461.

228. Bradford, J.M. Penetrability. In: *Methods of Soil Analysis. Part 1. Physical and Mineralogical Properties, Including Statistics of Measurement and Sampling*. (Eds). C.A. Black *et al.* ASA-SSSA, Madison, Wisconsin, 1986, pp. 463-477.

229. Gee, G.W. and Bauder, J.W. Particle-size analysis. In: *Methods of Soil Analysis, Part 1. Physical and Mineralogical Methods*, Second Edition. (Ed.) A. Klute. ASA-SSSA, Madison, WI, 1986, pp. 383-412.

230. Gardner, W.H. Water content. In: *Methods of Soil Analysis, Part 1. Physical and Mineralogical Methods*, Second Edition. (Ed.) A. Klute. ASA-SSSA, Madison, Wisconsin, 1986, pp. 493-544.

231. Brady, N.C. (Ed.). *The Nature and Properties of Soils*. Ninth Edition. Macmillan Publishing Company, New York, 1984.

232. Ledieu, J., DeRidder, P., DeClerck, P., and Dautrebande, S. A method of measuring soil moisture by time-domaine reflectometry. *Journal of Hydrology*, 1986, **88**, 319-328.

233. Topp, G.C. and Davis, J.L. Measurement of soil water content using time-domaine reflectometry (TDR): a field evaluation. *Soil Science Society of America Journal*, 1985, **49**, 19-24.

234. Cassell, D.K. and Klute, A. Water potential: tensiometry. In: *Methods of Soil Analysis, Part 1. Physical and Mineralogical Methods*, Second Edition. (Ed.) A. Klute. ASA-SSSA, Madison, WI, 1986, pp. 563-596.

235. Rawlins, S.L. and Campbell, G.S. Water potential: Thermocouple psychrometry. In: *Methods of Soil Analysis, Part 1. Physical and Mineralogical Methods*, Second Edition. (Ed.) A. Klute. ASA-SSSA, Madison, WI, 1986, pp. 597-617.

236. Campbell, G.S. and Gee, G.W. Water potential: miscellaneous methods. In: *Methods of Soil Analysis, Part 1. Physical and Mineralogical Methods*, Second Edition. (Ed.) A. Klute. ASA-SSSA, Madison, WI, 1986, pp. 619-634.

237. Reeve, R.C. Water potential: Piezometry. In: *Methods of Soil Analysis, Part 1. Physical and Mineralogical Methods*, Second Edition. (Ed.) A. Klute. ASA-SSSA, Madison, WI, 1986, pp. 545-562.

238. Bruce, R.R. and Luxmoore, R.J. Water retention: field methods. In: *Methods of Soil Analysis, Part 1. Physical and Mineralogical Methods*, Second Edition. (Ed.) A. Klute. ASA-SSSA, Madison, WI, 1986, pp. 663-686.

239. Klute, A. and Dirksen, C. Hydraulic conductivity and diffusivity: laboratory methods. In: *Methods of Soil Analysis, Part 1. Physical and Mineralogical Methods*, Second Edition. (Ed.) A. Klute. ASA-SSSA, Madison, WI, 1986, pp. 687-734.

240. Amoozegar, A. and Warrick, A.W. Hydraulic conductivity of saturated soils: field methods. In: *Methods of Soil Analysis, Part 1. Physical and Mineralogical Methods*. Second Edition. (Ed.) A. Klute. ASA-SSSA, Madison, WI, 1986, pp. 735-768.

241. Green, R.E., Ahuja, L.R., and Chong, S.K. Hydraulic conductivity, diffusivity, and sorptivity of unsaturated soils: field methods. In: *Methods of Soil Analysis, Part 1. Physical and Mineralogical Methods*, Second Edition. (Ed.) A. Klute. ASA-SSSA, Madison, WI, 1986, pp. 771-798.

242. Powers, R.F. Nitrogen mineralization along an altitudinal gradient: Interactions of soil temperature, moisture, and substrate quality. *Forest Ecology and Management*, 1990, **30**, 19-29.

243. Hanks, R.J. and Ashcroft, G.L. *Applied Soil Physics*. Advanced Series in Agricultural Sciences 8. Springer-Verlag, Berlin, 1980.

244. Jackson, R.D. and Taylor, S.A. Thermal conductivity and diffusivity. In: *Methods of Soil Analysis, Part 1. Physical and Mineralogical Methods*, Second Edition. (Ed.) A. Klute. ASA-SSSA, Madison, WI, 1986, pp. 945-956.

245. Husch, B., Miller, C.I., and Beers, T.W. *Forest Mensuration*. J. Wiley & Sons, New York, 1982.

246. Avery, T.E. and Burkhart, H.E. *Forest Measurements*. McGraw-Hill, New York, 1983.

247. Telewski, F.W. and Lynch, A.M. Measuring growth and development of stems. In: *Techniques and approaches in forest tree ecophysiology*. (Eds.) J.P. Lassoie and T.M. Hinckley. CRC Press, Boca Raton, FL, USA, 1991, pp. 503-555.

248. Solomon, D.S. and Brann, T.B. (Eds.). *Environmental Influences on Tree and Stand Increment*. Proceedings of the IUFRO Mensuration, Growth and Yield (S4:01-00), and Instruments and Methods in Forest Mensuration (S4:01-06) Workshop, Durham, NH, USA. University of Maine, Agricultural Experiment Station, Miscellaneous Publication 691, 1986.

249. Will, G.M. and Hodgkiss, P.D. Influence of nitrogen and phosphorus stresses on the growth and form of radiata pine. *New Zealand Journal of Forestry Science*, 1977, **7**, 307-320.

250. Walker, R.B. Measuring mineral nutrient utilization. In: *Techniques and Approaches in Forest Tree Ecophysiology*. (Eds.) J.P. Lassoie and T.M. Hinckley. CRC Press, Boca Raton, FL, USA, 1991, pp. 183-206.

251. Valentine, H.T., Tritton, L.M., and Furnival, G.M. Subsampling trees for biomass, volume, or mineral content. *Forest Science*, 1984, **30**, 673.

252. Neter, J., Wasserman, W., and Kutner, M.H. *Applied Linear Statistical Models*, Second Edition. R.D. Irwin, Homewood, IL, USA, 1985.

253. Madgwick, H.A.I. Estimating the above-ground weight of forest plots using the basal area ratio method. *New Zealand Journal of Forestry Science*, 1981, **11(3)**, 278-286.

254. Madgwick, H.A.I. and Satoo, T. On estimating the aboveground weights of tree stands. *Ecology*, 1975, **56**, 1446-1450.

255. Mead, D.J. Diagnosis of nutrient deficiencies in plantations. In: *Nutrition of Plantation Forests*. (Eds.) G.D. Bowen and E.K.S. Nambiar. Academic Press, London, 1984, pp. 259-291.

256. Timmer, V.R. Interpretation of seedling analysis and visual symptoms. In: *Mineral Nutrition of Conifer Seedlings*. (Ed.) R. van den Driessche. CRC Press, 1991, pp. 113-134.

257. Will, G.M. Nutrient deficiencies and fertilizer use in New Zealand exotic forests. FRI Bulletin No. 97. Forest Research Institute, Rotorua, NZ, 1985, 53 p.

258. Lambert, M.J. The use of foliar analysis in fertilizer research. In: *Symposium on Site and Productivity of Fast Growing Plantations*. (Eds.) D.C. Grey, A.P.G. Schonau, and C.J. Schutz. IUFRO Symposium Proceedings Vol. 1, Pretoria and Pietermaritzburg, South Africa, 1984, pp. 269-291.

259. Smith, C.T., Dyck, W.J., Beets, P.N., Hodgkiss, P.D., and Lowe, A.T. Nutrition and productivity of *Pinus radiata* following harvest disturbance and fertilization of coastal sand dunes. *Forest Ecology and Management*, 1993, (in press).

260. Olykan, S.T., Adams, J.A., and Nordmeyer, A. Effect of urea, ulexite, and rainfall on foliar boron levels of young *P. radiata* in Ashley Forest, New Zealand. In: *Productivity in Perspective*. (Ed.) P.J. Ryan. Third Australian Forest Soils and Nutrition Conference, Melbourne, 1991, p. 36.

261. Valentine, D.W. and Allen, H.L. Foliar responses to fertilization to identify nutrient limitation in loblolly pine. *Canadian Journal of Forest Research*, 1990, **20**, 144-151.

262. Raison, R.J., Khanna, P.K., Connell, M.J., and Falkiner, R.A. Effects of water availability and fertilization on N cycling in a stand of *Pinus radiata*. *Forest Ecology and Management*, 1990, **30**, 31-43.

263. Birk, E.M. and Vitousek, P.M. Nitrogen availability and nitrogen use efficiency in loblolly pine stands. *Ecology*, 1986, **67**, 69-79.

264. Brix, H. Effects of fertilization on photosynthesis and respiration in Douglas-fir. *Forest Science*, 1971, **17**, 407-414.

265. Brix, H. Effects of nitrogen fertilizer source and application rates on foliar nitrogen concentration, photosynthesis, and growth of Douglas-fir. *Canadian Journal of Forest Research*, 1981, **11**, 775-780.

266. Hillerdal-Hagstromer, R., Mattson-Djos, E., and Hellkvist, J. Field studies of water relations and photosynthesis in Scots pine. II. Influence of irrigation and fertilization on needle water potential of young pine trees. *Physiologia Plantarum*, 1982. **54**, 295-301.

267. Linder, S. and Rook, D.A. Effects of mineral nutrition on carbon dioxide exchange and partitioning of carbon in trees. In: *Nutrition of Plantation Forests*. (Eds.) G.D. Bowen and E.K.S. Nambiar. Academic Press, London, 1984, pp. 211-236.

268. Sheriff, D.W., Nambiar, E.K.S., and Fife, D.N. Relationships between nutrient status, carbon assimilation and water use efficiency in *Pinus radiata* D. (Don) needles. *Tree Physiology*, 1986, **2**, 73-88.

269. Brix, H. and Ebell, L.F. The effect of nitrogen fertilizer on growth, leaf area, and photosynthesis rate in Douglas-fir. *Forest Science*, 1969, **15**, 189-196.

270. Bouma, D., Dowling, E.J., and Wahjoedi, H. Some effects of potassium and magnesium in the growth of subterranean clover (*Trifolium subterraneum*). *Annals of Botany*, 1979, **43**, 529-538.

271. Terry, N., Waldron, L.J., and Taylor, S.E. Environmental influences on leaf expansion. In: *The Growth and Functioning of Leaves*. (Eds.) J.E. Dale and F.L. Milthorpe. Cambridge Press, 1983, pp. 179-205.

272. Leverenz, J.W. and Hallgren, J.-E. Measuring photosynthesis and respiration of foliage. In: *Techniques and approaches in forest tree ecophysiology.* (Eds.) J.P. Lassoie and T.M. Hinckley. CRC Press, Boca Raton, FL, USA, 1991, pp. 308-328.

273. Whitehead, D., Grace, J.C., and Godfrey, M.J.S. Architectural distribution of foliage in individual *Pinus radiata* D. Don crowns and the effects of clumping on radiation interception. *Tree Physiology*, 1990, **7**, 135-155.

274. Gower, S.T., Vogt, K.A., and Grier, C.C. Carbon dynamics of Rocky Mountain Douglas-fir: influence of water and nutrient availability. *Ecological Monographs*, 1992, **62(1)**, 43-65.

275. Brix, H. Effects of thinning and nitrogen fertilization on growth of Douglas-fir: relative contribution of foliage quantity and efficiency. *Canadian Journal of Forest Research*, 1983, **13**, 167-175.

276. Binkley, D. and Reid, P. Long-term responses of stem growth and leaf area to thinning and fertilization in a Douglas-fir plantation. *Canadian Journal of Forest Research*, 1984, **14**, 656-660.

277. Miller, H.G., Cooper, J.M., and Miller, J.D. Effect of nitrogen supply on nutrients in litter-fall and crown leaching in a stand of Corsican pine. *Journal of Applied Ecology*, 1976, **13**, 233-248.

278. Gholz, H.L., Vogel, S.A., Cropper, W.P. Jr., McKelvey, K., Ewel, K.C., Teskey, R.O., and Curran, P.J. Dynamics of canopy structure and light interception in *Pinus elliottii* stands of north Florida. *Ecological Monographs*, 1991, **61**, 33-51.

279. Linder, S. Responses to water and nutrients in coniferous ecosystems. In: *Potentials and limitations of ecosystem analysis. Ecological Studies, Volume 61.* (Eds.) E.-D. Schulze and H. Zwolfer. Springer-Verlag, Berlin & Heidelberg, 1987, pp. 180-202.

280. Gower, S.T., Grier, C.C., Vogt, D.J., and Vogt, K.A. Allometric relations of deciduous (*Larix occidentalis*) and evergreen conifers (*Pinus contorta* and *Pseudotsuga menziesii*) of the Cascade Mountains in central Washington. *Canadian Journal of Forest Research*, 1987, **17**, 630-634.

281. Mead, D.J., Draper, D., and Madgwick, H.A.I. Dry matter production of a young stand of *Pinus radiata*: some effects of nitrogen fertiliser and thinning. *New Zealand Journal of Forestry Science*, 1984, **14(1)**, 97-108.

282. Snowdon, P. and Benson, M.L. Effects of combinations of irrigation and fertilization on the growth and above-ground biomass production of *Pinus radiata*. *Forest Ecology and Management*, 1992, **32**, 87-116.

283. Axelsson, E. and Axelsson, B. Changes in carbon allocation patterns in spruce and pine trees following irrigation and fertilization. *Tree Physiology*, 1986, **2**, 205-214.

284. Birk, E.M. Poor tree form of *Pinus radiata* D. (Don) on former pasture sites in New South Wales. *Australian Forestry*, 1990, **53**, 104-112.

285. Santantonio, D. and Grace, J.C. Estimating fine-root production and turnover from biomass and decomposition data: a compartment-flow model. *Canadian Journal of Forest Research*, 1987, **17**, 900-908.

286. Ingestad, T. and Kahr, M. Nutrition and growth of coniferous seedlings at varied relative nitrogen addition rate. *Physiologia Plantarum*, 1985, **65**, 109-116.

287. Keyes, M.R. and Grier, C.C. Above- and below-ground net production in 40-year-old Douglas-fir stands on low and high productivity sites. *Canadian Journal of Forest Research*, 1981, **11**, 599-605.

288. Linder, S. and Axelsson, B. Changes in carbon uptake and allocation as a result of irrigation and fertilization in a young *Pinus sylvestris* stand. In: *Carbon uptake and Allocation: Key to Management of Subalpine Forest Ecosystems.* (Ed.) R.H. Waring. IUFRO Workshop Proceedings, Forest Research Laboratory, Oregon State University, Corvallis, Oregon, USA, 1982, pp. 38-44.

289. Payn, T.W. The effects of magnesium fertiliser and grass on the nutrition and growth of *P. radiata* planted on pumice soils in the Central North Island of New Zealand. Ph.D. Thesis. School of Forestry, University of Canterbury, Christchurch, NZ, 1992.

290. Jackson, D.S. and Chittenden, J. Estimation of dry matter in *Pinus radiata* root systems. 1. Individual trees. *New Zealand Journal of Forestry Science*, 1981, **11**, 164-182.

291. Santantonio, D. and Santantonio, E. Effect of thinning on production and mortality of fine roots in a *Pinus radiata* plantation on a fertile site in New Zealand. *Canadian Journal of Forest Research*, 1987, **17**, 919-928.

292. Joslin, J.D. and Henderson, G.S. Organic matter and nutrients associated with fine root turnover in a white oak stand. *Forest Science*, 1987, **33(2)**, 330-346.

293. Reid, J.B. and Petrie, R.A. Effects of soil aeration on root demography in kiwifruit. *New Zealand Journal of Crop and Horticultural Science*, 1991, **19**, 423-431.

294. Hendricks, J.J., Nadelhoffer, K.J., and Aber, J.D. The role of fine roots in energy and nutrient cycling. *Trends in Ecology and Evolution*, 1993, **8**, 174-178.

295. Singh, J.S., Lauenroth, W.K., Hunt, W.H., and Swift, D.M. Bias and random errors in estimators of net root production: a simulation approach. *Ecology*, 1984, **65**, 1760-1764.

296. Santantonio, D. Dry-matter partitioning and fine-root production in forests – new approaches to a difficult problem. In: *Biomass Production by Fast-Growing Trees.* (Eds.) J. Pereira and J. Landsberg. Kluwer Academic Publication, 1989, pp. 57-72.

297. Vogt, K.A. and Persson, H. Measuring growth and development of roots. In: *Techniques and Approaches in Forest Tree Ecophysiology.* (Eds.) J.P. Lassoie and T.M. Hinckley. CRC Press. Boca Raton, FL, USA, 1991, pp. 477-501.

298. Raison, R.J. and Myers, B.J. The biology of forest growth experiment: linking water and nitrogen availability to the growth of *Pinus radiata. For. Ecol. Manage.*, 1992, **32**, 279-308.

299. Brix, H. and Mitchell, A.K. Thinning and nitrogen fertilization effects on soil and tree water stress in a Douglas-fir stand. *Canadian Journal of Forest Research*, 1986, **16**, 1334-1338.

300. Linder, S., Benson, M.L., Myers, B.J., and Raison, R.J. Canopy dynamics and growth of Pinus radiata. I. Effects of irrigation and fertilization during a drought. *Canadian Journal of Forest Research*, 1987, **17**, 1157-1165.

301. Squire, R.O., Farrell, P.W., Flinn, D.W., and Aeberli, B.C. Productivity of first and second rotation stands of radiata pine on sandy soils. I. Height and volume growth at five years. *Australian Forestry*, 1985, **48(2)**, 127-137.

302. Bradford, K.J. and Hsiao, T.C. Physiological responses to moderate water stress. In: *Encyclopedia of Plant Physiology, Physiological Plant Ecology II: Water Relations and Carbon Assimilation (New Series).* Vol 12B. (Eds.) O.L. Lange, P.S. Nobel, C.B. Osmond, and H. Ziegler. Springer-Verlag, Berlin, 1982.

303. Pallardy, S.G., Pereira, J.S., and Parker, W.C. Measuring the state of water in tree systems. In: *Techniques and Approaches in Forest Tree Ecophysiology.* (Eds.) J.P. Lassoie and T.M. Hinckley. CRC Press, Boca Raton, FL, USA, 1991, pp. 27-76.

304. Scholander, P.F., Hammel, H.T., Hemmingsen, E.A. Hydrostatic pressure and osmotic potential in leaves of mangroves and some other plants. *Proceedings of the National Academy of Science (U.S.A.)*, 1967, **52**, 119.

305. Ritchie, G.A. and Hinckley, T.M. The pressure chamber as an instrument for ecological research. *Advances in Ecological Research*, 1975, **9**, 165.

306. Myers, B.J. Water stress integral - a link between short-term stress and long-term growth. *Tree Physiology*, 1988, **4**, 315-323.

307. Morris, L.A. Long-term productivity research in the U.S. Southeast: experience and future directions. In: *Research Strategies for Long-term Site Productivity.* Proceedings, IEA/BE A3 Workshop, Seattle, WA, August 1988. (Eds.) W.J. Dyck and C.A. Mees. IEA/BE A3 Report No. 8. Forest Research Institute, Rotorua, NZ FRI Bulletin No. 152, 1989, pp. 221-235.

308. Dyck, W.J. and Mees, C.A. (Eds.). *Research Strategies for Long-term Site Productivity.* Proceedings, IEA/BE A3 Workshop, Seattle, WA, August 1988. IEA/BE A3 Report No. 8. Forest Research Institute, Rotorua, NZ, FRI Bulletin No. 152, 1989.

309. Dyck, W.J. and Mees, C.A. (Eds.). *Long-term Field Trials to Assess Environmental Impacts of Harvesting.* Proceedings, IEA/BE T6/A6 Workshop, Florida, USA, February 1990. IEA/BE T6/A6 Report No. 5. Forest Research Institute, Rotorua, NZ, FRI Bulletin No. 161, 1991.

310. Draper, N.R. and Smith, H. *Applied Regression Analysis.* 2nd Edition. J. Wiley & Sons, New York, 1981.

311. Kimmins, J.P. *Forest Ecology.* Macmillan Publishing Company, New York, 1987, 531 p.

312. Vose, P.B. and Victoria, R.L. Re-examination of the limitations of the nitrogen-15 isotope dilution technique for field measurement of dinitrogen fixation. In: *Field measurement of dinitrogen fixation and denitrification.* (Eds.) R.D. Hauck and R.K. Weaver. ASA Spec. Publ. 18. ASA, CSSA, and SSSA, Madison, Wisconsin, 1986, pp. 23-41.

313. Larsen, D.R. and Kershaw, J.A. Jr. Leaf area measurements. In: *Techniques and Approaches in Forest Tree Ecophysiology.* (Eds.) J.P. Lassoie and T.M. Hinckley. CRC Press, Boca Raton, FL, 1991, pp. 465-475.

CHAPTER 11

IMPACTS OF HARVESTING ON LONG-TERM SITE QUALITY: FUTURE RESEARCH

N.B. COMERFORD

Department of Soil and Water Sciences, University of Florida
Gainesville, FL 32611, U.S.A.

D.W. COLE

College of Forest Resources, University of Washington
Seattle, WA 98195, U.S.A.

W.J. DYCK

New Zealand Forest Research Institute Ltd.
PO Box 31-011, Christchurch, New Zealand

INTRODUCTION

Review of Terminology

The term "long-term site productivity" has been loosely used by forest scientists and forest managers to denote the potential future capacity of a site to produce biomass, usually wood. It has often been confused with the more general term "productivity", which, as explained in Figure 2.1, is a function of both the inherent properties of a site (site quality) including climate, and the impacts of management on the site. In terms of the long-term productive capacity of forest sites, research needs to focus on the more specific term of site quality and, in particular, on the soil component (referred to as "soil capacity" in Chapter 3).

"Long-term" conjures up different time horizons depending on one's perspective; often related to the average crop rotation length of the region of interest. For example, researchers in Scandinavian coniferous forests may be most concerned about the long-term impacts of harvesting successive crops of spruce grown on 100-year rotations, whereas in Brazil, nutrient-drain induced by 8-year rotations of exotic eucalypts might mean that a long-term time frame is less than 25 years. In site productivity research the main objective is to better understand the impact of management practices on the capacity of the soil to produce the crop of interest. Obviously, the impact of harvesting is going to be greater on a site where the crop is harvested at a greater frequency and at a higher level of biomass utilization. Therefore, regardless of region, long term should be considered to be in the order of three rotations. For Douglas-fir forests in western Washington, for example, this means that long-term site productivity research should focus on being able to make predictions 150 years or more into the future.

The main impact of "harvesting" on site quality is through nutrient removal in biomass and soil physical damage by heavy machinery. Harvesting may also have off-site impacts as discussed in Chapter 4. Other forest management practices, including mechanical site preparation, burning, and fertilizer application, also affect site quality, often to a greater extent than harvesting (see Chapter 3). Thus, any future research on long-term site productivity must address these other operations and should also consider the more subtle effects of tree growth and nutrient cycling on site quality.

Although the premise throughout this book is that long-term site productivity research focus on wood production, it is realized that other forest values may be of greater importance in some situations. In these situations, site quality needs to be redefined in terms of the values of primary interest because the production of these non-timber values may be in direct conflict to the production of wood.

State of Knowledge

Previous chapters in this book reviewed our current understanding with regard to being able to predict the impacts of harvesting on long-term site productivity. From this we can conclude that there is very limited evidence of site quality degradation due to harvesting alone or as a result of subsequent site preparation practices (Chapters 1, 2, and 3). This is not necessarily because site quality decline has not occurred, but because it has been difficult to differentiate between the impacts of forest management operations on site quality versus the effects on productivity. When using the biota as the measure of response, we must realize that tree growth is affected not only by the factors of site quality, but also by the level of competition from both crop and non-crop vegetation, changed genetics, and variable re-establishment practices (e.g., variable planting stock quality). The effect of weed control alone will not affect inherent site quality but has a potentially enormous negative impact on productivity, especially in the shorter term (Chapters 3 and 9). Therefore, observed negative impacts of a practice are not necessarily evidence of reduced site quality, only of lower productivity.

The problem of predicting the impacts of harvesting is aggravated because we do not know how to measure site quality independent of the target crop. The key processes defining site quality are thought to be site dependent (Chapter 5) and the methodology to measure the fluxes resulting from these processes is currently inadequate (Chapter 10). Furthermore, site productivity studies outlined in previous chapters have often either not been designed with the objective of long-term results or have not been provided with adequate financial support to keep them active in the long-term (Chapters 2, 3, and 9).

An interesting point is that the long-term time frame of the questions asked in this book (e.g., 3 rotations) is often beyond the time frame of our research history. Therefore the question of a long-term effect from harvesting practices, at least when there is no short-term negative response, is unresolvable with just field trials. Likewise short-term responses (negative or positive) are not necessarily representative of long-term trends (Chapter 3). These limitations have led to the development of long-term predictive models. While being the most appropriate way to proceed, the models that have been developed have been found wanting (Chapter 6). Their major limitation is that they are nearly impossible to validate in the long-term time scale.

Harvesting and reestablishment operations often increase productivity, at least in the short-term. This is exemplified in Chapter 3 where the effects of stimulating nutrient cycles, changing understorey plant structure, adding fertilizers, and introducing improved genetic stock have more often resulted in increased productivity than the reverse. However, without continued attention to the factors of the site that have been altered (e.g., continuous fertilizer additions), these short-term trends of improved productivity may not continue into the long-term.

FUTURE RESEARCH

Given the above statements, the goals of future research should be to predict the consequences of today's harvesting/regeneration actions on *site quality* and to assure these predictions are valid in the long-term. We also want the option of modifying today's actions so we can control the long-term consequences. The intent of this section is not to list specific studies, but to outline the information needs.

Predicting the Consequences of Today's Actions

The needs in this area have been addressed by a number of chapters (Chapters 3, 4, 5, 9). First and foremost is the need to develop an experimental base that is grounded in well-designed and adequately supported field trials (Chapter 9). Through these field trials, and associated short-term process level studies, we need to develop a much better understanding of the key processes controlling site quality. In order to do so, we also need to develop a variety of techniques so we can finally measure how some of these processes work. The next challenge is to cross spatial and time scales with relevant processes (Chapter 5). The reliance on "indices" in the absence of adequate methodology should be reduced (mineral-ization indices, extractable nutrient indices, foliar tissue analysis for nutrition inference) (Chapter 10).

To make data and results from various trials comparable, a minimum data set should be developed. The minimum data set approach, which standardizes important environmental and site variables, has been useful in the CRISAT program [1] for modelling production of agricultural crops worldwide and the United States based Integrated Forest Studies (IFS) program (Electrical Power Research Institute) investigating the effects of atmospheric inputs on forest ecosystems [2].

Coincident with the field trials, there needs to be the continuous development *and* evaluation of predictive models. The questions to be asked are: Do these models work? What is the predictive accuracy of the output? What is the risk of accepting the model output? Since the evaluation is long-term, the first question is probably unanswerable, but continued re-evaluation of model predictions with results of field trials should provide the opportunity to upgrade the concepts of the model and build confidence in the long-term predictions.

Long-term predictions then need to be coupled with landscape classification schemes (Chapter 7). This step marries the research to the land base, extrapolates the results to the mixture of soils and climates, and gives the manager a tool to evaluate the consequences of his actions. This is the ultimate step to using the research and its predictions. With this capacity in hand it should be possible to (i) avoid mistakes in harvesting activities that could have a detrimental effect on long-term productivity, and (ii) correct mistakes made by management.

Avoiding or Correcting the Negative Consequences of Today's Actions

Avoiding degradation of site quality is achieved by accepting management techniques that do not adversely affect site quality (Chapter 8). If a negative effect is an unavoidable consequence of the management goal, then future long-term productivity problems need to be averted by incorporating the appropriate ameliorative technique into the silvicultural prescriptions for the site. This can be a very logical and cost-effective way of proceeding. To do so however, requires a full understanding as to what has been changed by the harvest practice, the effect of these changes on long-term productivity, and the correct ameliorative practice needed to minimize any long-term consequences to the quality of the site. From a research perspective, we are a long way from such an understanding of our forest systems and the modelling tools necessary to accurately predict the consequences of alternative corrective measures that might be used.

LITERATURE CITED

1. International Symposium on Minimum Data Sets for Agrotechnology Transfer. Proceedings of the International Symposium on Minimum Data Sets for Agrotechnology Transfer, March, 1983, ICRISAT (International Crops Research Institute for the Semi-Arid Tropics Center, Patancheru, Andhra Pradesh, India, 1984.

2. Johnson, D.W. and Lindberg, S.E. (Eds). *Atmospheric Deposition and Forest Nutrient Cycling.* Springer-Verlag, New York, 1992.

INDEX